Practical Motorsport Engineering

This guide and textbook on motorsport engineering is written from a practical point of view. It offers a wide-ranging insight into the nuts and bolts technology of practical car racing from saloons and sports cars to open wheelers. It gives the aspiring race engineer the tools to do the job by explaining all aspects of race car technology and offering crucial insight into the essentials of the motorsport engineering industry.

For motorsport engineering students at all levels, this book particularly covers the examination syllabuses for IMI (the Institute of the Motor Industry), EAL and BTEC, and meets the CPD requirements of most engineering institutions. Each aspect of the race car is covered in a separate chapter with test questions and suggestions for further study at the end. Combining the key points from his previous publications *Basic Motorsport Engineering* and *Advanced Motorsport Engineering*, the author draws on a career in teaching and industry to create the must-have, all-in-one reference. It is an ideal companion for the practising owner, driver or race engineer (whether amateur or professional), a suitable introductory text for HND and degree students and a great point of reference for any other keen fans with an interest in motorsport.

Andrew Livesey has set up a Kart Engineering Academy at Buckmore Park Kart Circuit, is a member of the Institute of the Motor Industry and the Institution of Mechanical Engineers, and has been an examiner for degrees in both Motorsport Engineering and Automotive Engineering. He is former Head of Motorsport on the Oxford Brookes University Brooklands HND and degree programmes, has worked for McLaren F1 and super car production, and has run a company building race and rally cars. He was also technical editor of *Cars and Car Conversions* and *Custom Car* magazines. He teaches part-time and runs an independent consultancy in Kent.

Practical Motorsport Engineering

Andrew Livesey

LONDON AND NEW YORK

First published 2019
by Routledge
2 Park Square, Milton Park, Abingdon, Oxon OX14 4RN

and by Routledge
52 Vanderbilt Avenue, New York, NY 10017

Routledge is an imprint of the Taylor & Francis Group, an informa business

British Library Cataloguing-in-Publication Data
A catalogue record for this book is available from the British Library

Library of Congress Cataloging-in-Publication Data
Names: Livesey, Andrew, author.
Title: Practical motorsport engineering / Andrew Livesey.
Description: First edition. | Abingdon, Oxon : Routledge/Taylor & Francis, 2019. | Includes bibliographical references and index. |
Identifiers: LCCN 2018038946 (print) | LCCN 2018040523 (ebook) | ISBN 9781351239165 (ePub) | ISBN 9781351239172 (Adobe PDF) | ISBN 9781351239158 (Mobipocket) | ISBN 9780815375692 (hardback) | ISBN 9780815375685 (pbk.) | ISBN 9781351239189 (ebook)
Subjects: LCSH: Automobiles, Racing—Design and construction—Textbooks. | Motorsports—Textbooks.
Classification: LCC TL236 (ebook) | LCC TL236 .L5825 2019 (print) | DDC 629.228/5—dc23
LC record available at https://lccn.loc.gov/2018038946

ISBN: 978-0-8153-7569-2 (hbk)
ISBN: 978-0-8153-7568-5 (pbk)
ISBN: 978-1-351-23918-9 (ebk)

Typeset in Times New Roman
by Apex CoVantage, LLC

Printed and bound in Great Britain by
TJ International Ltd, Padstow, Cornwall

This book is dedicated to the memory of
JOHN SURTEES CBE
The only person to be world champion on both two wheels and four wheels. He taught me how to go around corners and confirmed that it is important to support those less fortunate by providing accessible help and training.

Contents

Preface

We are now living in the Golden Era of Motorsport. It has become accessible to all. There are now more opportunities for motorsport than ever before, and the costs are the cheapest that they have ever been. The growth of track days, and clubman motorsport, in both two-wheel and four-wheel formats, give exciting opportunities for practical involvement either as an owner driver or small-scale business. Along with this are glorious professional opportunities in motorsport engineering as the big names fight out million-pound (dollar) battles around the world in events such as F1, DTM, Indy Car and Superbike.

I hope that you find this book useful whatever your involvement in the sport. If you see me at events please say hello, I'm always happy to chat.

Andrew Livesey MA CEng MIMechE FIMI
Herne Bay, Kent
Andrew@Livesey.US

Abbreviations and symbols

The abbreviations are generally defined by being written in full when the relevant technical term is first used in the book. In a very small number of cases, an abbreviation may be used for two separate purposes, usually because the general concept is the same, but the use of a superscript or subscript would be unnecessarily cumbersome; in these cases, the definition should be clear from the context of the abbreviation. The units used are those of the internationally accepted *System International* (SI). However, because of the large American participation in motorsport, and the desire to retain the well-known Imperial system of units by UK motorsport enthusiasts, where appropriate Imperial equivalents of SI units are given. Therefore, the following is intended to be useful for reference only and is neither exhaustive nor definitive.

α	(alpha) angle – tyre slip angle
λ	(lambda) angle of inclination; or measurement of air–fuel ratio
μ	(mu) co-efficient of friction
ω	(omega) rotational velocity
ρ	(rho) air density
a	acceleration
A	area – frontal area of vehicle; or Ampere
ABS	anti-lock braking system; or acrylonitrile butadiene styrene (a plastic)
AC	alternating current
AED	automatic enrichment device
AF	across flats – bolt head size
AFFF	aqueous film forming foam (fire fighting)
ATC	automatic temperature control
bar	atmospheric pressure
BATNEEC	best available technique not enabling excessive cost
BCF	bromochorodifluoromethane (fire fighting)
BS	British Standard
BSI	British Standards Institute
C	Celsius; or Centigrade
CAD	computer-aided design
CAE	computer-aided engineering
CAM	computer aided manufacturing
C_D	aerodynamic co-efficient of drag

CG	centre of gravity
CI	compression ignition
CIM	computer-integrated manufacturing
C_L	aerodynamic co-efficient of lift
cm	centimetre
cm^3	cubic centimetres – capacity; engine capacity also called cc. 1000 cc is 1 litre
CO	carbon monoxide
CO_2	carbon dioxide
COSHH	Control of Substances Hazardous to Health (Regulations)
CP	centre of pressure
CR	compression ratio
D	diameter
d	distance
dB	decibel (noise measurement)
DC	direct current
Deg	degree (angle or temperature)
dia.	Diameter
DTI	dial test indicator
EC	European Community
ECU	electronic control unit
EFI	electronic fuel injection
EPA	Environmental Protection Act; or Environmental Protection Agency
EU	European Union
F	Fahrenheit
ft	foot
ft/min	feet per minute
FWD	front-wheel drive
g	gravity; or gram
gal	gallon (USA gallon is 0.8 of UK gallon)
GRP	glass-reinforced plastic
GRP	glass-reinforced plastic (glass fibre)
HASAWA	Health and Safety at Work Act
HGV	heavy goods vehicle (used also to mean LGV – large goods vehicle)
hp	horse power (CV in French, PS in German)
HSE	Health and Safety Executive; also health, safety and environment
HT	high tension (ignition)
HVLP	high volume low pressure (spray guns)
I	inertia
ICE	in-car entertainment
ID	internal diameter
IFS	independent front suspension
IMechE	Institution of Mechanical Engineers
IMI	Institute of the Motor Industry
in^3	cubic inches – measure of capacity; also cu in. Often called 'cubes' – 61 cu in is approximately 1 litre
IR	infra red
IRS	independent rear suspension

ISO	International Standards Organization
k	radius of gyration
kph	kilometres per hour
KW	kerb weight
l	length
L	wheelbase
LH	left hand
LHD	left-hand drive
LHThd	left-hand thread
LPG	liquid petroleum gas
LT	low tension (12 volt)
lumen	light energy radiated per second per unit solid angle by a uniform point source of 1 candela intensity
lux	unit of illumination equal to 1 lumen/m^2
M	mass
MAG	metal active gas (welding)
MAX	maximum
MDRV	mass driveable vehicle
MI	Motorsport Institute
MIA	Motorsport Industry Association
MIG	metal inert gas (welding)
MIN	minimum
MoT	Ministry of Transport; also called DTp – Department of Transport and other terms depending on the flavour of the Government, such as the Department of the Environment Transport and the Regions (DETR), not to be confused with DOT which is the American equivalent
N	Newton; or normal force
Nm	Newton metre (torque)
No	number
OD	outside diameter
OL	overall length
OW	overall width
P	power, pressure or effort
Part no	part number
PPE	Personal Protective Equipment
PSV	public service vehicle (also used to mean PCV – public carrying vehicle in other words a bus)
pt	pint (UK 20 fluid ounces, USA 16 fluid ounces)
PVA	polyvinyl acetate
PVC	polyvinyl chloride
r	radius
R	reaction
Ref	reference
RH	right hand
RHD	right-hand drive
rpm	revolutions per minute; also RPM and rev/min
RTA	Road Traffic Act

RWD	rear-wheel drive
SAE	Society of Automotive Engineers (USA)
SI	spark ignition
std	standard
STP	standard temperature and pressure
TE	tractive effort
TIG	tungsten inert gas (welding)
TW	track width
V	velocity; or volt
VIN	vehicle identification number
VOC	volatile organic compounds
W	weight
w	width
WB	wheelbase
x	longitudinal axis of vehicle or forward direction
y	lateral direction (out of right side of vehicle)
z	vertical direction relative to vehicle

Superscripts and subscripts are used to differentiate specific concepts.

SI units

cm	centimetre
K	Kelvin (absolute temperature)
kg	kilogram (approx. 2.25 lb)
km	kilometre (approx. 0.625 mile or 1 mile is approx. 1.6 km)
kPa	kilopascal (100 kPa is approx. 15 psi, that is atmospheric pressure of 1 bar)
kV	kilovolt
kW	kilowatt
l	litre (approx. 1.7 pint)
l/100 km	litres per 100 kilometres (fuel consumption)
m	metre (approx. 39 inches)
mg	milligram
ml	millilitre
mm	millimetre (1 inch is approx. 25 mm)
N	Newton (unit of force)
Pa	Pascal
ug	microgram

Imperial units

ft	foot (= 12 inches)
hp	horse power (33,000 ftlb/minute; approx. 746 Watt)
in	inch (approx. 25 mm)
lb/in^2	pressure, sometimes written psi
lbft	torque (10 lbft is approx. 13.5 Nm)

Chapter 1

Power unit – engine

There ain't no substitute for cubes.

Whether it's cubic inches or cubic centimetres, the more of them that you have in your engine the more power you can develop. That's how it is for petrol and diesel engines – with electric motors it volts and amps you need.

Motorsport is often grouped into engine sizes, so the competitor is challenged to get the most power out of the engine. There are also usually regulations on what is, and is not, allowed to improve the power output.

Terminology

*The term **horse power** – **HP** – comes from steam engine sales agents of about two hundred years ago saying how many horses their engines could replace. In French this is **Cheval Vapour** – **CV**; in German this is **Pferde Starke** – **PS**.*

1 HP, in any language, that is CV, or PS is equal to 746 watts.

When we are talking about power output we must be careful to compare like for like. When we say **BHP** brake horse power we are talking about the engine power measured on an engine brake or dynamometer – **dyno**. But rolling road dynamometers measure power at the wheels – this is after the frictional losses in the transmission system – typically around 10–15%. Also, there are a number of different standards for measuring power outputs set by different organisations, the most popular are the American **SAE** standard and the German **DIN** standard. Both measure power but under different conditions – variations like the use of air filters and the way that the cooling system is connected.

Power is about doing work in amount of time – mathematically work done per unit time.

Identification

Identification of the engine before working on it is very important. The **VIN** number will help identify the type, or classification, of the engine. The detail of the engine will be given in a separate engine number, the prefix will identify the engine type, and the serial number will identify the exact engine.

Figure 1.1 Lola T70B V8 engine

With motorsport engines the build may be completely different to standard; for this reason you should keep a log of the engine build, detailing all the components including part numbers, sizes, and any other variants.

Engine performance

The two common terms used in motorsport are:

Power – this is **work done** in unit time.
Torque – turning moment about a point.

Let's discuss them for clarity then look at the calculations. When we are using the term power we are referring to how much energy the engine has. A big heavy car needs a big powerful engine. Power is about doing work in a time period, it means burning fuel in the time period. We can make a small four-cylinder engine – say one from a motorcycle like a Kawasaki ZX6R produce over 100 BHP from its 600 cc; but we need it to revving at about 12,000 rpm. Such engines are used in clubman cars; but starting from rest necessitates slipping the clutch until the car reaches about 30 mph. The torque produced by such an engine is very low. On the other hand, a large slow revving 2 or 3 litre diesel engine as used in many commercial vehicles will set off at about 500 rpm and pull a big payload; it is designed to have lots of torque.

For a mathematical definition of these terms we need to start with work done. Work done is the amount of load carried multiplied by the distance travelled. The load is converted into force, the force needed to move the car for instance in Newtons (N). The distance is measured in metres. That is:

$$\text{Work done}\ (\text{Nm}) = \text{Force}\ (\text{N}) * \text{Distance}\ (\text{m})$$

As we also express torque in Nm, so it is common to use the term joule (J) for work done.

Racer note

Joule is a term for energy. 1 J = 1 Nm

If we use a force of 10,000 N to take a dragster down a 200 m drag strip then we have exerted 2,000,000 Nm, or 2,000,000 J. We'd say two mega joules (2 MJ). We'd need to get this amount of energy out of the fuel that we were using.

The force is generated by the pressure of the burning gas on top of the piston multiplied by the area of the top of the piston. So, the work done is the mean (average) force of pushing the piston down the cylinder bore multiplied by the distance travelled.

Example

The work done during the power stroke of an engine where the stroke is 60mm and the mean force is 5kN:

$$\begin{aligned}\text{Work Done} &= \text{Force} * \text{Distance}\\ &= 5\text{kN} * 60\text{mm}\\ &= 5{,}000\ \text{N} * 0.06\ \text{m}\\ &= 300\text{J}\end{aligned}$$

Racer note

The mathematical symbols used in this book are those found on your calculator or mobile phone,

** is multiply and / is divide*

The same mean force is going to create the torque; this time we are going to use the crankshaft throw – this is half of the length of the stroke.

Example

Using the same engine:

$$\begin{aligned} \text{Torque} &= \text{Force} * \text{Radius} \\ &= 5\text{kN} * 30\text{mm} \\ &= 5{,}000\text{N} * 0.03\text{m} \\ &= 150\text{Nm} \end{aligned}$$

The work done by a torque for one revolution is the mean force multiplied by the circumference. The circumference is $2\Pi r$ so:

$$\text{Work Done} = F * 2\,\Pi\, r$$

As $$Fr = T$$

So, $$\text{Work Done} = 2\,\Pi\, T$$

That is for one revolution. For any number of revolutions, where n is any number, the formula is:

$$\text{Work Done in n revolutions} = 2\,\Pi n T$$

Example

Using the same engine of the previous examples.
The work done in 1 minute at 6,000 rpm will be:

$$\begin{aligned} \text{WD in n revolutions} &= 2\,\Pi n T \\ &= 2 * \Pi * 6{,}000 * 150 \\ &= 5657\text{ kJ} \end{aligned}$$

Power is, as we said; work done in unit time, which is:

$$\text{Power} = \text{Work Done} / \text{Time}$$

The motorsport industry uses a number of different units and standards for power, from our calculations we can use watts (W) and kilowatts (kW) and then convert:

Racer note

1 kW = 1,000W

$$1 \text{ Watt} = 1\text{J} / \text{second}$$

And $$1 \text{ kW} = 1\text{kJ} / \text{s}$$

Example

Following on from our engine in the previous calculations and examples:

$$\text{Power} = \text{Work Done} / \text{Time}$$
$$= 5657\text{kJ} / 60$$
$$= 94.3\text{kW}$$

The term **horse power** (HP or hp) was derived by James Watt as the average power of a pit pony. These were small horses used to turn pulleys to draw water from Cornish tin mines (pits) before steam power became more popular. He equated the power of his steam engines to a number of these pit ponies. For our purposes 1 HP equals 33,000 ftlb/minute.

In French, horse power is *cheval vapour* (CV); in German it is *Pferde Stracker* (PS).

For conversion purposes 1 HP is equal to 746 W.

When talking about power and doing work the weight of the vehicle comes into play. A vehicle like a clubman's car such as Caterham or a Westfield with a small engine can accelerate almost as fast as a Bugatti with a very large engine because it is lighter in weight. So, the criterion becomes BHP per tonne.

Terminology

One metric tonne is 1,000 kg. As a kilogramme is equivalent to 2.25 pounds, a metric tonne is the equivalent of an Imperial ton – 2250 lb.

The aim of any race car designer is to get the maximum BHP per ton, not forgetting to have enough torque to get off the starting line – especially for hill climb cars. See Table 1.1.

Table 1.1 Typical BHP per ton figures

Vehicle	*Capacity*	*BHP*	*Weight – kg*	*BHP per ton*
Typical BTCC car	2 litre turbo	350	1000	350
Kawasaki H2 motorcycle	1 litre supercharged	310	215	1,442
Clubman's car	1.6 litre turbo	200	500	400
German touring car – DTM	4 litre supercharged	500	1,122	445
Bugatti 16/4	8 litre with 4 turbos	1,200	1,990	603

Figure 1.2 JCB Diesel World land speed record holder

Fuel energy output measurement

The energy output of fuel varies (see Table 1.2). Some classes of racing use control fuels – that is only one type or brand of fuel is allowed, the idea being that all the competitors are restricted to the same and cannot vary the fuel. Others are not restricted or may have different classes for different fuels. Whatever the case, it is worth having a knowledge of the fuel that you are using.

There are two ways of measuring the quantity of fuel: this may be by the kilogram or by the litre. Energy companies tend to give outputs for kilograms, this is a more stable method of measurement and is traditionally used for ships and aeroplanes where mass is a more important factor. Typical values for consumer fuels are:

- Petrol 45.8 MJ/kg
- Diesel 45.5 MJ/kg

However, diesel is about 15% more dense than petrol, when measured at 15.55 °C this figure is used as it is the equivalent of 60 °F – the temperature scale used in non-European countries. The specific gravity (also called relative density) of consumer fuels is:

- Petrol 0.739
- Diesel 0.82–0.95

The range of sg values for diesel is because of the variety of products for commercial use and summer and winter products. Winter diesel is lighter than summer diesel to prevent waxing. Waxing, when diesel turns solid, takes place at between 14 °C and 18 °C – the latter for winter use.

So, taking typical consumer fuels – pump fuels, we get the following values per litre:

- Petrol 33.7 MJ/litre
- Diesel 36.9 MJ/litre

If you are not able to measure the calorific value of the fuel it is a good idea to record the sg every time that you refuel, especially if you are not able to refuel from a regular source – as when rallying. The sg is measured using a standard type of hydrometer or refractometer – or for laboratory use more complex digital machines are available.

Your log should include: date, mileage, quantity, source of fuel, sg and temperature.

Alternative fuels – certain classes of racing allow, sometimes insist on, the use of other fuels. **Methanol** is a clean burning fuel, it is used in Indy cars and midgets in the USA. **Biodiesel**, like methanol, is made from renewables and is therefore environmentally friendly. **Liquified natural gas** (LNG) is a product of shale – bored or fracked from the ground – not environmentally friendly, there are various form of gas products – they all burn dry and limit engine life even with special operating procedures. LNG/LPG is used by cost-conscious operators as it attracts a lower level of duty or tax.

Bomb calorimeter – this is used to measure the energy output of fuel. It is shaped like you would imagine a bomb to look and works by burning – exploding – a spoon full of fuel to see how much energy it will generate. You do this by measuring the temperature change brought about by burning the fuel under controlled conditions. Bomb calorimeters are readily available and allow you to check the fuel that you are using. Remember that for maximum power you want the fuel that produces the highest energy output for a given volume, so it must be high density, high calorific output.

Table 1.2 Typical energy outputs of alternative fuels

Fuel	*MJ/kg*	*MJ/litre*
Methanol	19.9	15.9
Biodiesel	37.8	33.3
LNG	38	25.5

Table 1.3 Data for bomb calorimeter

No	*Name*	*Reading*
1	Density of fuel – density	0.845
2	Water equivalent of calorimeter – mfg figures	652 g
3	Mass of water	200 g
4	Mass of fuel	1 g
5	Temperature of water before combustion	19.26 °C
6	Temperature of water after combustion	23.84 °C
7	Temperature rise, No 6 – No 5	4.58 °C
8	Specific heat of water	4.19 kJ/kg deg K

A typical calculation using a bomb calorimeter

The formula:

$$\text{Calorific value} = \text{total water mass} \times \text{Temperature rise} \times \text{specific heat of water}$$
$$= 2.652 * 4.58 * 4.19$$
$$= 50{,}900\,\text{kJ / kg}$$

Engine construction

Key points

- The engine burns fuel at a very high temperature to force the piston down the cylinder bore
- The connecting rod and crankshaft convert the reciprocating motion of the piston into rotary motion
- Engine size is calculated as a product of the cylinder bore and the piston stroke
- Engines may be petrol (SI) or diesel (CI)
- Two-stroke petrol engines are used in small motorcycles and garden machinery

There are several different types of engines in use; the most common type in the UK and the USA is the **four-stroke petrol engine**. The second most common in the UK is the **four-stroke diesel engine**; the diesel engine is much more popular in France and other countries in continental Europe as diesel fuel is cheaper there than in the UK. Petrol engines are also called spark ignition (**SI**) engines; diesels are called compression ignition (**CI**) engines. Both types of engines are very similar in appearance and construction; the main components are as follows.

> **Racer note**
>
> Race engines are developed to produce the maximum power with a specific capacity to conform with the regulations of the racing classification – this usually means high engine speed with a narrow power band

Cylinder block (block) – this forms the main part of the engine and carries the other engine parts. The cylinder block is made from either cast iron or, on high-performance engines, aluminium alloy. Aluminium alloy is both lighter and a better thermal conductor. The cylinder head fits onto the top of the block; the crankshaft fits into bearing housings in the lower part of the block. The pistons run in the cylinder bores, which are at right angles to the crankshaft. The block must be accurately machined, and very rigid, so that the components are held in exact positions relative to each other.

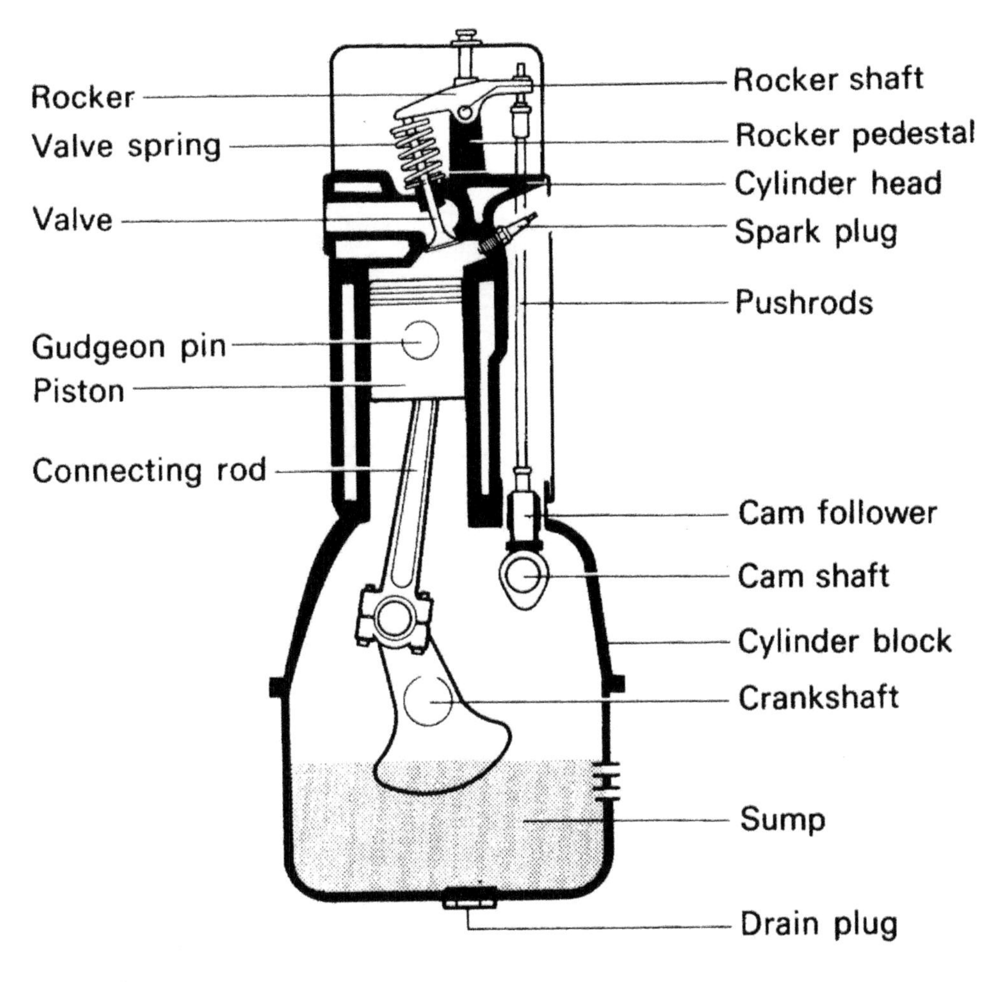

Figure 1.3 Engine components

Points to be noted when preparing a race engine:

- Manufactures sometime use different blocks for the same engine in different applications, variations include materials and manufacturing processes.
- Sometimes one block is used for a variety of different engine capacities – the capacity is changed by using a crankshaft with a different stroke.
- There are several different types of block construction.
- The same material can have different properties varying with manufacturing process, for instance aluminium can be sand-cast or die-cast. The former is more resilient for race engines, although the later might be lighter in mass.

Pistons – these move up and down in the cylinder bores. This up and down movement is called **reciprocating motion**. The piston forms a gas-tight seal between the combustion chamber and the crankcase. The burning of the fuel and air mixture in the combustion chamber forces the piston down the cylinder to do useful work. The pistons are usually made from aluminium alloy for its lightweight and excellent heat conducting ability. The top of

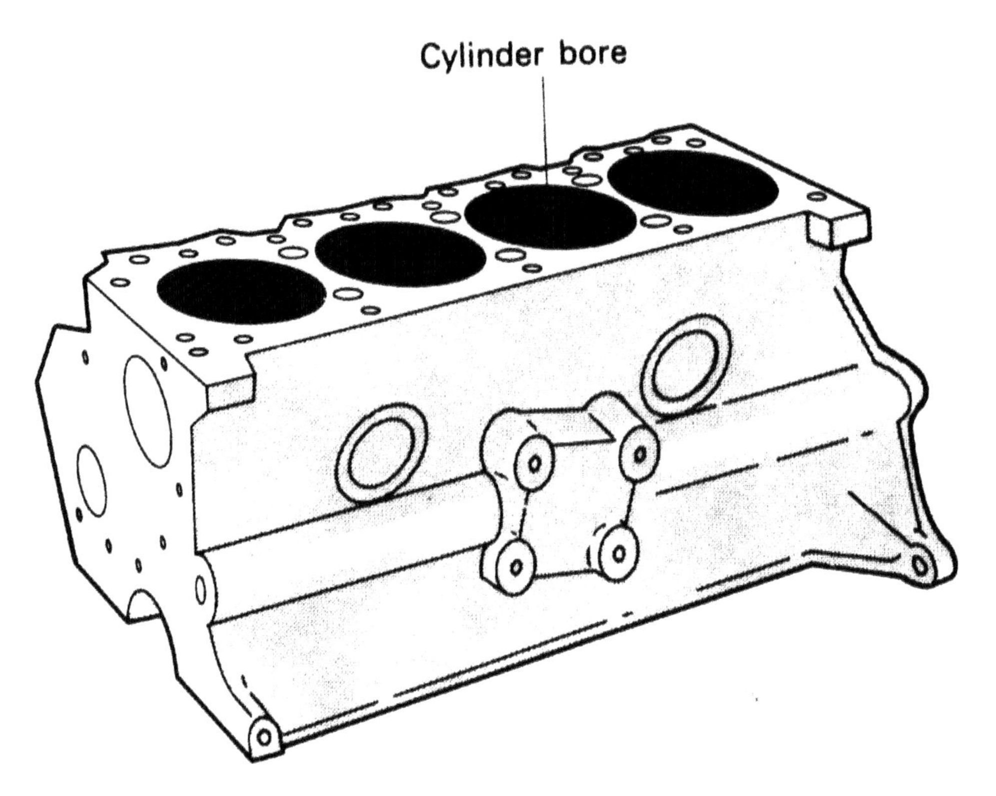

Figure 1.4 Cylinder block

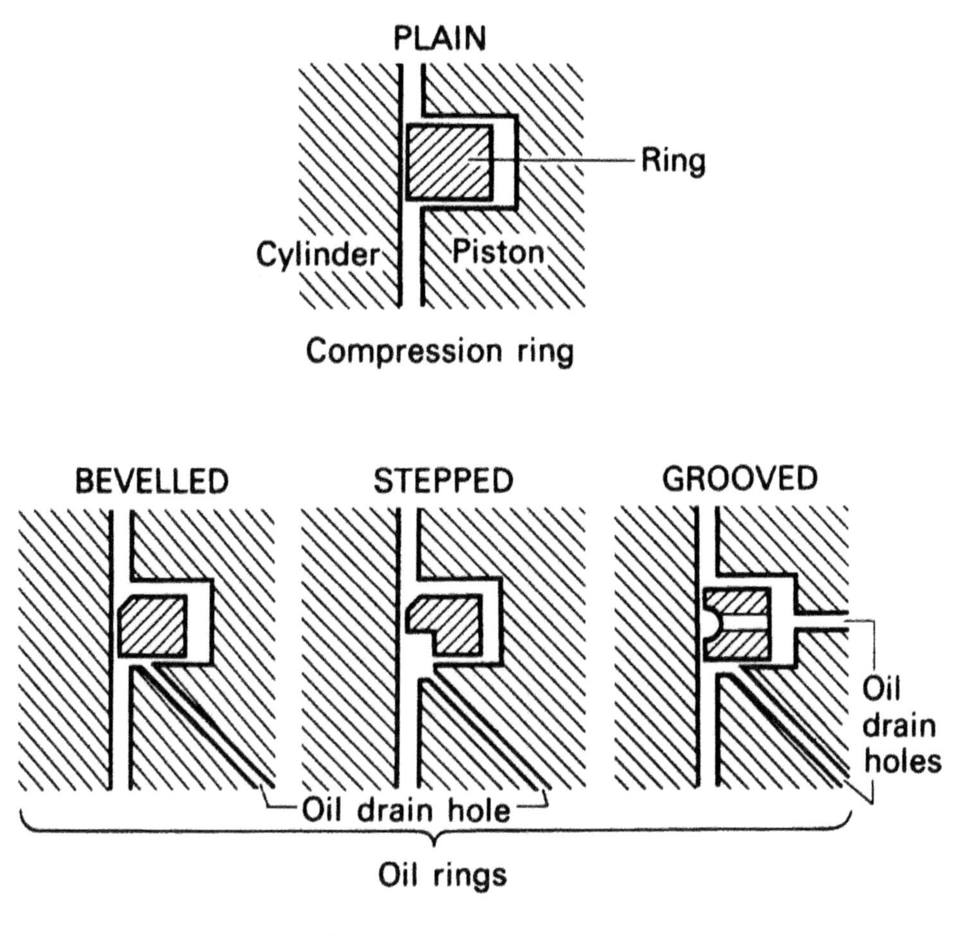

Figure 1.5 Types of piston rings

the piston is called the crown; the lower part is called the skirt. The pistons must be perfectly round to give a good seal in the bore when the engine is at its normal running temperature. However, aluminium expands a lot when it is heated up. The pistons have slits in their skirts to allow for their expansion in diameter from cold to their normal operating temperature. When cold, pistons may be a very slightly oval shape, so that when at running temperature they are a perfect fit in the cylinder bore.

The pistons are fitted with piston rings to ensure a gas-tight seal between the piston and the cylinder walls. This is needed to keep the burning gases in the combustion chamber. The piston rings are made from close grain cast iron, a metal that is very brittle. But the piston rings are slightly springy because of their shape. Usually there are three piston rings. The top two are compression rings to keep the gases in the combustion chamber. The bottom one is an oil ring; its job is to scrape the oil off the cylinder walls. The oil is returned to the sump by passing through the slots in the piston rings and running down inside the pistons. As piston rings are made from cast iron – this is very brittle; so when piston rings are being fitted great care must be taken not to break them.

Racer note

Pistons on race engines are as light as possible to give very high-speed running – to reduce the weight of them the skirts are made very short and cut away to a *slipper* shape – referred to as slipper pistons.

FAQs What is meant by bore and stroke?

The bore is the cylindrical hole, or cylinder, in which the piston runs. The bore must be perfectly smooth, round and parallel. It is also the term used to describe the diameter of the cylinder; this is usually expressed in millimetres (mm) or inches (in). The stroke is the distance the piston travels from the bottom of the cylinder – called bottom dead centre (BDC) – to the top of the cylinder – called top dead centre (TDC). The stroke may be measured in millimetres or inches. The surface inside the cylinder is called the cylinder wall.

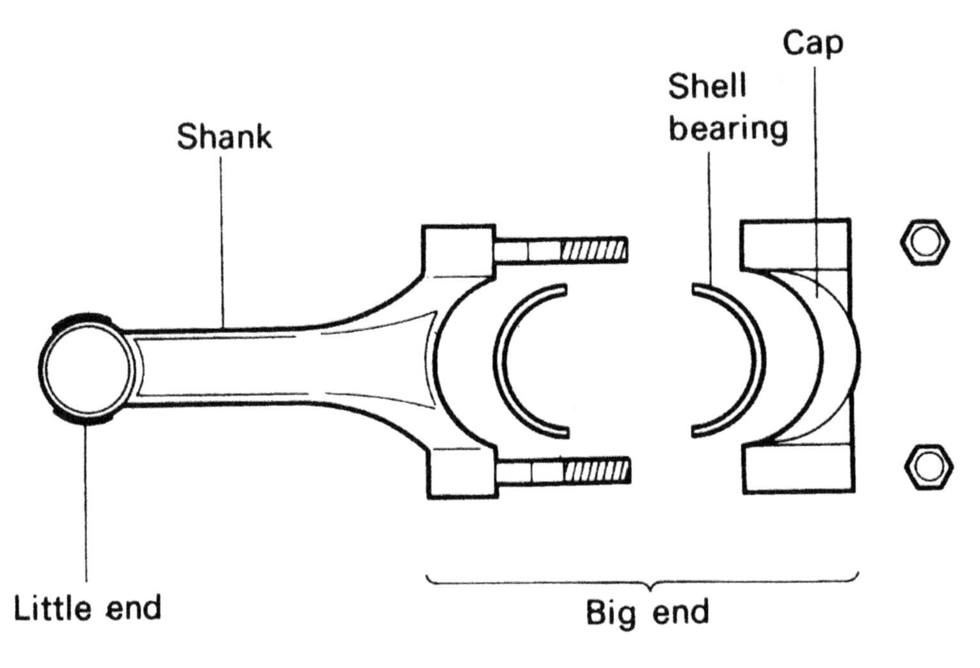

Figure 1.6 Connecting rod

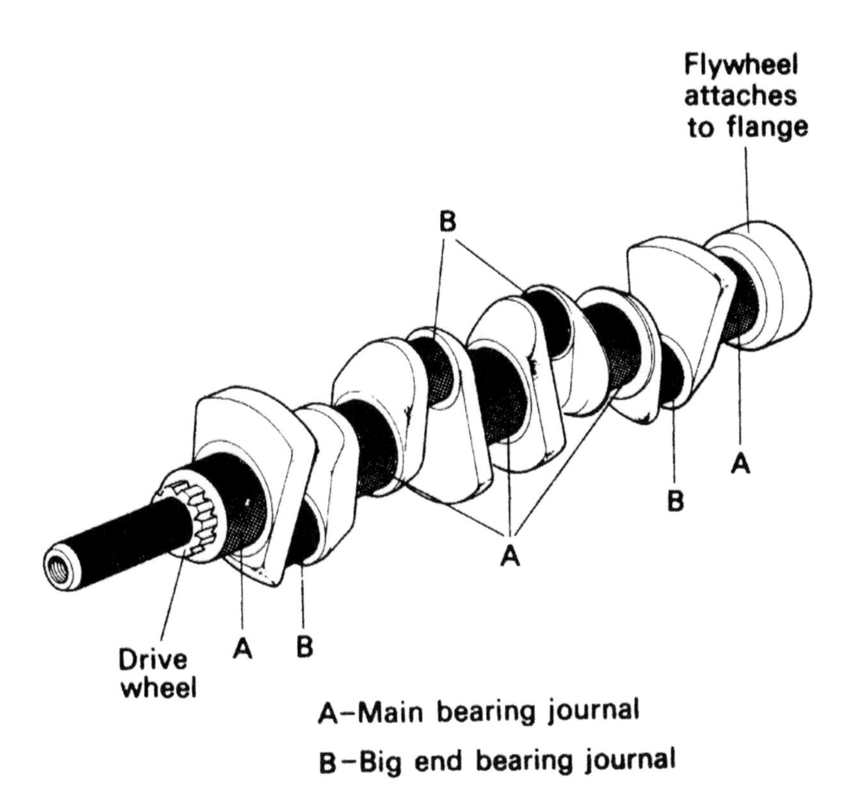

Figure 1.7 Crankshaft for a four-cylinder engine

Connecting rod (con rod) – this connects the piston to the crankshaft. The con rod has two bearings; the **little end** connects to the piston and the **big end** to the crankshaft. The con rod is made from either cast iron or forged steel. The big-end bearing is a shell bearing; this allows for easy replacement and cheap manufacture.

Crankshaft – this, in conjunction with the con rod, converts the reciprocating motion of the pistons into the rotary motions which turns the flywheel. The crankshaft is located in the cylinder block by the main bearings. The big-end bearings are attached to the crank pins; the crank pins are at the ends of the crank webs. The distance between the centre of the **crank pin** and the centre of the **main bearing** is called the **throw**. The throw is half of the **stroke**.

Key points

From TDC to BDC is the stroke. For the piston to travel from TDC to BDC the crankshaft rotates 180 degrees – half a revolution. The crank pin is moved from being above the main bearing – the length of one throw – to being below the main bearing – the length of another throw. That is, two throws are equal to the length of the stroke.

Figure 1.8 Cylinder head

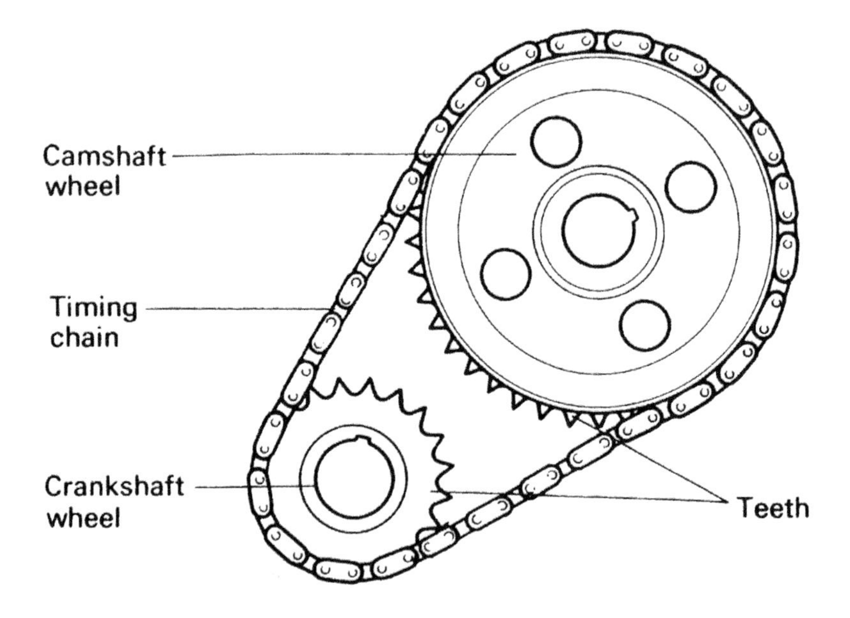

Figure 1.9 Timing chain

Cylinder head (head) – the head sits on top of the cylinder block. The head contains the combustion chambers and valves. Between the head and the block is a cylinder head gasket. The cylinder head gasket allows for the irregularities between the block and the head and keeps a **gas-tight seal** for the combustion chamber. A SI engine cylinder head locates the spark plugs; a CI engine cylinder head locates the injectors.

Valve cover and sump – the cylinder head is fitted with a valve cover (also called **rocker box**, or **cam box**.) The valve cover encloses the valves and their operating mechanism, forming an oil-tight seal for the engine oil. The bottom of the block is fitted with a sump. The sump has two jobs; it is a store for the engine lubricating oil and forms an oil-tight seal to the bottom of the engine. Both the valve cover and the sump are usually made from thin pressed steel.

Timing mechanism – at the front of the engine is the timing mechanism. That is either a belt, or a chain, which connects the crankshaft to the camshaft. A plastic casing covers the timing mechanism. The timing end of the engine is also called the free end. **Cylinder numbers** always start from the free end.

Flywheel – the flywheel is attached to the crankshaft. The flywheel end of the engine is the drive end. That is, the flywheel turns the clutch and the gearbox to move the vehicle.

Figure 1.10 Flywheels

Racer note

As a race mechanic, it is good to understand the different metals that are used in engines. The different metals must be considered when you are handling the part, and particularly when tightening up nuts and bolts.

- Cast iron is used for many cylinder blocks and heads; it is very heavy and brittle, so it will break if dropped.
- Aluminium is light in weight and expands a great amount when heated up. It is also soft, so it is easily scratched. You must be careful not to over tighten spark plugs in aluminium cylinder heads or you will damage the threads.
- Pressed steel is used for sumps and valve covers; this is easily bent. A bent sump may leak around the joints.
- Hardened steel is used for the crankshaft; this is both heavy and expensive.

Four-stroke petrol engine

The four-stroke petrol engine works on a cycle of four up and down movements of the piston. These up and down movements are called strokes. The piston moves down from top dead centre (TDC) to bottom dead centre (BDC), then up to TDC again. Each stroke corresponds to half of a turn of the crankshaft; therefore, the complete cycle of four strokes takes two revolutions of the crankshaft.

The petrol and air mixture are burnt in the combustion chamber during one of the strokes. The heat from the burning fuel causes a pressure increase in the combustion chamber. This pressure forces the piston down the bore to do useful work. The mixture is ignited by the spark plug, hence the term spark ignition (SI).

The cylinder head is fitted with inlet valves; these open and close to control the flow of the petrol and air mixture from the inlet manifold into the combustion chamber. The cylinder head is also fitted with exhaust valves to control the flow of the spent exhaust from the combustion chamber into the exhaust manifold and exhaust system. The passage in the cylinder head, which connects the manifold to the combustion chamber, is called the port. There are inlet ports and exhaust ports. The valves are situated where the ports connect into the combustion chamber. The valves are operated by the camshaft; this is discussed later in this chapter.

Induction stroke

The piston travels down the cylinder bore from TDC, drawing in the mixture of petrol and air from the inlet manifold. This is like a syringe drawing up a liquid. The downward movement of the piston has caused a depression above the piston; this depression, or partial vacuum, is satisfied by the air coming into the inlet manifold through the air filter. The air mixes with the petrol that is supplied either from the injectors or a carburetter.

Compression stroke

When the piston reaches BDC it starts to return up the bore. At about BDC the inlet valve is closed by the camshaft; the exhaust valve was already closed. The mixture of petrol and air,

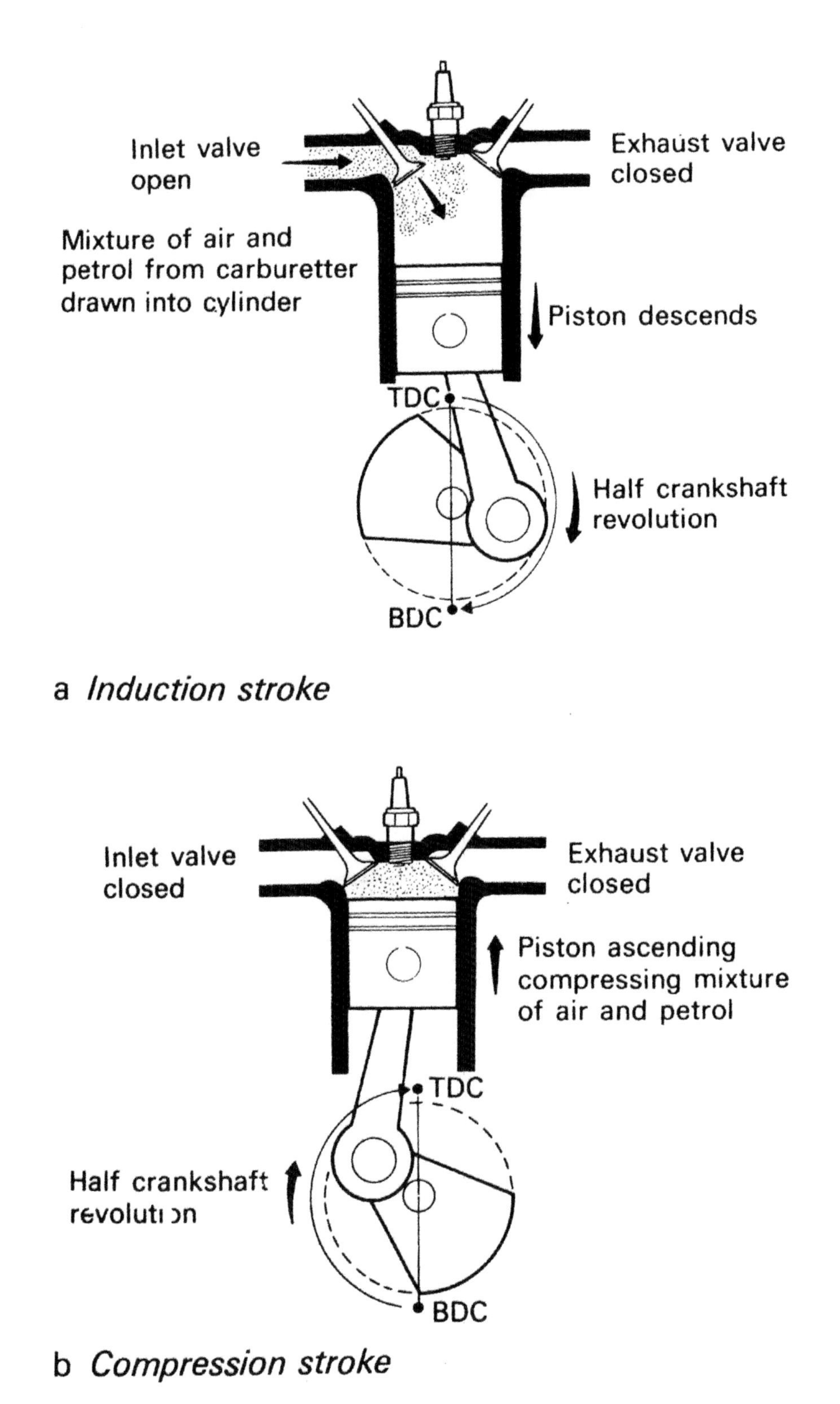

Figure 1.11 Four-stroke Otto cycle

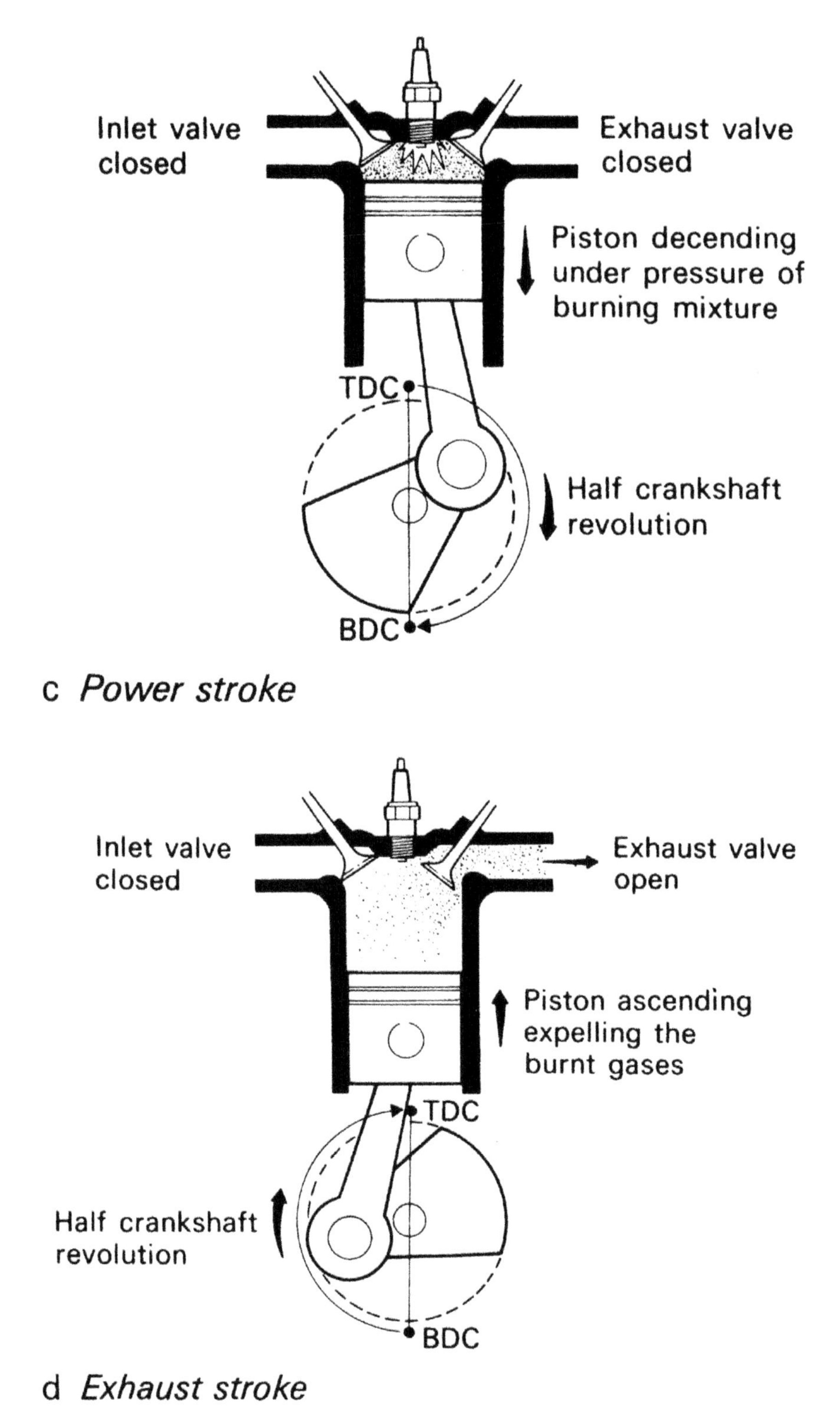

Figure 1.11 (Continued)

which was drawn in on the induction stroke, is now compressed into the combustion chamber. This increases the pressure of the mixture to about 1250 kPa (180 psi). The actual pressure depends on the compression ratio of the engine, on race engines it is typically between 10:1 and 16:1. On formula cars this figure may exceed 20:1. Increasing the compression ratio within limits will increase the power output.

Nomenclature

The mathematical sign : means to – signifying a ratio

Power stroke

As the piston reaches TDC on the compression stroke, the spark occurs at the spark plug. This spark, which is more than 10 kV (10,000 volts), ignites the petrol–air mixture. The mixture burns at a temperature of over 2,000 degrees Celsius and raises the pressure in the combustion chamber to over 5,000 kPa (750 psi). The pressure of the burning petrol–air mixture now starts to force the piston back down the cylinder bore to do useful work. The piston rings seal the pressure of the burning mixture in the combustion chamber, so that the pressure exerts a force on the piston, the gudgeon pin and then the con rod, which converts this downward motion into rotary motion at the crankshaft. It's good to remember that a modest speed engine will fire on each cylinder over 50 times every second.

FAQs How much force does the burning mixture actually exert on the con rod or crankshaft and why do they not bend?

The amount of force depends on the size of the engine; but, as a rough guide, imagine an elephant sat on the top of the piston every time it goes down. The components will not bend as long as the engine is rotating and the force is being passed on to the transmission to move the vehicle.

Exhaust stroke

At the end of the power stroke the exhaust valve opens. When the piston starts to ascend on the exhaust stroke, this is the last stroke in the cycle; the burnt mixture is forced out into the exhaust. The mixture of petrol and air has been burnt to change its composition. Its energy has been spent. The temperature of the exhaust gas is about 800 to 1,200 °C. The petrol–air mixture has been burnt to become carbon monoxide (CO), carbon dioxide (CO_2), water (H_2O), nitrogen (N) and free carbon (C). The exhaust gas is passed through the exhaust system to the catalytic converter to be cleaned and made non-toxic. The exhaust gas exits the engine in waves – it is not a continuous stream, nor in parcels like a sausage machine. The speed of sound at 20 °C is 343 m/s – the different exhausts give different pitches, compare the exhaust noise of a Ferrari V12 with a Ducati twin.

Factors affecting the exhaust efficiency and sound are:

- Manifold design and subsequent system
- Material used for exhaust – some vintage cars use copper – this produces a delightful resonance
- Use, or otherwise, of a catalytic converter – cats operate best at about 900 °C
- Use of a turbo-charger

Flywheel inertia

There is only one firing stroke for each cycle. The flywheel keeps the engine turning between firing strokes. Single cylinder engines need a bigger flywheel in proportion to their size than those with more cylinders do. The flywheel on a large V8 engine is smaller than one on a four-cylinder engine. The flywheel's desire to keep rotating is called inertia; it is inertia of motion.

Valve operation

The inlet and the exhaust valves each open once every two revolution of the crankshaft. The mechanism for opening the valves is a camshaft, which is either direct acting on the valves; or operates them through a pushrod and a rocker shaft assembly.

If the camshaft is situated above the valves the engine is referred to as overhead cam (OHC). Other types are overhead valve (OHV) and side valve. A rubber toothed belt or a chain may drive the camshaft. As the camshaft must rotate at half the speed of the crankshaft, to open the valves once for every two revolutions of the crankshaft, the drive is through a two to one (2:1) gear ratio. The cam gear wheel has twice as many teeth as the crankshaft gear wheel.

The actual point at which the valves open and close depends on the engine design. The vehicle's workshop manual will give the valve timing figures; these are usually expressed in degrees of the crankshaft relative to TDC and BDC.

It is essential that the valves close firmly against their seats to give a good gas-tight seal and allow heat to conduct from them to the cylinder head so that they may cool down. The valves are held closed by springs. To ensure that the valves close firmly, even when the components have expanded because of the heat, the valve mechanism is given a small amount of clearance. The valve clearance is measured with a feeler gauge; a typical figure is 0.15 mm (0.006 in). If the valve clearance is too great then a light metallic rattling noise will be heard.

Two-stroke petrol engine

The two-stroke petrol engine is used mainly in small motorcycles, although they have been used in some cars. It operates on one up-stroke and one down-stroke of the piston; that is one revolution of the crankshaft. The most common type is the Clerk Cycle engine; this has no valves, just three ports. The three ports are the inlet port, the transfer port and the exhaust

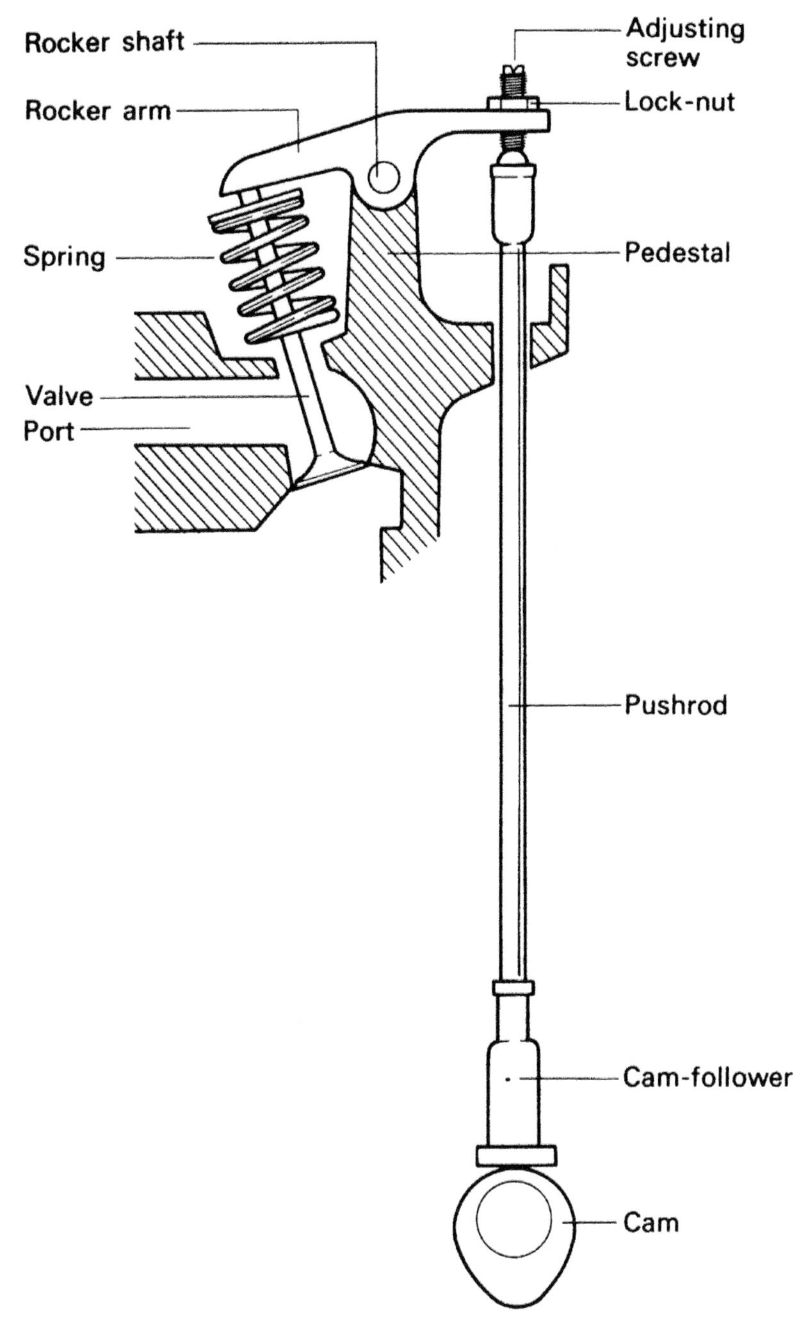

Figure 1.12 OHV arrangement

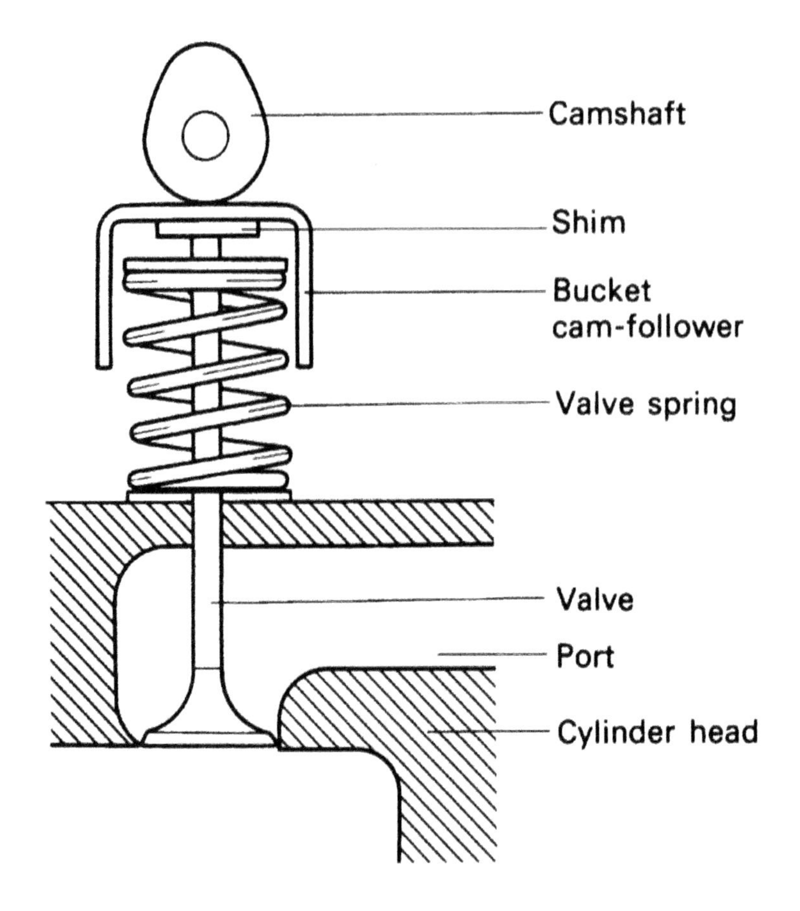

Figure 1.13 OHC arrangement

port. The flow of the gas through these ports is controlled by the position of the piston. When the piston is at TDC both the transfer and the exhaust ports are closed. When the piston is at BDC the piston skirt closes the inlet port.

The piston travels up the bore; as it reaches TDC it closes the both the transfer and the exhaust ports. At the same time, the piston is compressing the charge of petrol and air above it into the combustion chamber. At about TDC the spark plug ignites the petrol–air mixture. The burning of the petrol–air mixture increases the temperature and the pressure, so that the burning gas pushes the piston down the bore. The downward force of the piston is passed through the gudgeon pin to the con rod and crankshaft to drive the vehicle.

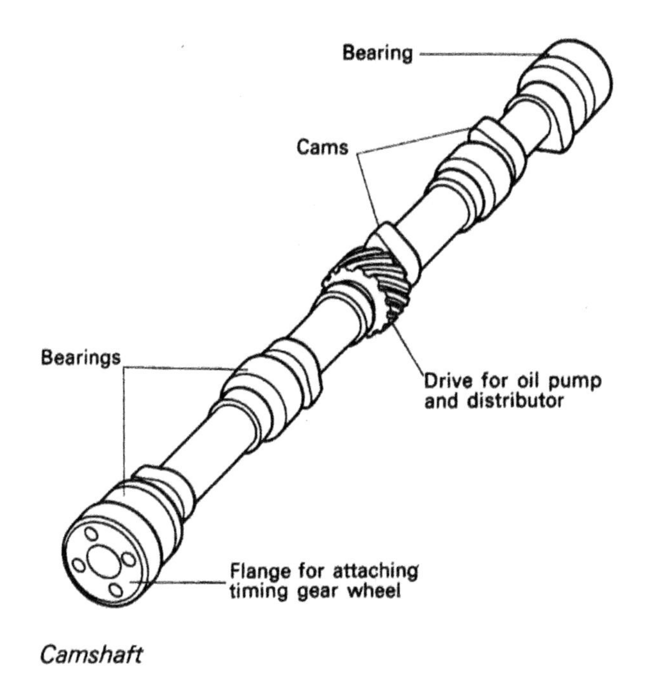

Figure 1.14 Camshaft

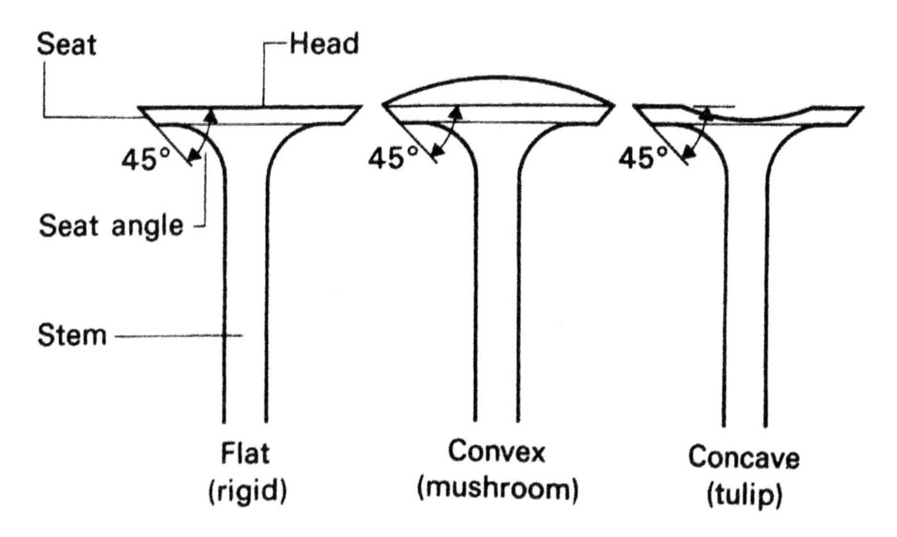

Figure 1.15 Valve shapes

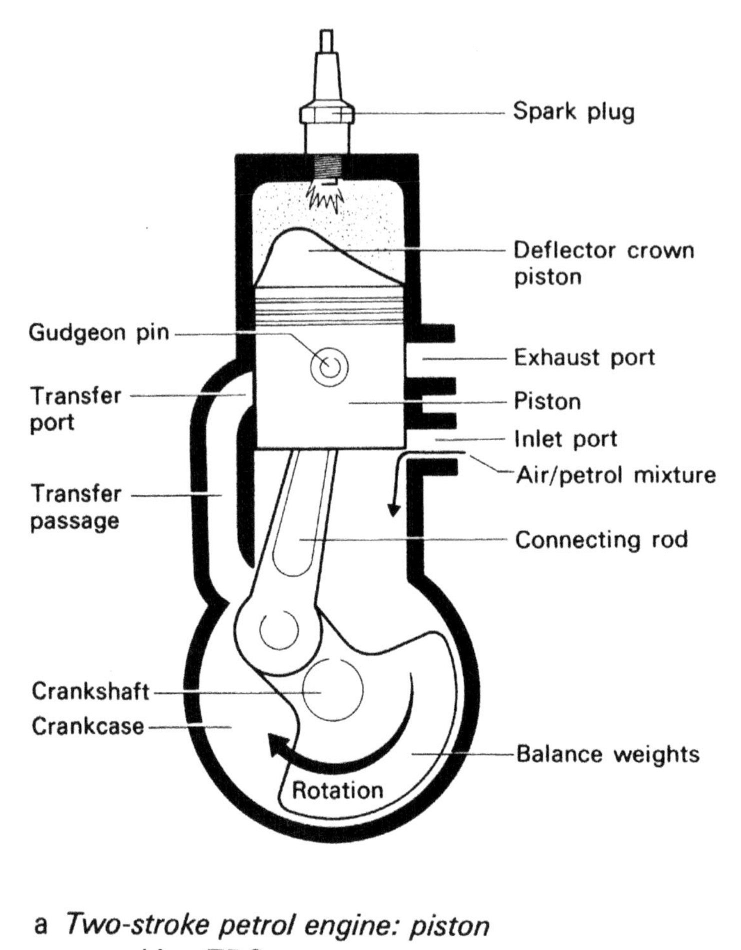

a *Two-stroke petrol engine: piston approaching TDC*

Figure 1.16 Two-stroke cycle

Whilst the piston is ascending, its skirt uncovers the inlet port. The upward motion of the piston causes a vacuum in the crankcase, which is satisfied by the petrol–air mixture from the carburetter entering through the now open inlet port.

When the piston is travelling downwards – being forced down by the burning mixture – the piston crown first uncovers the exhaust port. This allows the spent gas to escape into the exhaust system. The skirt of the piston covers the inlet port at the same time as the piston crown uncovers the transfer port at the top of the cylinder. The underside of the piston therefore acts like a pump plunger, forcing the fresh charge of petrol–air that is in the crankcase up and through the transfer port into the cylinder so that another cycle is started.

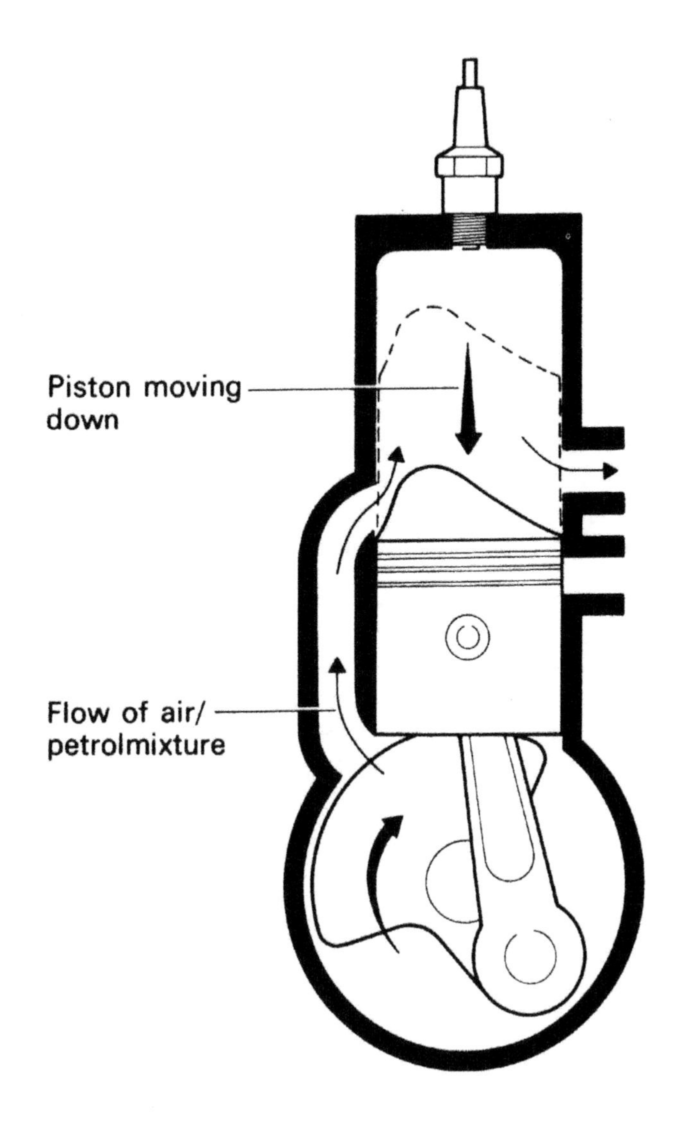

b *Two-stroke petrol engine: piston approaching BDC*

Figure 1.16 (Continued)

The two-stroke petrol engine is much lighter and simpler than the four-stroke petrol engine and has fewer moving parts. It has neither valves nor valve-operating mechanism. Two-stroke petrol engines usually run at high speeds. The problem is that oil must be mixed with the petrol for lubrication; this causes the exhaust to smoke.

Racer note

Two-stroke engines were used in the early karts and there are racing classes for these; however because of emission regulations no new ones are being made. Old racing two-stroke motorcycles were loved by many for the smell of the burnt oil – even though it could be a health hazard. The oil, *Castrol R,* was vegetable based – it is currently available for historic applications

Four-stroke diesel engine

The operation of the four-stroke diesel engine is very similar to the four-stroke petrol engine. The diesel engine draws in air only, and then compresses this to a very high pressure and temperature to cause the fuel to combust and burn. The fuel is injected directly into the engine at a very high pressure. Vehicles with diesel engines are very economical and they produce lots of pulling power at low engine speeds.

Racer note

There's lots of arguments about diesel cars; it is worth keeping an open mind about them and the use of them in competition. Most manufacturers produce a high-performance diesel variant: BMW, Mercedes Benz, Audi and Porsche as examples. The author uses one, and Audi won Le Mans with one.

Flywheel inertia

Because of their high compression ratios, and heavy moving parts, diesel engines usually have large flywheels.

Performance tuning diesel engines – diesel engines can be tuned in the same way as petrol engines – changing the inlet and exhaust and other parts; but such parts are not usually available as they are for petrol vehicles. The most cost-effective way of tuning a diesel is by ECU remapping. An example (Table 1.4) is that by Quantum Tuning, increasing BHP by 20% on a Mercedes Benz 2.2 litre diesel.

Table 1.4 Mercedes Benz 2.2 litre performance increase by ECU remapping. Courtesy of Kent.Quantum.co.uk

Original power	150 BHP
Tuned power	180 BHP
% increase in power	20%
Original torque	340 Nm
Tuned torque	410 Nm
% increase in torque	20.6%

Firing order

By increasing the number of cylinders, the engine becomes more compact for its size and smoother running. Smoothness of running is further improved by setting the sequence in which the cylinders fire; this is called the firing order. The normal firing orders for four-cylinder engines are 1-3-4-2 and 1-2-4-3.

Engine capacity

To find the capacity of an engine, first you need to find the size of each cylinder. This is called the swept volume – the volume that is displaced when the piston goes from BDC to TDC. The swept volume of each cylinder is the product multiplying the cross-sectional area and the stroke.

$$\text{Swept Volume} = \Pi\ D^2L / 4$$

$\Pi = 3.142$
D = Diameter of bore
L = Length of stroke
All divided by 4

The engine capacity is the product of the swept volume and the number of cylinders.

$$\text{Capacity} = \text{Swept Volume} * \text{Number of Cylinders}$$

You will find the following abbreviations:

V_s = Swept volume
N = Number of cylinders

The engine capacity is usually measured in cubic centimetres (cc).

> **Nomenclature**
>
> There are 1,000 cc in 1 litre. A 1,000-cc car is therefore referred to as one litre. American cars are sized in cubic inches (cu in). 1 litre is equal to 62.5 cu in.

Compression ratio

The compression ratio is the relationship between the volume of gas above the piston at BDC compared to that at TDC. You need to know the swept volume (Vs) and the volume of the combustion chamber that is referred to in this case as the clearance volume (Vc).

$$\text{Compression Ratio} = \left(Vs + Vc\right) / Vc$$

Road version petrol engines have compression ratios of between about 9:1–11:1. Race engines are up to about 20:1. Diesel engines are usually up to about 22:1.

Volumetric efficiency – in other words the efficiency of the engine of getting in fuel and air to fill the cylinder. The fact that an engine has cylinders of 1 litre (61 cu in) does not mean

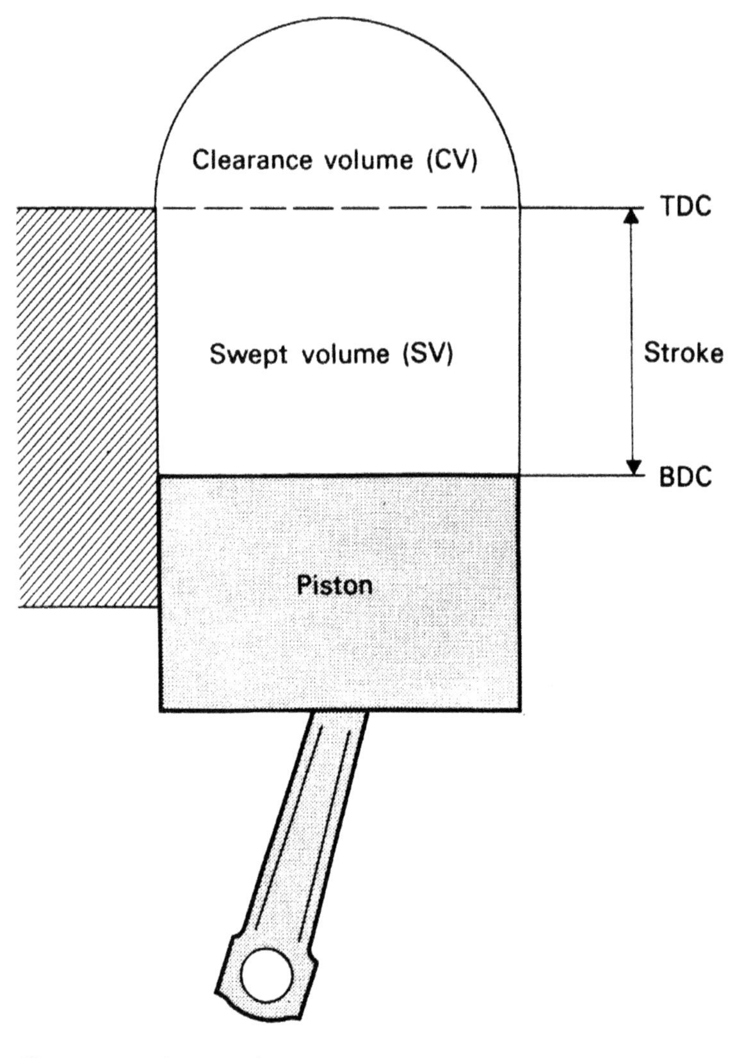

Figure 1.17 Compression ratio

that you are getting that amount of air and fuel into the cylinders. The flow of gas is affected by a number of factors, mainly:

- Size, shape and number of valves
- Valve timing
- Size and shape of the inlet and exhaust ports
- Shape and location of combustion chamber

- Bore to stroke ratio
- Engine speed
- Type of induction tract – air filter, carburetter or throttle body, inlet manifold
- Type of exhaust system – manifold, silencers, cat and pipe layout

The volumetric efficiency is calculated by measuring the amount of air entering the engine and finding this as a percentage of the actual volume of the engine. The formula is:

Volumetric Efficiency = Actual Air Flow / Capacity of Engine

The airflow can be measured on the test dynamometer by using either an airflow meter on the inlet or calculating the airflow from the pressure drop across an orifice. You would normally do this over a set time period so that the airflow is calculated in litres per minute (cu in/minute or cu ft /minute).

A typical engine might have a volumetric efficiency of 80%, a well-tuned naturally aspirated engine may have an efficiency approaching 90% with throttle bodies and all the right manifolds. With a turbo-charger or supercharger fitted that is pressurising the air into the cylinder, the figure will be over 130%.

The greater the amount of air that can be packed into the cylinder, the more power the engine is going to give out. The saying is, ***there ain't no substitute for cubes***. That is, the more cubic inches (litres) of air the more power the engine will develop. An example of an affordable engine with a high volumetric efficiency is the Kawasaki HS2; this gets 310 bhp from 1 litre.

Thermal efficiency – from volumetric efficiency you saw that you need to get the engine efficient in getting the air (actually air and fuel) into and out of the engine. That is one thing. Now when the air and fuel is in the engine it needs to be burnt efficiently to produce as much energy as possible to be turned into power and that conversion of burning gas into turning the crankshaft needs to be as efficient as possible to get the maximum power out of the engine:

Thermal Efficiency = Power Output / Energy Equivalent of Power Input

The power output is easy to measure on a dynamometer, either as an engine in a test rig, or the complete vehicle on the rolling road.

Racer note

Lots of drivers know the power output of their engines; very few know the efficiency of the same engine – greater efficiency means fewer refuelling stops and hence faster race times – in other words more wins

To calculate the energy input, you are going to need to know:

- Mass of air used – that is volume of air multiplied by the density. The density depends on altitude, temperature and weather conditions
- Mass of fuel
- Specific calorific value of the fuel – how much energy it produces for a given amount

The calculation for the mass of air follows on from the measurements of airflow from the volumetric efficiency calculations. The calculations for density may be made by either sampling or using tables. When carrying out calculations or comparing figures it is worth checking whether they are at STP or NTP.

STP – Standard temperature and pressure

STP – Standard temperature and pressure – is defined as air at 0 °C (32 °F) and 1 bar (14.7 psi).
Also STP – commonly used in the Imperial and USA system of units – as air at 60 °F and 14.696 psi.

NTP – Normal temperature and pressure

NTP – Normal temperature and pressure – is defined as air at 20°C (68°F) and 1 bar (14.7 psi).
Air density at sea level at NTP is 1.2 kg/m^3 (0.075 lb/ft^3).

Racer note

Have you ever noticed that on a cold, damp day the engine runs well and sounds well – this is because the air is at maximum density and the water cools the cylinder to prevent detonation.

The energy from fuel is discussed under fuel composition separately in this chapter. Examples of calculations are also given.

Racer note

Some simple changes, like chipping, may increase the power by 5% or 10% but increase fuel consumption by 40% – customers may have a shock at the increase in fuel costs.

Valve timing – is very important on a competition engine. The basic principles are to open the valves as quickly as possible, as great a distance as possible and for as long as possible. The problem is that if you do this, then the engine will need to run fast to maintain gas velocity, so it is likely to stall at low engine speeds, or at least produce low power. To this end many vehicles use variable valve timing.

If you fit a high-performance camshaft you need to ensure that the timing is exact – especially on engines where fully open valves may touch the piston crown. To this end a **vernier adjustable timing pulley** may be needed.

Cam design – there are several methods of cam design. These are: **constant velocity**, **constant acceleration** and **simple harmonic motion**.

Combustion

Flame travel – the combustion of the air and fuel mixture is a burning action – we often refer to it as an explosion because it is very fast; however the fuel burns with a flame that is travelling very quickly. At 6,000 rpm each cylinder fires 50 times each second; in a Formula 1 engine at 22,000 rpm this is 183 times per second. In this time the piston has completed four strokes, so the stroke time is a quarter of this. So the maximum duration of the burning, if it were between TDC and BDC, would be 0.005 s (5 ms) at 6,000 rpm and 0.0016 s (1.6 ms) at 22,000 rpm. On race engines the spark plug needs to be as near to the centre of the combustion chamber as possible to give even burning of the gas. The faster the engine speed is the more important this is. Flame travel should not be confused with spark plug burn time, which is the length of time that the plug is sparking.

Pre-ignition, **post ignition**, **pinking** and **detonation** are often confused as their symptoms and effects are almost identical; that is the air and fuel are burnt in a noisy manner, there is usually a knocking noise and there is a loss of power or uneven running. **Running-on** is associated with these symptoms too. Pre-ignition is when the fuel is ignited before the spark occurs – this is usually caused by something burning, or overheated in the combustion chamber. Often this is the spark plug insulator, a carbon deposit or a section of damaged cylinder head gasket. Post-ignition is when the fuel burns late, or more likely combustion continues after the engine is switched off – the engine continues to run for a while when the ignition is switched off – this is running-on. Pinking, also called **knock**, is the noise made by the mixture burning too quickly, or on two flame fronts, caused by the ignition timing being too advanced or too low octane rating fuel being used. **Detonation** is when pockets of fuel burn in an irregular way, usually because of poor air–fuel mixing, incorrect ignition timing or poor combustion chamber design.

Octane rating – there are two methods of measuring fuel anti-knock ability. These are:

- **research octane number** (RON) a measure of anti-knock during acceleration under medium load
- **motor octane number** (MON) a measure of anti-knock during acceleration under heavy load

As neither are perfect measures, an average of the two is often used. This is called the **anti-knock index**:

$$\text{Anti-knock Index} = (\text{RON} + \text{MON}) \ / \ 2$$

Cetane rating – is the diesel equivalent of octane rating.

Think safe

Petrol is highly flammable – race petrol is especially volatile.

Cylinder head

Valve layout – the valve layout for high-performance engines is designed to get the maximum amount of air and fuel into the engine, and then out again as exhaust gas. The most common design uses four valves, the inlet valves being larger than the exhaust valves as the exhaust gas is forced out by the piston under high pressure. The gas flows from one side of the engine to the other to reduce the time taken; this allows an increase in engine speed.

Cooling – for good cooling, aluminium cylinder heads are almost universal – they dissipate the heat very quickly. Coolant (water) jackets may be opened out to improve coolant flow.

Valves – the use of a variety of materials is used to give longer valve life and better cooling. Typically, high chromium alloy steel to resist corrosion and give wear resistance. Valve head shapes tend to be with thin heads – often called penny on a stick – for minimum weight and less mass for heat build-up. Variations of seat angles and seat materials are used to give good sealing.

Racer note

The thinner the valve seat the tighter will be the seal; conversely, the wider the seat the better the heat transfer from valve to cylinder head.

With aluminium alloy cylinder heads, seat inserts are needed to give the required hardness. Iron cylinder heads may also have inserts for competition use – these are often fitted at the same time that the cylinder head is machined to fit larger valves. Valve seat inserts may be screwed – now rare – or press fit into the cylinder head.

To improve thermal efficiency, two factors of combustion chamber design should be considered. These are: **swirl ratio** and **surface to volume ratio**. The formulas are quite self-explanatory. They are:

Swirl Ratio = air rotation speed / crankshaft speed

The higher this figure is the better the engine will run – look for figures over 5.

Surface to volume ratio = surface area of combustion chamber/volume of Combustion chamber

The lower this number is the better the engine will run – a sphere will give the best ratio.

Valve springs – double or triple valve springs may be used to enable high revs.

Camshaft – When building a competition cylinder head, you should pay attention to the fit of the camshaft(s). If possible use white metal bearings and line bore to ensure perfect straightness.

Racer note

Turn the camshaft like the crankshaft when setting it up; that is, ensure that it turns freely at each step of tightening the caps, or when inserted – you will need to do this before fitting the valves.

Standard camshafts tend to be cast iron or low-grade steel – for competition use you will need a high-grade alloy steel. These can be hardened by **nitriding** – lower quality competition camshafts may be **tockle hardened**, **tufftrided**, or simply coated to give a better wearing surface.

Camshaft drive – for competition use high tensile bolts will be used.

Gas flow – this is a commonly misused term, or used as a catch-all for anything to do with cleaning and polishing cylinder heads – gas flowing. The volume and velocity of gas flow through a cylinder head is measured on a **flow bench**. The cylinder head is attached to the flow bench so that the air flowing through it can be measured. The exhaust valve is kept firmly shut whilst the inlet valve is opened in small increments and the flow recorded. The same can then be done with the exhaust valve. The modifications are then made to the relevant part of the cylinder head. Typically the inlet port is smoothed and polished, then the flow is measured again. It is not the absolute flow rate that is important – this may be difficult to measure in a finite way – the percentage improvement is the indicator looked for. A 5% improvement is very good. The factors effecting gas flow are:

- Port shape
- Port finish
- Manifold to port fit – do they align?
- Gasket fit – does it over lap the port?
- Valve seat angle
- Valve size
- Valve opening

Figure 1.18 Stones for polishing cylinder head ports

Short block assembly

Materials – for lightness and cooling dissipation, aluminium alloy is the obvious choice – but this will depend on the racing class. To improve aluminium block strength and improve wear resistance **cryogenic treatment** is often carried out – check the regulations to see if it is allowed for the class. This involves taking the temperature of the block down to a temperature of about −180 °C (−300 °F) for a period of about 48 hours. This smooths out the grain structure of the metal making it stronger.

> ### Racer note
>
> You can cryogenically treat any part of the car – brake discs (rotors) and pads are a common and economical choice.

Liners are used in aluminium blocks to give the essential resistance to wear. They may be **dry liners**, or **wet liners**. Wet liners are the most common to give easy-to-make open blocks. If you are cytogenetically treating the liners you will need to bore them out about 0.002 in (two thou) to return them to true round – you must bear this in mind if you are blue printing the engine.

Figure 1.19 Cylinder block – open type

Racer note

Never hit liners, even with something soft – they easily distort.

Drillings – when building a race engine check the drillings and passage ways to ensure that they are clear and the correct size to give sufficient water and oil flow.

Crankshaft – made from high-grade alloy steel with a high nickel content so that is can be nitrided, this involves leaving it in a bath of ammonia at 500 °C (900 °F) for about 24 hours.

When fitting the crankshaft ensure that it turns after each bearing cap is tightened. A very small amount of very light (5 SAE) oil should be placed on each journal and bearing shell; but ensure that there is no oil behind the shell. Also check for crankshaft end float after fitting the thrust bearings. End float must be the minimum to ensure con-rod alignment and clean clutch operation – remember the clutch thrusts against the crankshaft thrust bearings.

Con rods – increasingly made from carbon fibre composite as this method allows relatively low-cost batch production. The alternatives are high strength aluminium and titanium. Consider the speed at which they are travelling and you will appreciate the need for aerodynamics in their design. Changing the engine stroke or the piston height may mean changing the con rods:

Bore – stroke ratio – this is simply the ratio of the two dimensions

$$= \text{bore} / \text{stroke } (\text{mm or inches})$$

If the bore is larger than the stroke, the engine is said to be an over square engine; if bore and stroke are the same it is said to be to be a square engine; if the stroke is greater than the bore then it is called a long stroke engine. Generally, race engines are over square and rev at high speeds – this can be over 20,000 rpm. Older engines tend to be long stroke, being lower running and producing more torque than ultimate power.

Balancing – pound (dollar) per horse power, balancing an engine gives the best return. Not only does it ensure smooth operation, and longer life, it also allows higher revs at the top end. Balancing breaks down into two parts – these are **static** and **dynamic**.

Static balancing can be done on the bench with a simple weigh scale. Dynamic balancing needs a specialised machine. To carry out static balancing you start with the pistons. Weigh each piston and note its weight. Now remove material from the heavier ones until they all weigh the same. Reduce the weight of the heavy pistons by removing material from the bottom inside of the skirt or a relatively unstressed part of the boss.

Racer note

If you have access to more than one box of pistons of the same size weigh them all – you may find the number that you want that are all the same to save machining.

Next are the con rods. Support the little end on a piece of wire (or similar) and weigh the big ends on the scale. Do the same supporting the big end. Remove metal by drilling small holes until they are all the same weight. Weight can be added on steel rods by drilling small holes and filling them with lead – make sure that they are very clean and use flux so that there is no risk of the filler coming out.

You can statically balance the flywheel and clutch by assembling them onto the crankshaft, supporting the crankshaft in vee blocks and spinning, then wait until it stops. The heavy part will be at the bottom. Again, remove weight by drilling small holes in unstressed parts. When it is balanced you should be able to stop the assembly at any point without it rotating.

Properly statically balanced – dynamic balance should be easy – indeed if on a budget, static balance only will give very good results.

Surface finish – surface finish is important for mating surfaces and surfaces which are subject to air or other fluids flowing over them. On a race engine, good surface finish makes the engine look good – a very important point. Nicely polished parts make the engine attractive.

Blue printing – this is another often misused term. It simply means building the engine to the manufacturer's specification – in other words the blue print – or drawing. Final designs were printed on large sheets of paper in blue colour. As we now use CAD this means that we no longer print drawings in blue. When blue printing you should pay detailed attention to each part fit, and ensure that it is the correct size within design tolerance.

Induction

Gas laws – air is not always what you think it is. You may think that air goes through the engine like sausages through a sausage machine. You may think that a chunk of it goes into the cylinder to be processed, and then it comes out as exhaust gas. Certainly, air is mixed with petrol; it does go into the cylinder; it is burned and it does come out as exhaust. However, the gas is travelling at several hundred miles per hour (hundreds of kilometres per hour). And it travels in waves, not in sausage like lumps.

Induction systems fall into three categories. These are:

- **Open induction**, usually the carburetter, or the throttle body has an open ram pipe
- **Closed induction**, this uses a plenum chamber with air filter
- **Forced induction**, as with a turbo-charger, or supercharger

The **ram pipe** length can be changed to suit the engine operating speed – this is very much a trial and error activity. The plenum (Latin for full) chamber is designed to give the engine a supply of still air. With forced induction air is in effect pumped into the cylinder under pressure.

The ram pipes are guiding the air into the cylinder, and helping to control the waves.

Racer note

Changing from plenum chamber to ram pipes can add 5% to the power output of some engines.

Exhaust

Exhaust gas travels in the same way – that is it come out in waves; the waves are generated by the piston, in the same but opposite way to the induction waves. The piston going up the bore pushes the gas out, not as a sausage, but as a pulse. The pulse velocity depends on a number of factors, which we will look at later in this section. First of all, let us look at the speed of sound – exhausts as you know make a sound, so the pulse wave is travelling at the speed of sound. The speed of sound varies with a number of variables.

Historic racer note

Race cars and bikes at Brooklands had to have their exhaust tail pipe extending past the rear wheel spindle, to satisfy the complaining residents of nearby St Georges Hill – this led to fancy fish tail silencers.

The **speed of sound** in air (C) is calculated from the heat capacity (γ) of air (1.4) the air pressure (p) and the air density (ρ):

$$C = \sqrt{\gamma\ p\ /\rho}$$

Example

At NTP

$$C = \sqrt{1.4\ *\ 101000 / 1.2} = 343\ \text{m/s}$$

The speed of sound varies with temperature (T) in the ratio:

$$C2 / C1 = \sqrt{\ T2 / T1}$$

Taking T1 as NTP of 20 °C and C1 as 343 m/s and remembering that this should all be in absolute temperature form Kelvin (K). At 800 °C the speed of sound will be:

$$C2 / 343 = \sqrt{\ 800 + 273 / 273 + 20}$$
$$= 656\ \text{m/s}$$

Now we have the speed of sound in the very hot exhaust. This allows us to work out the length of the exhaust **primary pipe** to suit the engine speed. You'll see lots of variations of exhaust manifolds and systems. The primary pipe length from the exhaust valve to the 'Y' joint into the rest of the system; or the open pipe on a dragster is important to give maximum power at any particular speed. The reason for this is that as the exhaust valve opens it sends a pulse wave at the speed of sound – we calculated that – down the primary pipe. When it gets to the end of the primary pipe and the gas can expand, the sound wave travels back down the pipe to the exhaust valve. This then returns back to the end of the primary pipe.

Try this

Think of the exhaust gas as a wave on the sea shore – it goes in and out even though the tide is coming in all the while.

If you can get the primary pipe to be the length that will be just right for gas to go up and down between the exhaust valve opening and closing, it will leave a negative pressure (vacuum) at the exhaust valve as it closes. This will scavenge (clean out) the exhaust gas from the engine most efficiently. So, let's look at how we do this. We know that time taken to cover a distance is that distance divided by the velocity:

$$\text{Time}\left(t\right) = \text{distance}\left(d\right) / \text{velocity}\left(v\right)$$

The distance is from the exhaust valve to the end of the primary pipe and back – we'll call this 2L. The velocity is the speed of sound (C). The time varies with engine speed and valve period – the number of degrees the exhaust valve is open, for example 120 degrees. 120 degrees is 0.333 of a revolution (360 degrees). At 6,000 rpm this equates to 100 rev/sec, so 0.333 of a revolution will take 0.00333 seconds, that is: 0.00333 / 100 = 0.00333. We calculated that the speed of sound at 800 °C is 656 m/s. So the distance travelled by the sound wave at 656 m/s in 0.00333 s is:

$$0.00333 * 656 = 2.18 \text{ m}$$

The length of the primary pipe is half of this – remember it is down and back – so it is 1.09 m.

Try this

Calculate the length of the primary pipe if the engine speed is 10,000 rpm.

Historic racer note

The same principle applies to the induction, which is why you see old racing cars with long inlet manifolds and long ram pipes.

Helmholtz theory – if you look at the inner tube in a silencer you see that it is drilled with lots of holes. Helmholtz worked out that if you chop the end off a wave it will reduce its energy, in other words the noise that it makes. So, as the sound wave passes each hole in a silencer, a part of it goes through the hole reducing the energy and hence the noise.

Think safe

Exhaust gas is both hot and poisonous.

Modifications

Caution

The first four paragraphs in this section apply to all the topics in this book. If you carry out any modifications to any vehicle, whatever happens is your responsibility.

Health and safety and environment

As with any other task, you must be aware of health and safety and environmental issues.

Legal implications

It is important to be aware of the legal implications of any modifications that you make to a vehicle. It is not illegal to sell many parts that when they are fitted to a road-going vehicle could be illegal. Also, on competition vehicles, the parts must comply with the requirements of the racing regulations relating to that particular type of racing, or specific class. As a technician, under corporate law, which is vicarious by its nature, if a vehicle to which you have fitted a part is involved in an incident, and the part that you fitted might have been causal towards the damage, then you may be held wholly, or partially, liable for the damage caused. That damage could be the death of an innocent person, in which case you could be charged with manslaughter.

So, think carefully about the modifications that you are making. Do not just do it because a customer asked you to do it. Customers are often unaware of the laws, and indeed racing regulations. It is your duty to advise and guide them; you have a duty of care to your customers. Table 1.5 lists some of the commonly broken regulations or problems.

Table 1.5 Common mistakes in modifications

No	*Area*	*Modification*	*Mistake*	*Comment*
1	Tyres	Racing tyres	Slicks with no tread	Illegal on road
2	Tyres	Asymmetric tyres	Wrong direction of rotation	Unsafe in wet
3	Wheels	Wide wheels	Extending beyond wheel arches	Illegal on road or track
4	Lighting	Fitting spot or fog lamps	Incorrect height or position	Illegal on road

(Continued)

Table 1.5 (Continued)

No	*Area*	*Modification*	*Mistake*	*Comment*
5	Lighting	Any lamps moved by changes to bodywork	Moved or restricted vision	Check that they comply with position on vehicle and for angle of vision
6	Exterior fillings	Bolts or screws on exterior of bodywork	Must have a minimum of 2 mm radius and no sharp edges	Common reason for Kit cars failing SVA
7	Exhaust	Change system, or parts	Noise limits	Especially appropriate to rally cars
8	Ignition	Change parts	May alter engine operation	Could affect emissions
9	Fuel system	Change parts	May change fuel emissions	Check within limits
10	Brakes	Changing pads	Could alter braking characteristics	Lots of drivers are surprised to find the extra effort needed with competition brake pads
11	Engine	Fit a bigger/more powerful one	Need to upgrade the brakes and suspension too	I admit to doing this – wondered why it took a long time to stop – very dangerous
12	Suspension	Fit lowered springs	Need to fit shock absorbers to suit	Suspension will bottom
13	Tow bar	Incorrect attachment	Need appropriate fitting kit	Don't forget the electrical connections too

Some of those points should make you smile – but they are all potentially dangerous and most seriously to be avoided; think the modification through, think safe.

Reasons for modifications

The reasons why people modify, or enhance, their vehicles are many and varied; but usually they can be classified as one of the following:

- Making the vehicle faster
- Making the vehicle more powerful
- Making the vehicle lighter
- Making the vehicle handle better
- Making the vehicle stop more quickly
- Making the vehicle more comfortable to drive – especially under specific conditions, such as extra lights for night driving
- Making the vehicle look good – attractive
- Making the vehicle comply with specific racing regulations

A good motorsport technician will know all the right tweaks to enable the car to perform better than the other competitors.

Cylinder head

Cylinder heads may be made from cast iron (CI) or aluminium alloy. The aluminium alloy ones are about a third of the mass (weight) of the equivalent CI ones. Aluminium is a better conductor of heat than CI and therefore more suitable for high-performance engines that produce more heat for the equivalent capacity. Remember that when petrol and air, or diesel and air, is burnt it produces heat. About 40% of the heat does useful work in pushing the piston down the bore; 30% goes to the coolant jacket and 30% to the exhaust. If you increase the power output of the engine, then you are also going to increase the heat to the coolant and the exhaust. Changing to an aluminium cylinder head will improve the rate of heat conduction and so help to prevent engine failure through overheating. If you are changing a cylinder head for this reason, it is also worth checking and if necessary changing the:

- Hoses – high pressure
- Hose clips – high clamping pressure
- Coolant – inhibitor for use with aluminium
- Use of wetting agent
- Radiator flow capacity
- Coolant (water) pump – flow rate
- Coolant pump drive belt and velocity ratio
- Cylinder head gasket – check material, thickness and use of rings
- Cylinder head studs or bolts – need high tensile strength ones

Both types of cylinder heads may be modified to accommodate larger valves by machining; the valve seat angles changed, the compression ratio increased by skimming; and the ports smoothed and polished to improve gas flow.

Valves

The purpose of the valve is to open and close to control gas flow. It's opening and closing speed is limited by its mass – remember Newton's Second Law, $F = Ma$. So the valve needs to be as light in weight as possible; for this reason titanium is frequently used. The higher the engine output the higher the amount of heat that is generated. So the valves need to be able to dissipate heat – for this reason sodium filled valves are sometimes used as the latent heat of the sodium filling serves to reduce the valve temperature.

Valve shape is also of consideration, in terms of valve seats and head shape to improve the flow of the gases both into and out of the cylinder head.

Valve springs

The valve springs close the valves. The camshaft opens the valves, so the faster the engine revs, the faster they will open. Remember that if we can make the engine rev faster it will probably be able to develop more power – within certain restrictions. So, we are now dependent on the springs closing the valves – if we are increasing the engine speed we will need to fit stronger valve springs. Otherwise the springs will not close the valves fast enough and the engine will suffer from valve bounce – indeed the valves may touch the piston crown. To modify the valve spring closing rate there are usually three options – stronger single springs, double springs or, if double springs are already fitted, triple springs can be fitted.

Fitting double or triple springs may only be possible if the valve retaining caps are changed to match and new springs and the spring seat on the cylinder head may need machining to accommodate the new springs.

Camshafts

The camshaft controls the valve opening in terms of both lift and period. That is:

- Valve lift – the distance the valve is lifted form its seat. Like opening a door – the greater the lift the more gas that can be put through at any one point of time. The valve lift may be the same as the lift at the cam, or there may be a ratio so that the valve lift is greater than the cam lift – when using a rocker mechanism. Direct acting cams do not have this option.
- Period (also called duration) is the number of degree that the valve is open – measured between the opening and the closing points. The number of degrees will correspond to a period of time for any given engine speed. Again – like a door – the longer it is open the more petrol and air that can get through it. If the valve period is increased from say 100 degrees to 110 degrees it will be open 10% longer, so 10% more gas can pass through it. This is often simply referred to as valve timing or just timing.

Figure 1.20 Adjustable camshaft drive wheels

There are three main options:

1. Change the camshaft for one with different opening periods and/or lift.
2. Change the rocker mechanism ratio.
3. Change the cylinder head and camshaft for one giving a different location – for instance fitting a double overhead cam (DOHC) cylinder head in place of an overhead valve (OHV) arrangement.

Also, to be considered is the material from which the camshaft is made. Many popular vehicles use cast iron because it is easily made, cheap and sufficiently hard for normal usage. For competition use it is normal to use a high grade of steel with sufficiently high carbon content so that it can be induction hardened.

Camshaft drive

Camshaft drive is usually toothed belt or chain with a small number of vehicles using gear drive. Table 1.6 compares the different types.

Cylinder block

Cylinder blocks are usually the subject of detailed regulation for most competition classes – because the block is usually the largest mechanical component and limits the power through its capacity (swept volume), configuration and general structure.

Material – aluminium alloy is the best choice for low weight and thermal conductivity. This may be either die-cast or sand-cast. The grain structure of aluminium alloy engines can be improved by head treatment – this can include deep freezing for several days.

Capacity – changing the bore and/or the stroke will alter the engine capacity. Usually there are competition limits on over boring, and over boring will weaken the structure of the engine.

Configuration – that is number and layout of cylinders. Now before you say that not much can be done about this point, it must be remembered that the first Cosworth V8 was made from two four-cylinder engines put together, and SAAB made a range of engines using the same components in different configurations.

Table 1.6 Camshaft drive comparisons

No	*Type*	*Advantages*	*Disadvantages*	*Comment*
1	toothed belt	simple, no lubrication needed, adjustable, quiet in operation	may break unexpectedly	most popular
2	chain	long lasting, adjustable, wear clearly visible	needs constant lubrication, expensive	
3	gear (metal to metal)	long life	noisy, expensive, needs constant lubrication	
4	gear (metal to fibre)	quieter than (3) and does not need lubrication	fibre gears need frequent replacement	used on racing motorcycle engines

Liners – these may be of the wet type or the dry type. The liners may be coated or plated to reduce friction or increase life. One common material is chrome – its shiny surface reduces friction and therefore can increase power output. The liners may also be made out of different materials; steel is normal for motorsport vehicles. If a cast iron cylinder block is being over bored – taken to a size above normal rebore limits – then it is often good practice to machine it out even more, then to add steel liners. This gives better durability and ensures a minimum cylinder wall thickness.

When new dry liners are fitted into a block, the fitting is usually an interference fit, that is, a press fit. This may cause irregularities on the inside of the cylinder – these are removed by honing with a very fine grade oil stone. Of course, this procedure cannot be carried out with coated liners.

Deck height – the height of the top of the cylinder block in relation to the piston crown. The deck height affects the compression ratio. A minimum deck height of about 0.2 mm (0.010 inch) (always check manufacturer's figures) is needed to prevent the piston crown touching the cylinder head due to the dynamic forces at maximum engine speed.

Sealing – increasing the engine's power output will put a greater pressure load on the engine's seals. Some engines are built without gaskets; as gaskets are used to allow for uneven surfaces these engines are machined very accurately. A sealant is used between the mating surfaces.

Cylinder head gaskets may be replaced on motorsport engines by ones made from different materials – frequently this is one sheet of malleable metal. Where the compression ratios are very high the block may be machined with annular rings around the cylinder bores to accommodate sealing rings.

The sump and other gaskets may be made from materials that are more able to cope with the high pressures and temperatures of competition engines. They will also be made to more exacting levels of accuracy than standard road versions.

Bearings – the crankshaft main and big-end journals are likely to be harder than those of road cars, so the bearing surfaces must be able to work with the materials. Normally the bearings are harder, containing more tin and less lead to make up the white metal alloy.

The bearing caps may also be upgraded by either fitting steel caps that are then line bored to match the crankshaft, or fitting steel straps over the existing caps to increase their strength.

Pistons

For high performance purposes, the piston is a very important part of the modification process. Table 1.7 sets out the characteristics of piston design.

As an overview, there are only a small number of companies that produce pistons in the UK and USA and, notwithstanding the Asian market, they tend to produce for the worldwide motorsport range too. This is because piston production is very specialised precision engineering requiring expensive machine tools and highly skilled staff. The motorsport piston supply is largely related to modification of standard piston – although this is done at the manufacturing stage, not as a post-production change; or, the application of a piston to one engine, that is used as standard in another engine. If you think of the Ford range of engines that come in various capacities and are used in different models you may get a better picture.

Connecting rods

The most common modification to connecting rods (con rods) is balancing them. There are a number of different ways of balancing con rods. The basic principle is to ensure that they all

Table 1.7 Characteristics of piston design

No	*Part of piston*	*Design characteristic*	*Reason*	*Comment*
1	Crown	Height above gudgeon pin	Alter compression ratio and relation to deck height	
2	Crown	Shape	Changes to swirl and cut outs for valve pockets	Valve pockets may need to be machined to cope with extra valve lift
3	Gudgeon Pin	Diameter	To cope with extra load	This will also involve modifications to the con rod
4	Gudgeon pin	Type of fit	To suit con rod	Types of fit include fully floating and interference
5	Skirt	Shape	Slipper skirt to reduce friction with cylinder wall and reduce weight	
6	Skirt	Design	Solid skirt to increase strength	
7	Skirt	Length	Short skirt to reduce friction with cylinder wall and reduce weight	
8	Rings	Number	Fewer rings – two or three – to reduce friction with cylinder wall	
9	Rings	Type	To enhance combustion sealing	Ring material may also be changed
10	Material	May be forged, billet or cast from a variety of aluminium alloys	will affect weight, strength and machining processes	Pistons can be made any size or design; but if non-standard will cost accordingly

weigh the same amount, and that the weight of the top (little end) and the bottom (big end) is also equal on each con rod. This is done by supporting one end and weighing the other. To lighten the con rod, metal is moved by drilling holes in lesser stressed section. Weight can be added by drilling holes and filling them with lead, which is much denser.

The con rods offer resistance to the engine, both from their mass – resistance to accelerate – and the air resistance moving inside the crankcase. The acceleration is improved by using lighter materials. In ascending order of cost the options are: aluminium alloy, titanium and carbon fibre. The aerodynamics are improved by making them an aerodynamic shape.

An economical improvement can be made to the original con rods by balancing, shot blasting smooth then polishing.

If the stroke length is being changed by changing the crankshaft – a very popular modification on many engines, then it may be necessary to change the con rod so ensure that the piston operates in the correct area of the cylinder bore.

When building engines, no matter which type of con rod is being used – you should:

- Check the oil way drilling
- Check the little end fastening

- Check for straightness
- Check the big-end fitting, especially the cap and retaining nuts and stud or bolts

More big-end caps come loose than con rods break in the middle. Use the correct fastenings and torque to the correct setting. As each piston and con rod is added to the block, check that the engine turns freely. Also ensure that the little end fit is correct – as if it is not the piston is likely to break at this point. That is, the piston boss breaks away from the piston skirt and the engine is likely to be wrecked.

On very high-performance engines the con rods should be seen as a service item – that is replaced at each rebuild along with the bearings.

Crankshaft

Materials – the crankshaft is one of the main components in limiting the engine speed and power output. Standard crankshafts may be made from either cast iron (CI) or steel. The steel is likely to be a forged medium carbon variety. For high-performance applications the steel is usually upgraded to one that will accept surface hardening readily. This usually means a slight increase in carbon content and the addition of other metals such as chromium, silicone, copper and aluminium.

The two main types of surface hardening, to increase crankshaft life by reducing wear, are:

- **Nitriding** – that is immersing in a *salt bath* – that is a solution of nitric acid (very hazardous) for a controlled (long) time period at a pre-set temperature. This is usually applied to more expensive steel alloys and increases the strength of the crankshaft.
- **Tufftririding** – a cheaper option for use with cheaper steels and only giving a surface hardness. This involves immersion of the crankshaft into a bath of (very hazardous) sodium cyanate at 570 °C for two hours.

Design – the design of the crankshaft has a number of factors for consideration when modifying the engine. Looking at the main ones:

- Configuration – this affects the firing order and balance; on six- and eight-cylinder engines there are a number of options
- Webs and counter balance – for high-speed engines these are removed to make the crankshaft light; balance becomes dynamically controlled
- Balance – done on a balancing machine both statically and dynamically, usually in conjunction with the flywheel, con rods and pistons
- Thrust races – to prevent longitudinal movement, thin needle roller bearings may replace the plain metal ones to reduce friction
- Oil ways – the oil passage ways are cross-drilled, that is extra holes are drilled across the crankshaft to ensure a good oil supply to the big-end bearings

Flywheel

As the purpose of the flywheel is to keep the engine turning between firing strokes, the faster the engine runs the less the flywheel inertia is needed. So faster running engines, and ones with more cylinders, require flywheels with less mass (weight). The less the mass of the flywheel the faster the engine will be able to accelerate. However, this will lead to uneven running at low engine speeds – not a problem on race cars.

The flywheel also locates the starter ring gear and provides one surface of the clutch assembly.

Modified road cars – the process is to maintain the original flywheel, lightening it by removing surplus material from less stressed areas like the back of the flywheel towards the outer edge.

Full race cars – the flywheel can be replaced by a thin sheet of plannished plate (usually hammered or rolled to be staff and flat) with a flanged edge to accept the starter ring gear, and a small diameter multi-plate clutch pack used that does not need the flywheel as a surface for friction.

Balance – the flywheel will be balanced both statically and dynamically in conjunction with the crankshaft and other rotating parts.

Be aware – changes to the flywheel may have an effect on dynamic balance and it may also be necessary to look to the crankshaft front pulley, TV damper and any counter-balance shaft mechanism.

Questions and skills

1 Draw up a detailed specification table for your favourite car, including items like power, torque, BHP per litre and BHP per tonne.
2 Make a list of what you consider to be the most important modifications needed to a standard road car.
3 Calculate the bore/stroke ratio of an engine.
4 Calculate the compression ratio of an engine.
5 Use a workshop manual to carry out any service/repair/modification on an engine.

Chapter 2

Ignition and fuel

The ignition system and the fuel system are separate systems, but they work closely together, almost as one. We are starting off with the systems found on classic cars for two reasons: classic car racing is very popular and accessible and they are a good way of understanding the practicalities of what goes on.

The purpose of the ignition system is to provide a **spark** in the combustion chamber of the SI engine which will **ignite the mixture** of petrol and air whilst it is under pressure. As the piston compresses the petrol–air mixture on the compression stroke the pressure may be increased to over 2,000 kPa (300 psi). The voltage needed for the spark to jump across the spark plug gap at this high pressure is about **40 kV** (40,000 volts).

Figure 2.1 Induction system

Key points

- There are several different types of ignition system; the Kettering System is the simplest to understand
- The main components of the Kettering System are the battery, coil, spark plugs and distributor
- Ignition timing can be checked with the use of a small lamp
- Electronic ignition does not use cb points, it is more reliable and gives better fuel economy

Kettering System

There are many different types of ignition system; however, the original system, which has been used for about 100 years, is the Kettering System, designed by Dr Kettering. The fully electronic and other systems are easier to understand if you first understand the mechanical-electrical Kettering system. The main components of the Kettering System are the **battery**, **coil**, **spark plugs** and **distributor**. Let's look at these components individually and see how the system works.

Racer note

Lots of kit cars and racing specials use the Kettering System.

Battery

The battery is the **source of electrical power** for the ignition system (and other systems too); it is usually located in the engine compartment. The battery supplies electrical power to the ignition switch. The battery has a nominal voltage of **12 volts**; the ignition circuit draws about 0.5 amp. The battery is usually connected negative side (-) to the chassis earth and positive side (+) to the main **fuse box**.

Racer note

Wiring on race cars and bikes is usually through a separate isolator switch, and may incorporate a connection for a trolley-mounted starter battery, or power pack.

Safety note

Battery acid is corrosive – mind that you do not spill it.

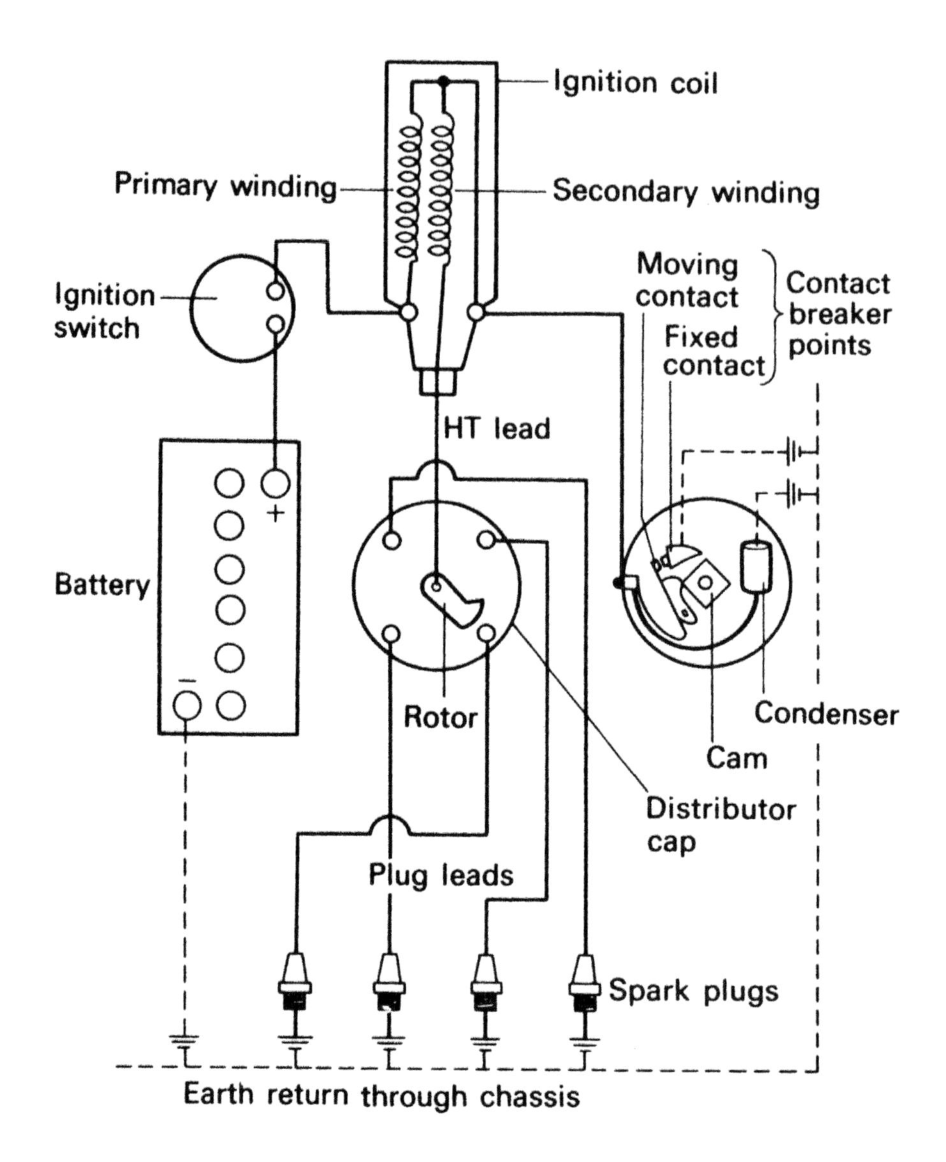

Figure 2.2 Ignition system

Ignition switch

This is connected to the battery through the main fuse box. The ignition switch makes and breaks the ignition circuit. When the engine is not running the switch must be off to disconnect the electrical power supply and so prevent the coil from overheating. Switching the ignition off also switches off other related circuits and prevents the battery from being **discharged**. The ignition switch is also a **security device** to prevent the vehicle from being stolen or used unlawfully. The ignition switch is combined with the steering lock, for added security and the convenience of needing only one key for both the steering lock and the ignition switch. The ignition switch/**steering lock** is situated on the side of the steering column near to the dashboard within easy reach for the driver. The ignition switch on road cars also incorporates the starter switch.

Racer note

Race cars usually have separate ignition switch and starter button which are covered when not in use to prevent accidental operation.

Ignition coil

The ignition coil is a kind of transformer. It changes the **low tension (LT)** 12 V from the battery to a **high tension (HT)** 10 kV at the spark plugs.

Nomenclature

Low tension (LT) refers to the components of the ignition system, which operate at a nominal voltage of 12 V. High tension (HT) is that part of the ignition system which operates at several thousand volts. The HT can range from 5 kV for an older car, to 40 kV for a motorsport vehicle. We refer to the battery as having a nominal voltage of 12 V. We say this is a 12 V battery though the actual voltage may be between 9 V and 16 V depending on a number of factors. A kilovolt is 1,000 volts.

The coil has three electrical terminals, one that goes to the ignition switch, one to the distributor body and one to the distributor cap. The ones to the ignition switch and the distributor body are low tension; the one to the distributor cap is high tension.

Safety note

The high voltage spark at the spark plugs, and associated leads, can kill you. If the system is wet, or you touch a bare cable or component, the shock will cause you to react and you may hit your head, or elbow, on another part of the car, which could result in a serious injury.

The connection to the ignition switch is the 12 V power supply. The connection to the distributor body goes to the contact breaker points, which carries out the switching action. The HT to the distributor provides the spark for the spark plugs.

The ignition coil operates on the principle of **difference of turns**. That is, it has two separate **windings** wound around a **soft iron core**. The **primary winding** is connected to the distributor LT; the **secondary winding** is attached to the distributor HT. The secondary winding has many more turns of wire than the primary winding; the increase in voltage from LT to HT is proportional to the difference in the number turns of the wire.

Ignition coils are usually filled with oil to improve cooling. Be careful not to damage the coil case as this is usually soft aluminium and, if damaged, may allow the oil to drain and the coil overheat.

Spark plugs

The metal end, or **body**, of the spark plug screws into the cylinder head so that the **electrode** protrudes into the combustion chamber. The screw thread with the washer and mating surfaces form a gas-tight seal, so preventing the loss of compression from the cylinder.

The spark plug has two electrodes; a **centre electrode** and a **side electrode**. The terminal end unscrews for use on motorcycles and agricultural machines, but on cars the **plug leads** are usually a push fit.

The diameter and the reach of the spark plug vary from engine to engine. The most common diameters are 10 mm, 14 mm and 18 mm. The common reaches are 3/8 inch, ½ inch and ¾ inch. It is important that the correct reach of spark plug is fitted to the engine, otherwise the electrode may foul against the piston crown. The diameter will only fit the tapping size in the cylinder head. The spark plug can be identified by the letter and number code that is printed on either the **insulator** or the body.

FAQs

How can I check that one make of spark plug matches another?

Spark plugs are made by a number of different companies, for example Champion, Bosch and NGK. Most motor factors and accessory shops have lists of equivalents. These lists show the various makes and model of cars and tabulate the code numbers for the different spark plug manufacturers.

When spark plugs are replaced, typically at between 16,000 km (10,000 miles) for old vehicles and 60,000 km (40,000 miles) for modern vehicles, they must be replaced with the correctly coded ones. The spark plug manufacturer's list, or the workshop manual, should be used to check the exact code for the model. It is common for different specification, or year models, of vehicle to use different spark plugs.

Servicing spark plugs is limited to cleaning and gapping between replacements. If cleaning is needed then a special machine is needed. Gapping the plug means setting the size of the **gap** between the fixed and the side electrode. Spark plug gaps are usually between 0.020 and 0.040 inch (0.5 and 1 mm). New spark plugs are usually ready gapped from the factory; but it is worth checking them before fitting. Plug condition is usually indicated by the engine analyser test.

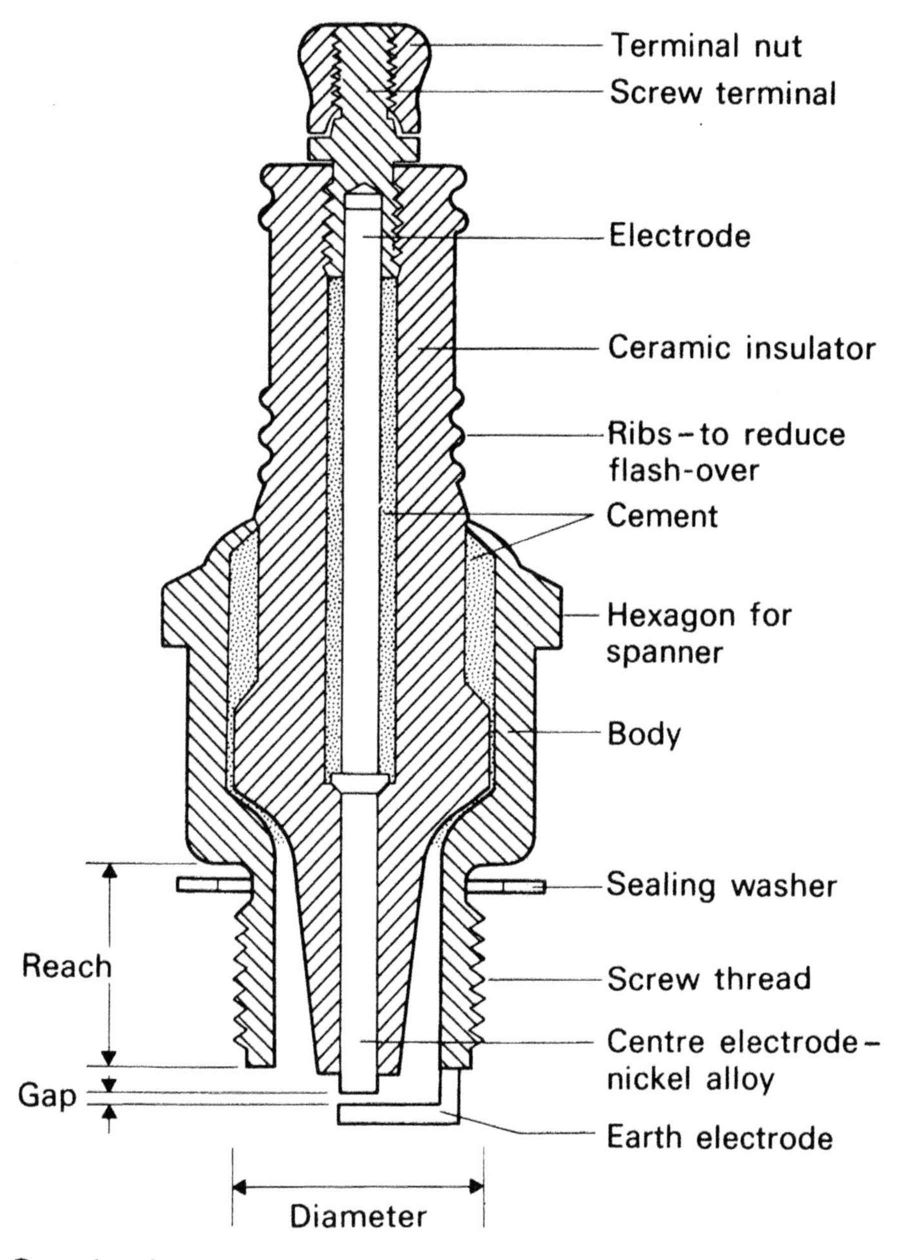

Figure 2.3 Spark plug

Racer note

Setting spark plug gaps takes skill and practice. Hold the feeler gauge on the flat surfaces between your first finger and thumb. The correct clearance can be felt as a faint touching drag when you move the feeler gauge through the gap. Close the gap by gently tapping the side electrode on a hard metal surface – such as a bench top. Do not use a screwdriver or pliers to alter the gap. Before fitting new spark plugs always check the gap settings. When tightening up sparking plugs always use a correctly set torque wrench. The taper fitting spark plugs must not be over tightened. If the spark plug thread in the cylinder head is slightly rusty, or, in an aluminium cylinder head, dry, apply a little light oil to the plug threads before fitting them.

Distributor

The inside of the distributor is divided into three parts. The lower part houses the mechanical components and the linkages; the middle part comprises the LT components and the upper part is the HT section. The HT components are mainly the **distributor cap** and the **rotor arm**. HT electricity is delivered from the coil to the centre of the cap.

Electricity flows from the cap to the rotor arm through a **brush** arrangement. The **rotor arm** is connected to the **distributor spindle** so that it goes around at the same speed as the **spindle**. As the rotor arm rotates, its free end aligns with **segments** in the cap, one at a time. The sequence is the firing order. As each segment is aligned the current is passed from the

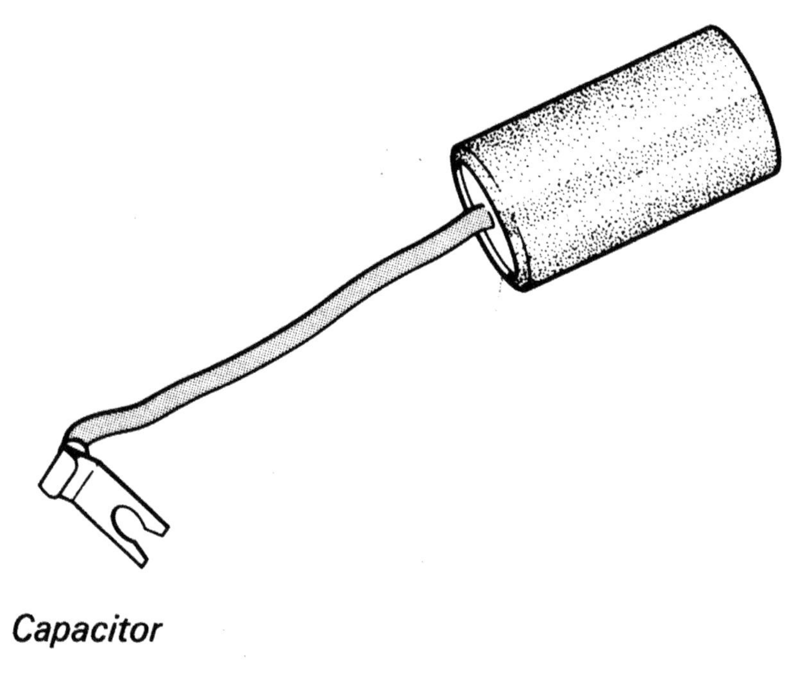

Figure 2.4 Capacitor

rotor arm to the segment. The segments each have a **plug lead** attached to carry the electricity on its way to the spark plug.

The low-tension part of the distributor comprises the **contact breaker (cb) points** and the **capacitor**. The **cam ring** which is formed on the outside of the distributor spindle rotates ate the same speed as the spindle. As the cam goes round it opens and closes the cb points. It is this action which causes the current to flow in the HT circuit by **induction** in the coil. The **gap** of the cb points, and their position in relation to the spindle, affects the **ignition timing** and general efficiency. The cb points should be checked for condition and size of gap every 8,000 km (5,000 miles). They should be replaced every 16,000 km (10,000 miles).

The **capacitor** (also called a **condenser**) is fitted to give a good quality spark by controlling the flow of electricity; this reduces **arcing** at the points and gives a longer life to the cb points.

The drive for the distributor is picked up from the camshaft by an angular gear called a **skew gear**. This skew gear is on the lower end of the distributor spindle. The spindle passes through the mechanical and the vacuum timing mechanisms. These mechanisms advance and retard the ignition timing to suit different engine speeds and load characteristics.

FAQs Does the distributor rotate at the same speed as the crankshaft, or that of the camshaft?

The distributor rotates at the same speed as the camshaft, that is, half the speed of the crankshaft. Therefore, on a four-cylinder engine the cb points open four times for each revolution of the distributor spindle.

Ignition timing

The ignition system is designed so that the spark occurs in the combustion chamber a small number of degrees **before top dead centre** (BTDC). Car manufacturers give specific figures for each of their different models. Typically, the static timing is 10 degrees BTDC, that is when the engine is rotated by hand. When the engine is running, dynamic timing, the timing will advance to about 30 degrees BTDC

Racer note

You will find the ignition timing settings in the workshop manual, or service data book. The static timing can be easily be checked using a small low wattage bulb (such as a side lamp bulb) in a holder with two wires attached. Attach one wire to the distributor terminal on the ignition coil – usually marked with a positive sign (+). Attach the other wire to a good chassis earth. Disconnect the HT king lead, or remove the distributor cap so that the engine will not start. Switch on the ignition and turn the engine by hand. The bulb should light just as the timing marks come into line. The timing can be adjusted by slackening the distributor clamp and moving the distributor gently. With the timing marks aligned move the distributor until the bulb just lights.

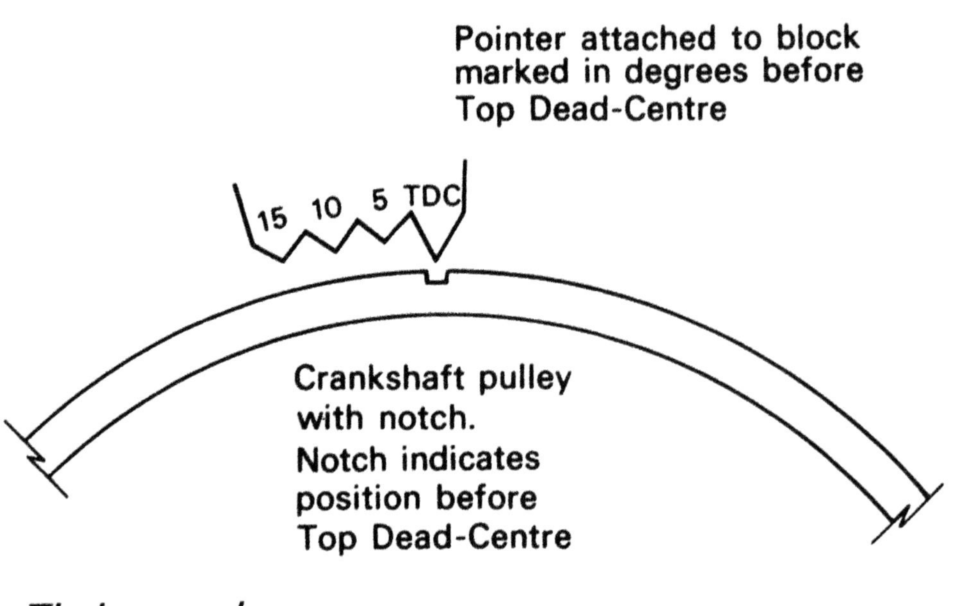

Figure 2.5 Timing marks

Dwell angle

This is the period for which the contact breaker points remain closed. When the cb points are closed the magnetic field is building up in the ignition coil. When the points open the spark is triggered at the plugs. The dwell angle is directly proportional to the number of cylinders and the cb points gap and as such is used as an indicator to the engine condition. For four-cylinder engines it is typically 60 degrees.

Electronic ignition

There are many different types of electronic ignition. The most common type has a distributor that does not have cb points. The distributor has an electronic trigger device instead of the cb points. On the outside of the distributor is an amplifier unit. The switch, coil and battery remain the same.

There are three main types of trigger devices used instead of cb points. These are:

- **Inductive type** – this uses a magnet attached to the distributor shaft to induce a small electrical current into a pick-up coil. The current produced is low voltage (typically 2v) alternating current (AC).
- **Hall Effect** – this uses a small integrated circuit (IC) as a switch which is turned on and off by the passing of metal drum shaped component. This works at low voltage (typically 5v).

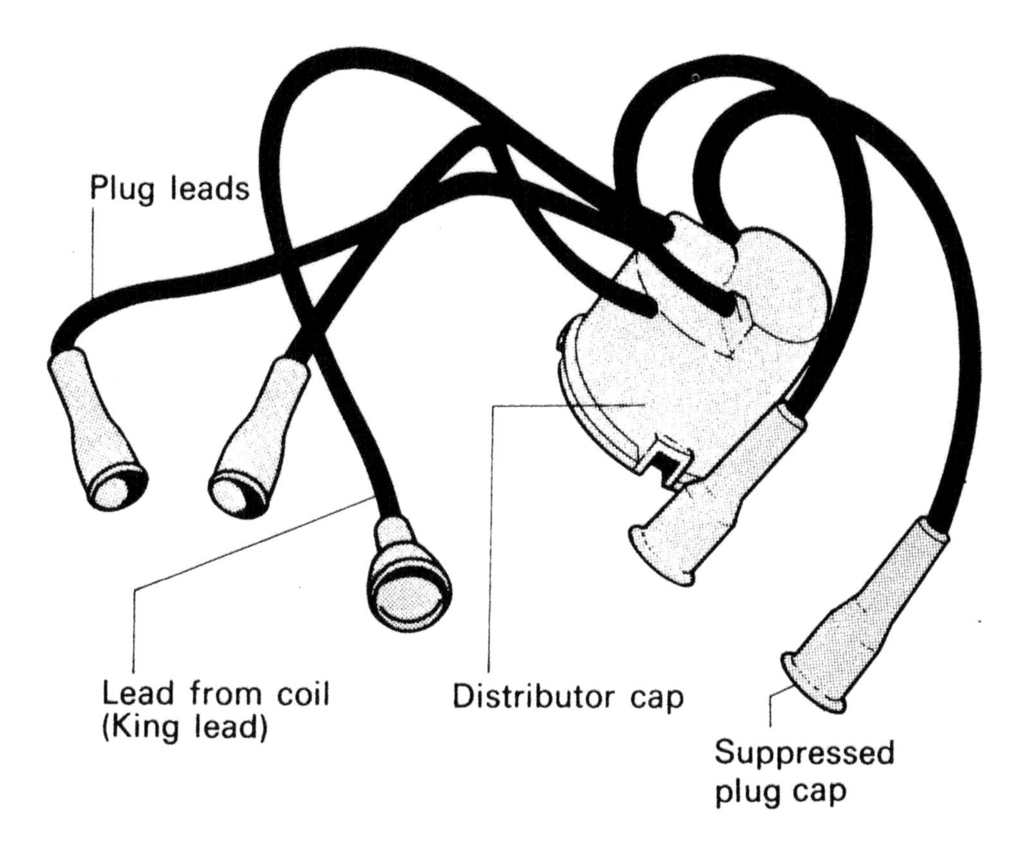

Figure 2.6 Distributor cap and plug leads

- **Optical type** – this uses an infra-red light emitting diode (LED) which shines on a phototransistor. The light is turned on and off by a Maltese Cross-shaped component usually referred to as a light chopper. Operating voltage is typically 9 volts.

As the electronic ignition system does not have cb points there are fewer moving parts, so reliability is increased along with engine economy and performance. The only ignition servicing that is required is changing the spark plugs. On electronic ignition cars the spark plugs may only need changing every 60,000 km (40,000 miles).

Other types of electronic ignition do not have a distributor and do not have a conventional coil. Also, the ignition lock and key have become a more complicated security device. The key incorporates a very small electronic device called a transponder; this transponder is a sort of electronic key that electronically unlocks the ignition at the same time that it is being mechanically unlocked.

Distributor-less ignition system (DIS)

The distributor-less ignition system (DIS) is usually part of an engine management system, which comprises two parts **ignition** and **fuel**. Looking at the ignition part.

DIS comprises of:

- **Ignition IC** (chip) in the Electronic Control Unit (ECU)
- **Sensors** for: **crankshaft position**, **TDC**, **knock** (pinking), **throttle position**, **engine temperature**
- **Spark plugs**
- **Ignition coils** for each spark plug

Racer note

Sensors on different makes of systems may vary and have different values for different engines.

The ignition IC is **programmed** to trigger a spark according to engine load and road conditions – it is fully self-regulating and needs no servicing apart from spark plug replacement at set intervals. In the event of a fault this is found using **electronic diagnostic equipment**.

Fuel system

Key points

- Both petrol and diesel fuel must be handled with care; they are highly flammable and can be a hazard to your health.
- Petrol engines mainly use fuel injection, but some older vehicles have carburetters.
- Diesel fuel is injected into the combustion chamber at a pressure of about 170 bars (2,500 psi).
- Air filters and fuel filters need changing regularly.
- Diesel fuel system components must be handled with care to prevent personal injury.
- The fuel system must be set up to supply the correct amount of fuel at the correct time.
- Correct servicing is needed to maintain both fuel economy and reduce environmental pollution.

The fuel system supplies the motor vehicle with the necessary amount of fuel for it to be able to do its work efficiently. The engine must receive the correct amount of fuel at the right time or else it will not run properly, if at all. Fuel is the food of the engine. Most vehicles run on either petrol or diesel fuel, which is bought in liquid form. Both petrol and diesel fuel are hydrocarbons, as they are made up of hydrogen and carbon atoms; but the similarity ends there. You cannot run a petrol engine off diesel fuel, or vice versa. Indeed, just putting the wrong fuel in the tank can cause a lot of expensive damage.

It is important to remember that burning fuel creates an exhaust hot gas, which, if the fuel system is not set up correctly, can cause illegal pollution.

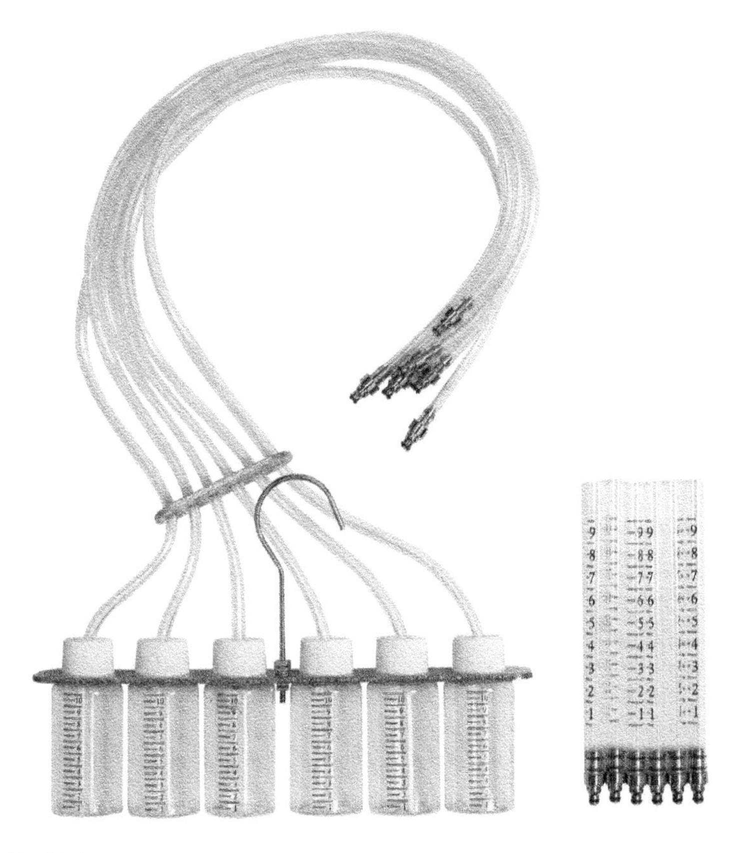

Figure 2.7 Fuel injector test

Working on a fuel system, either petrol or diesel, presents the mechanic with a number of dangers and hazards. Remember to follow the correct safety procedures and you will be quite safe. Let's look at the petrol system first.

Petrol supply system

Safety notes

Before working on a petrol supply system, you should be aware of the following hazards:

- Petrol is highly flammable; do not smoke or have any naked lights near petrol.
- Petrol dries the skin and can cause skin disease; always wear protective gloves and avoid direct contact.
- If your overalls are doused in petrol at all, change them; there is an extreme fire risk even after the petrol as dried off.
- If draining a fuel tank, a sealed and electrically earthed draining appliance must be used.
- Petrol must not be stored in the garage.

Basic carburetter petrol supply system

This section covers the basic carburetter system found on older vehicles, kit cars and special vehicles.

The basic system, which is fitted on older vehicles, uses the carburetter to mix the petrol and air in the correct proportions. The main components are the petrol tank, the petrol pump, the carburetter and the air filter. Let's briefly look at each of the components in turn.

Petrol tank – usually this is situated underneath the car and behind the driver and the passengers to reduce the risk of its contents spilling over them in the event of an accident. Petrol tanks may be made from either pressed low carbon steel or moulded plastic. Plastic petrol tanks are lighter and do not rust. The capacity of the petrol tank depends on the expected use of the vehicle and the average fuel consumption. For example, if the vehicle has a fuel consumption of 10 litres per 100 kilometres and its average daily usage is expected to be 500 kilometres, then:

$$\text{Fuel tank capacity} = 500 / 100 * 10$$
$$= 50\,\text{litres}$$

The petrol tank is fitted with a pick-up pipe that is just above lowest point of the tank so that it does not pickup any sediment. The filler cap must seal against the tank. The breather is from the highest part of the tank, so not to allow fuel to leak out; the open end of the breather is at a low point at the rear of the vehicle so that fumes do not enter the passenger compartment. The petrol tank is also fitted with a floating sensor to measure the amount of petrol in the tank. The sensor sends an electrical signal to the dashboard-mounted gauge. The gauge may record the level of the petrol in relative terms, such as full, half full and empty, or in actual litres.

Nomenclature

Fuel consumption for cars in SI units is given in litres per hundred kilometres (litres/100 km). In Imperial units it is in miles per gallon (mpg). For conversion purposes 9 litres / 100 km approximately equals 30 mpg.

Fuel (petrol) pump – this may be operated mechanically from the camshaft, or electrically through the ignition switch. The mechanical petrol pump is self-regulating. That is, it uses the force of the spring to increase the petrol's pressure. If the carburetter does not need petrol, the mechanism freewheels, the spring holding a constant pressure. The diaphragm-type electrical pump operates in the same way.

The petrol pump draws up petrol from the tank and sends it, under pressure, to the carburetter. The fuel lines, or pipes, may be made from steel or plastic. The pipes are usually a push fit onto the petrol tank pickup, the pump and the carburetter. Where the pipes run along the side of the chassis, or across parts of the body, they are clipped to hold them in place and stop chafing. If the petrol pipe is chafed through then the vehicle will not run, or worse still, the fuel that leaks out could cause a fire.

FAQs How can I tell if the petrol pump is faulty?

The pump supplies petrol to the carburetter at a pressure of about 30 kPa (5 psi); if the pump is suspected of being faulty, the pressure can be measured with a gauge.

Carburetter – this mixes the petrol and the air in the correct proportions for it to be burnt inside the combustion chamber.

The simple carburetter

The basic principles of carburation are embodied in the simple carburetter. The simple carburetter was used on the very first cars and motorcycles and can be found on lawnmowers and other single-speed garden equipment. The petrol is in the **float chamber**, which is connected to the **discharge jet** through a tube that contains the main metering jet. When the engine is spun over with the throttle valve and the **choke butterfly** open, air is drawn into the **venturi**. As the air passes through the narrow section of the venturi its speed increases. Increasing the air speed causes the air pressure to drop, so that the air pressure in the venturi is below normal atmospheric pressure. The petrol in the float chamber is at atmospheric pressure; it now flows through the main metering jet and out of the discharge nozzle to mix with the air. As the petrol leaves the discharge nozzle it breaks into very small droplets. This is referred to as **atomisation**.

Nomenclature

Venturi is the name given to the narrowing of an air passage; in a carburetter this is sometimes referred to as the choke tube, or simply choke. High-performance Weber carburetters usually have two chokes; American Holley carburetters have four chokes.

The **throttle flap** controls the flow of petrol and air through the carburetter; on a car this is operated by the throttle pedal. The **choke flap** controls the mixture strength by limiting the airflow into the carburetter. The choke is closed for cold starting to give a richer mixture.

The flow of petrol into the carburetter float chamber is controlled by the **needle valve** and float. As the **float chamber** fills up with petrol, the float rises. The rising float pushes the needle valve up against its seat and stops the flow of petrol.

Air:petrol ratio

For the complete combustion of petrol and air, approximately14 parts of air are needed to burn one part of petrol. However, when an engine is running under different conditions the mixture strength needs to be altered. Table 2.1 shows typical mixture strengths for different operating conditions.

Adjusting the carburetter – you can usually adjust both the **mixture strength** and the **idling speed** of the carburetter. The workshop manual will give information on both of these settings.

Table 2.1 Air:Fuel Ratios

Operating condition	*Air:petrol ratio*
Cold starting	9:1
Slow running	13:1
Accelerating	11:1
Cruising	19:1

FAQs How are carburetter adjustments checked?

The idling speed on most carburetter engines is between 600 and 900 rpm. Generally, if it sounds nice it is correct. Because of the exhaust emission regulations, you must set the mixture using an exhaust gas analyser – cheap toolbox ones are available for the home mechanic on a tight budget.

Air filter – the air filter assembly has three functions, namely:

1. To filter and clean the incoming air so preventing the entry of dust, grit and other foreign bodies into the engine; these could seriously damage the engine.
2. To silence the air movement, so making the engine less noisy.
3. To act as a flame trap, so preventing a serious under-bonnet fire should the engine backfire.

Most vehicles are fitted with paper element air filters. These must be replaced at set mileage intervals – usually about 20,000 miles. Air filters must be replaced regularly even if they look clean; their micropore surface may still be blocked.

Racer note

Replacing the standard paper air filter with a mesh type, such as the K&N version, may increase the engine power output.

Petrol injection

This section covers the petrol injection system used on most new vehicles – variations are used for specialist application.

All new cars are fitted with petrol injection (PI). Petrol injection cars have a similar petrol tank, petrol pipes and air filter. The carburetter is replaced by injectors and a high-pressure pump. PI systems are usually more efficient than carburetters in that they give more power, better fuel consumption and more controllable emissions; they also need less maintenance. Let's have a look at the main components.

Fuel (petrol) pump

The petrol pump on a PI engine is likely to be electrically operated and submerged in the petrol tank. It is submerged to keep it cool and prevent the entry of air bubbles into the petrol pipes. The PI engine petrol pump raises the petrol pressure to about 1,650 kPa (110 psi) and sends it to the injectors via a filter. The petrol pipes on a PI system are much heavier than those on a carburetter engine car; the pipe couplings are by screw threaded ends. The petrol pump is controlled electronically by the electronic control unit.

Electronic control unit (ECU)

The ECU is a sealed box containing a number of microprocessor integrated circuits – usually referred to as microchips – similar to those in a computer. The ECU controls both the idle speed and the mixture strength, so there is no need to adjust these settings. The ECU for the petrol injection equipment also controls the ignition system.

Injectors

There may be one centrally mounted injector, called single point injection, or one for each cylinder. The petrol under pressure is supplied to the injectors from the petrol pump. The ECU electronically controls the amount of fuel to be delivered by the injectors.

Injectors open and close about 50 times each second when the engine is running at full speed, on each occasion delivering petrol. Injectors can become blocked or worn. To maintain injectors in good working order they should be removed and checked at regular service intervals. Their spray pattern is observed and they may be cleaned or replaced as needed. Cleaning may be by ultrasonic vibrations; or the use of a chemical, cleaning agent. Injector cleaning chemicals are available to be added to the petrol in the tank.

Nomenclature

In this book we have used the term petrol injection and the abbreviation PI because it is a general term. You will come across electronic fuel injection (EFI), just injection (I, or i), gasoline injection (GI) and a range of equivalent terms in different foreign languages.

Air filter – on a PI engine this is similar to the one used on carburetter engine cars.

Airflow control

This controls the flow of air into the engine, and hence the engine speed. A throttle body similar to that in a carburetter is used. This is connected by the accelerator cable to the accelerator (or throttle) pedal. The throttle butterfly is connected electrically to the ECU.

Fuel cell

On race cars it is essential to store the fuel safely. The petrol used in race cars tends to be very volatile – so that it easily ignites; also, it burns to release more energy than that used in road cars. In other words, it is very hazardous and needs stringent controls to keep it safe. To this end a fuel cell is used instead of a normal tank. The fuel cell contains a foam material, which in effect absorbs the petrol and prevents is from spilling or leaking if the tank is inverted or damaged. The fuel cell is filled in the same way as a standard tank, though a large filler cap – usually these are connected to the filler hose with a sealed push and twist type of coupling.

Racer note

A number of different fuels are used for race cars – for certain series of racing a control fuel is used – that is one which is only available from one source.

Diesel fuel supply

The main difference between the petrol supply system and the diesel fuel supply system is that the petrol system takes in a mixture of petrol and air together, whereas the diesel draws only air into the engine and the diesel fuel is injected separately into the combustion chamber.

Safety notes

- Diesel fuel may look clean, but it contains materials that can cause dermatitis or other skin problems – always use barrier cream and rubber gloves when working on diesel engine components.
- The internal parts of the injector pump and the injectors will be damaged if touched directly by your hands – always handle the parts submerged under the special oil which is available for this – do not substitute other oils or diesel fuel as may damage both your skin and the parts.
- If you spill diesel fuel on your skin wash it off immediately and reapply your barrier cream.
- You must change your overalls regularly, and immediately if you spill diesel fuel on them.
- Do not keep rags that have diesel fuel on them in your pocket; this can lead to skin disease.

Diesel fuel tank

On a diesel engine car the fuel tank will be very much like the petrol tank on the petrol version of the same model. The filler neck is likely to be marked *Use Diesel Fuel Only*, or with similar words. Diesel fuel is sometimes referred to as DERV, or gas oil.

Diesel trucks have fuel tanks of at least 250 litres (50 gallons); this is because they have high fuel consumption and cover large distances. For instance if the truck's fuel consumption was 10 mpg, a 50-gallon tank would allow fewer than 500 miles between refills – you cannot completely drain the tank.

NB

You should never allow a diesel vehicle to run out of fuel; if you do this the system will need venting (bleeding) and the pumping components may be damaged.

Air filter

The air filter on a diesel engine is similar to that on a petrol engine. However, they are usually larger to allow in a greater amount of air. It is important that the air filter is replaced at the recommended service intervals.

Lift pump

The diesel engine has a lift pump to draw fuel from the tank to the injector pump via the fuel filter. The lift pump operates at about 30 kPa (5 psi), similar to the petrol pump on a carburetter engine. The diesel fuel lift pump is fitted with an external handle so that it can be operated manually when the engine is stationary. This is used for priming the fuel system before starting the engine. See the section on venting for more information.

Fuel filter

Because it is essential that the fuel is very clean before it goes to the injector pump and the injectors, it is passed through a very comprehensive filter. The filter uses a very fine paper element. The filter housing is also fitted with a mechanical trap to separate the fuel and any water and an area for sediment to settle into.

Injector pump

There are two main types of injector pumps:

- In-line injector pump
- Distributor, or rotary, injector pump – often called a DPA (distributor pump assembly) pump.

The in-line pump is used on larger diesel engines – those used in trucks and buses. The DPA pump is used almost universally on cars and light goods vehicles.

The injector pump sends fuel to the injectors at a pressure of about 170 bar (2,500 psi). Because of the extremely high pressure, special thick-walled pipes are needed; these are referred to as injector pipes. The injector pipes are connected to the pump at one end and the injectors at the other end, using special couplings called glands. It is usual to use special slotted hexagonal spanners to tighten up these coupling glands. These spanners will not slip on the gland nuts.

Injectors

The injector is fitted into the cylinder head. It projects into the combustion chamber so that the fuel can be injected into the hot compressed air just before TDC. The reason for injecting the fuel at 170 bar is so that it can enter into the compressed air and circulate a little before igniting. If the fuel were at a lower pressure than the compressed air it would not be able to enter the combustion chamber.

The injector breaks the fuel into a fine mist of almost atom-size particles – atomisation. You should remove and test the injectors at regular service intervals – typically every 24,000 miles. The test comprises of looking at the spray pattern made by the injector nozzle and checking the pressure at which the injector opens.

Safety notes

The injector spray pattern looks very attractive, like a shower head; but do not be tempted to place your hand under the spray. At 170 bar the finely atomised diesel fuel can go straight through your skin and inject a lethal dose into you blood stream.

Venting (also called bleeding). If you remove any of the diesel fuel system components you will need to vent out the air from the system before you can restart the engine. The fuel filter and the injector pump are both fitted with vent valves, small screws, which when undone, allow the air to escape.

Basically, you open the vent valves then operate the fuel pump primer lever manually so that fuel flows from the tank to the injector pump. Close the valves one at a time as the diesel fuel can be seen to squirt out. That is, as the fuel flows to prime the system the air is vented out, the fuel will first flow out of the filter inlet side, tighten that vent valve. Carry on pumping and the flow will move to the outlet side, tighten that valve and then you can move onto the injector pump.

Modifications

Carburetters

No new cars have been made with carburetters since the end of the 1990s because of the vehicle emission regulations. However, many old competition vehicles have carburetters and some special builders use them on track day and race vehicles because they are cheap, easy to understand and can be adjusted manually with the minimum equipment in the paddock. So, a motorsport technician could be expected to know a little of the background of the common motorsport carburetters.

SU – the oddly named Skinner's Union, designed by Army Colonel Skinner for his daughter's racing Austin 750. This variable choke, constant depression carburetter is a very popular choice for modifying engines. **Stromberg** later made a similar carburetter called the CD (constant depression), which uses a rubber diaphragm instead of the metal piston used in the SU.

Weber – designed by Edoardo Weber who first made carburetters for the racing FIATs of the 1920s; then they were fitted to the first Enzo Ferrari built car bearing his own name in 1948. The Weber carburetter's unique design point is that it is a fixed choke carburetter body on which every single part can be changed. **Dellorto** introduced a similar carburetter in the 1960s.

Zenith – a fixed choke carburetter made in a number of different styles, very popular on many cars and eventually combined with the similar operating **Solex**.

Holley – originally made in Manchester, England moved to America and now found on most American V8 engined race cars.

Amal – originally designed for motorcycle use; but used on many small-engined race cars such as the Mini. Its unique design point is the integration of the choke body concentric with the float chamber.

Minnow-Fish – a precursor of the fuel injection throttle body using the low pressure 3–5 psi (0.2–0.3 bar) fuel supply from a normal petrol lift pump.

Main modifications – the main modifications of carburetter arrangements are:

- Increasing the number of carburetters – ideally one per cylinder
- Increasing the size of the carburetters
- Changing the air filter (cleaner) for a high flow one – for example K&N
- Fitting air trumpets of different lengths
- Fitting water heated, or non-heated inlet manifolds
- Altering the carburetter setting to give different mixture strengths under different running conditions

Injector systems

Fuel injection systems are typically modified in two ways:

- The electronically programmable read only memory (EPROM) in the fuelling ECU is reprogrammed to give more fuel under particular conditions. This is called *chipping* as the EPROM is a type of micro-chip. Chips can readily be bought for most popular performance cars. On rally cars, which have off-road performance requirements separate to their on-road legal requirements, it is often the case to have two separate, but switchable ECUs.
- To deliver the extra fuel, high flow rate injectors are used when the engine is chipped. These are sometimes identified by the colour code green as against the normal grey.

Throttle bodies

Instead of the normal injectors and manifold arrangement, each cylinder is fitted with short individual stubs, which incorporate the injectors. This improves air and petrol flow.

Inlet manifold

These may be changed for ones, which give different airflow characteristics and carburetter or injector arrangements.

Air filter

The air filter (also called air cleaner) must be appropriate for the vehicle and its operating conditions. The air filter removes dirt from the incoming air to reduce the risk of engine damage, quietens the noise of the air intake and acts as a flame trap.

Figure 2.8 Petrol pump in boot

Figure 2.9 Plenum chamber

Figure 2.10 After market air filter

For track use an air filter may be thought unnecessary and intake trumpets used instead. For a road car a high flow system such as the popular K&N system may be used. For an off-road vehicle a high level – to keep out of the water when wading – system may be used with two air filters in series to reduce the risk of an ingress (taking in) of dirt or dust. One of the air cleaners may be a centrifugal one.

If twin turbo-chargers are used then two air filters may be appropriate.

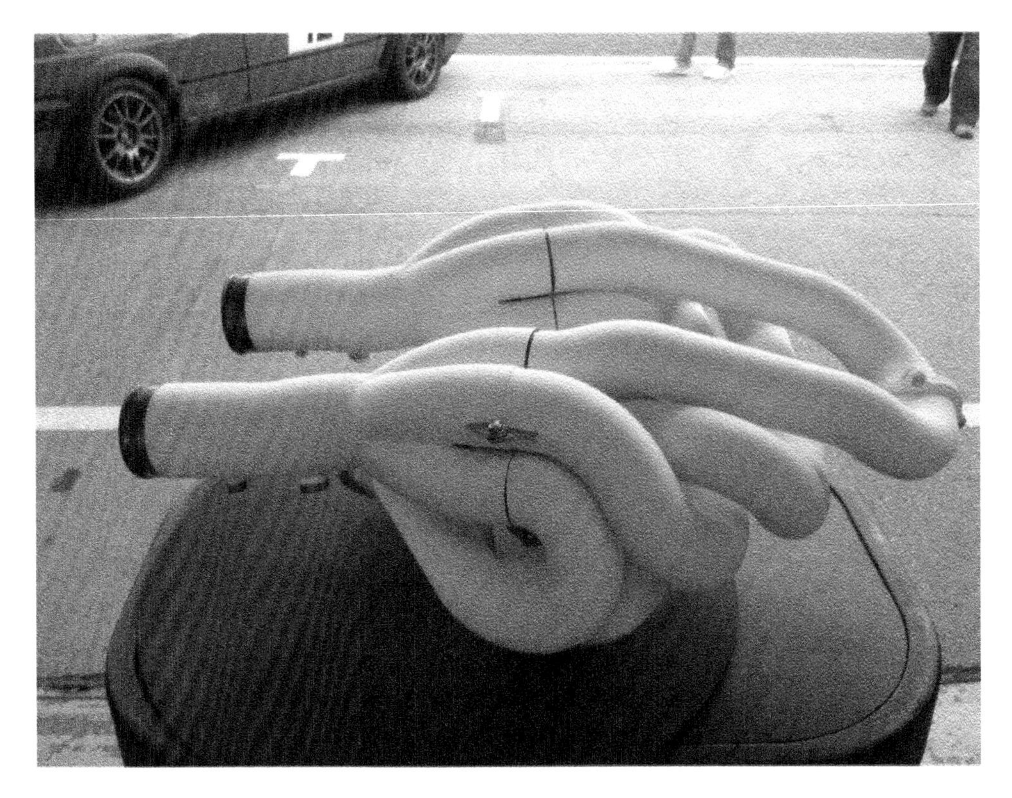

Figure 2.11 Coated exhaust system

Exhaust system

The exhaust system has the job of quietening the noise of the exhaust gas and cooling it before it leaves the vehicle. It will also, except on older vehicles, incorporate a catalytic converter (cat) and a lambda sensor. Most modifications relate to improving the flow of the exhaust gas from the engine to the outside of the vehicle. Some are solely about altering the exhaust note to make it sound *sporty*. The exhaust note is very important to some people – about impressions of power. It is worth mentioning that Alfa Romeo has a special department to ensure that their cars all have the correct corporate image exhaust note. It is good to listen to different exhausts and see if you can identify the cars by them. Typical exhaust modifications include:

- Change rear silencer (back box) for one with a larger diameter tail pipe – referred to as a *big bore* or *drain pipe*. This is for looks and noise
- Change the manifold (in USA called headers) for one to give a better gas flow – usually this means longer individual pipes before the cylinders are joined

- Complete system change – for serious performance increases, usually matching the inlet manifold and cylinder head and camshaft changes at the same time
- Change to a stainless-steel exhaust system – most appropriate to classic motorsport vehicles for longevity

Supercharger/turbo-charger

Superchargers and turbo-chargers are fitted to increase the amount of air and petrol (air only on diesel engines) going into the engine cylinder. The greater the amount of gas that goes into the cylinder the greater the power output will be. Superchargers and turbo-chargers typically increase engine output by 30%.

The supercharger is in effect an air pump driven by the crankshaft – they started life on aircraft where the air is less dense and needed to be compressed both for the engine and the passengers.

The turbo-charger is driven by the exhaust gas; it has two turbines. One turbine is driven by the exhaust gas; the other turbine is driven by the shaft from the first turbine and compresses the air into the cylinder. On high-performance and diesel vehicles, an intercooler may be used to cool the air between the turbo-charger and the cylinders.

Although superchargers and turbo-chargers do the same job, they do it in a very different way. Table 2.2 compares superchargers and turbo-chargers.

Questions and skills

1. Ignition system components cause many vehicle breakdowns, so it is important to be able to actually remove and refit all the components that we have discussed in this chapter. On a running engine, either in or out of a car, remove the ignition components and refit them using a workshop manual for guidance.

Table 2.2 Comparison of supercharger and turbo-charger

No	*Type*	*Drive*	*Characteristics*	*Advantages*	*Disadvantages*	*Comment*
1	Supercharger	Belt from crankshaft	Straight-line pressure increase dependent on engine speed	Immediate response to throttle	Needs drive from crankshaft	Ideal for rapid acceleration – dragsters
2	Turbo-charger	Exhaust gas	Delayed pressure increase dependent on exhaust gas pressure and A/R ratio	Exhaust drive giving better economy for given application	Delayed response to throttle – prompts use of twin turbo-chargers	Ideal for application combining power and economy – rally cars and diesels

2. Investigate an electronic ignition system on a car of your choice; sketch the wiring connections. Use a workshop manual for help.
3. Locate a spark plug identification table and identify the spark plugs that may be used in six vehicles of your choice.
4. Compare the fuel consumption figures of a range of petrol and diesel vehicles.
5. Examine the different types of fuel fillers – petrol and diesel – and report on how they differ.

Chapter 3

Lubrication and cooling

Lubrication system

Lubrication and cooling fit together like eggs and bacon. The lubrication system cools the inner parts of the engine as well as lubricates then; the cooling system cools the larger parts of the block and head. The radiator and the oil cooler usually occupy juxtapositions at the front of the cold air intake grille.

Key points

- Lubrication is needed to reduce friction and wear
- The two main types of lubrication are full-film and boundary
- There are many different types of oil; it is important to choose the correct one
- Most engines use a pressurised oil supply using an oil pump, sump, filter and pressure relief valve
- Engines and other units with rotating parts need oil seals to keep the oil in and the dirt out
- Two-stroke engines use total loss lubrication; this is only suitable for small motorcycles and garden equipment

Lubrication is needed to keep mating bearing surfaces apart; this reduces friction and wear. The lubricant also acts as a coolant, taking heat away from the bearing surfaces to maintain a constant running temperature. The liquid cooling system runs at about 85 °C; the lubrication system runs at a slightly higher temperature, 90–120°C being typical figures for road use; for racing engines this may exceed 200°C. The lubricant also picks up small particles of metal and carbon from the components that it passes over and deposits them in the oil filter. It is the particles that are too small to be filtered out that make the oil a dirty black colour.

Racer note

Oil temperature gauges are typically red lined at about 150 °C.

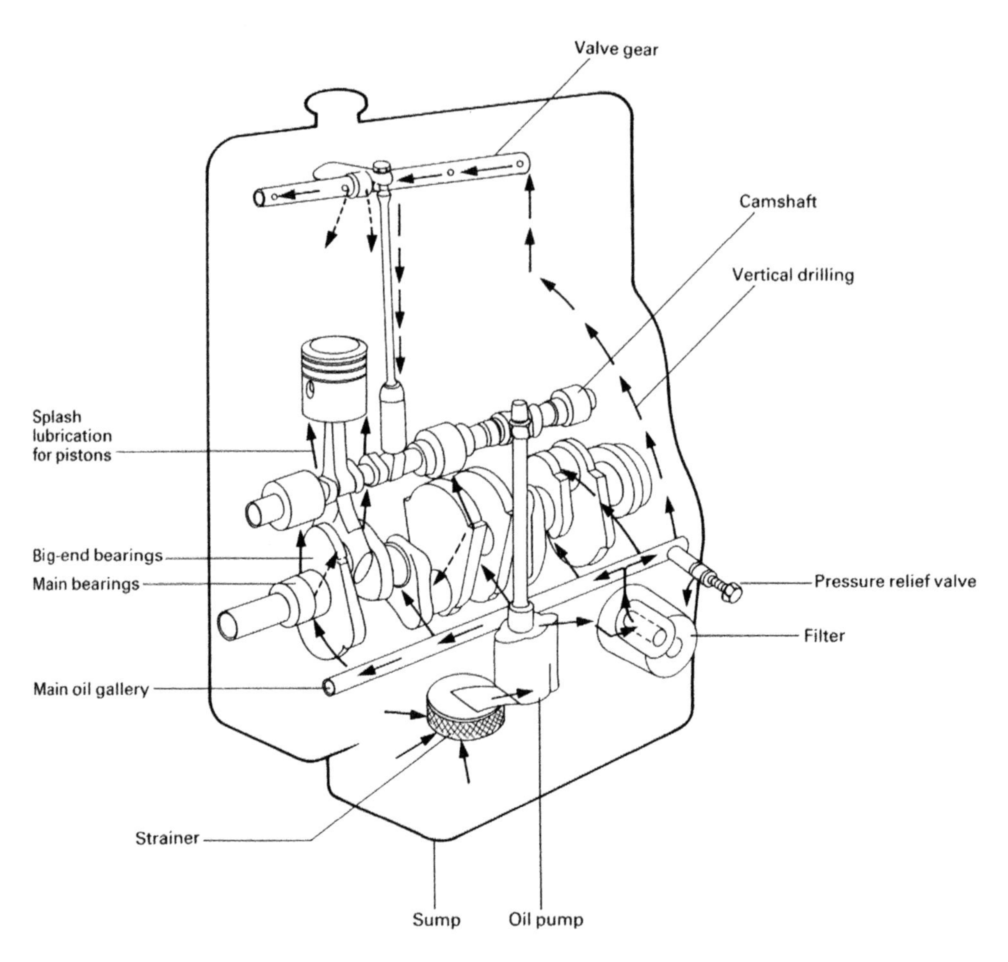

Figure 3.1 Lubrication system

Friction

Friction is the resistance of one surface to slide over another. The amount of friction is referred to as the co-efficient of friction and indicated by the Greek letter Mu that has the symbol μ.

To calculate the co-efficient of friction the formula used is:

$$\mu = F/W$$

where F is the force that is required to slide the object over the mating surface and W is the weight of the object. Both F and W are given in the same units, usually Newton (N). The value of μ is always less than unity; that is between 0 and 1.

Lubrication keeps the mating surfaces apart so that they can slide easily over each other. The simplest form of lubricant is water; think of how slippery a floor is when it is wet. Motor

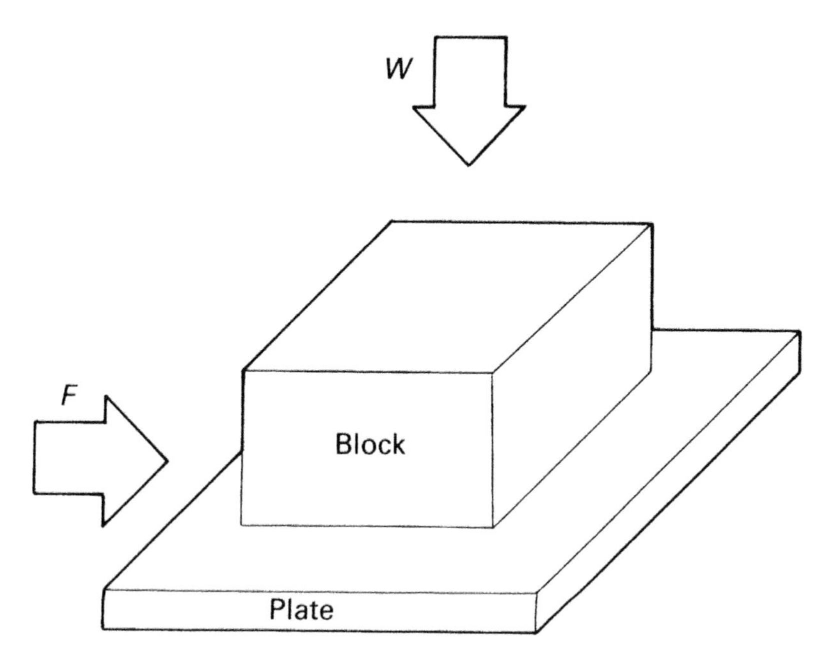

Figure 3.2 Friction equals F/W

vehicles use a variety of oils and greases in a range of different ways. The following sections looks at some of them.

> **Nomenclature**
>
> m almost always as a value of less than unity (in other words one) as it is usually easier to push something along than to lift it.

Types of lubrication

Lubricants are generally used in the form of oil, which is liquid, or grease, which is semi-solid. Oil in the engine is fed to the bearings under pressure; the film of oil keeps the bearing surfaces apart; this is known as **full-film lubrication**. Grease, which is used for steering and suspension joints, and wheel bearings, does not keep the bearing surfaces fully separated; this type of lubrication is known as **boundary lubrication**.

Viscosity

Oils are classified according to their viscosity. **Viscosity is the resistance of an oil to flow**; it is calculated by timing how long it takes a fixed quantity of oil to flow through a specific diameter aperture, at a pre-set temperature. The aperture is a hole in a piece of metal that looks like a washer. The time is in seconds; the convention is to call this amount of time Redwood Seconds, so that it is clearly seen that this process has been used. The longer the amount of time taken, the higher the oil's viscosity is said to be, and the higher the viscosity number it is given. In common language it is said to be thicker. Basic oils are:

5 SAE	– Cycle oil
30 SAE	– Straight engine oil (not commonly used)
90 SAE	– Gear/axle oil
140 SAE	– Truck axle oil

SAE is the abbreviation for the **Society of Automotive Engineers**; this is an American-based international organisation, which sets standard for many areas of automotive engineering.

When the temperature of oil is raised its viscosity usually decreases; that is to say when it gets hotter it gets thinner too. Therefore, most modern oils for vehicle engines are multigrade. That is to say they have two viscosity ratings, one for when they are cold and one for when they are hot. A typical oil is 15/40 SAE. This oil is rated as a thin 15 SAE when it is cold to give easy cold starting; it is rated a thick 40 SAE when it is hot to give good protection to the engine when it is hot – such as on the motorway. Other popular grades are 10/40 SAE and 5/30 SAE.

Types of oil

Oil can be classified in a number of different ways; it is important to choose the correct oil for the vehicle. Let's look at some of the common types:

- **Mineral oils** are those pumped from the ground, and **vegetable oils**, such as **Castrol R** oil, are those that are made from vegetable products. Castrol R oil is used for a small number of specialist applications such as racing motorcycle engines. Most cars and trucks use mineral oils. Mineral and vegetable oils cannot be mixed.
- **Synthetic oils** are chemical engineered oils; they are specially prepared mineral oils. Synthetic oils are very expensive. A cheaper alternative is **semi-synthetic** oil.
- Oil for diesel engines and ones with turbo-chargers are specially classified to withstand the very high temperatures and pressures of these engines – look for the marking on the container. Synthetic and other oils may be used for these special purposes.
- Extreme pressure (**EP**) oils are used in gearboxes; usually their viscosity rating is 80 or 90 SAE. They maintain a film of oil under a very heavy load between two gear teeth.
- **Hypoy** (or hypoid) oil is used for a special shape of gear teeth in rear axles on large vehicles and trucks. It is usually of 90 SAE or 140 SAE viscosity.
- Automatic transmission fluid (**ATQ** or similar) is an oil which is also a hydraulic fluid. It is used in power steering as well as the automatic gearbox.
- 3-in-1 oil, or **cycle oil**, is 5 SAE and is used because it cleans and prevents rusting as well as lubricating. On motor vehicles it is used for control cables, door hinges and electrical components.

Racer note

Because of the exacting nature of lubrication in race engines it is essential to use the right oil irrespective of cost – a race engine rebuild may cost more than the price of a small family car. Many of the specialist race lubrication suppliers and manufacturers sponsor race car teams because they enjoy the sport – not just to sell more oil.

Engine lubrication system

The most popular type of engine lubrication is the **wet sump** type. This is so called because the sump is kept wet by the engine oil, which uses the sump as a supply reservoir.

In operation the oil is drawn from the sump by the oil pump. To prevent foreign bodies from being drawn into the pump it passes through a gauze strainer at the lower end of the pick-up pipe. From the pump the oil is passed through a paper filter and then through a drilling in the cylinder block to the main oil gallery. The oil pressure in the filter and main gallery is about 400 kPa (4 bar or 60 psi). To ensure that the design pressure is not exceeded, and so to prevent damage to the oil seals, an oil pressure relief valve is fitted in the main gallery. The oil, which is still at a pressure of about 400 kPa, goes through drillings in the block to the main bearings. The oil then passes through drillings in the crankshaft to the big-end bearings. At one end of the engine there is a long vertical drilling to take the oil supply to the camshaft and valve gear.

Racer note

Oil pressure gauges typically read in psi but may also be written lb/in^2. During a race the oil pressure should normally remain constant within a given range – surges or drops outside this range indicate possible faults.

Gear-type oil pump

The function of the oil pump is to draw oil from the sump and send it under pressure to the filter and on to the main gallery from where it is distributed to other parts of the engine. The gear-type oil pump is used in a large number of vehicles. It is driven by the camshaft, usually by a skew gear mechanism. The skew gear drives only one of the two toothed gear wheels in the pump; the other follows the driven gear. Oil enters the pump through the inlet from the pick-up pipe. Oil is carried round the pump in the spaces between the gear teeth; it is then forced out of the pump into the filter. The meshing of the gear teeth on their inside prevents the oil returning to the sump. If the teeth wear so that the clearance between the gear teeth and the casing exceeds 0.1 mm (0.004 inch) the oil pressure will be reduced and the engine might become noisy, especially when it starts from cold when big-end bearing rattle might be heard.

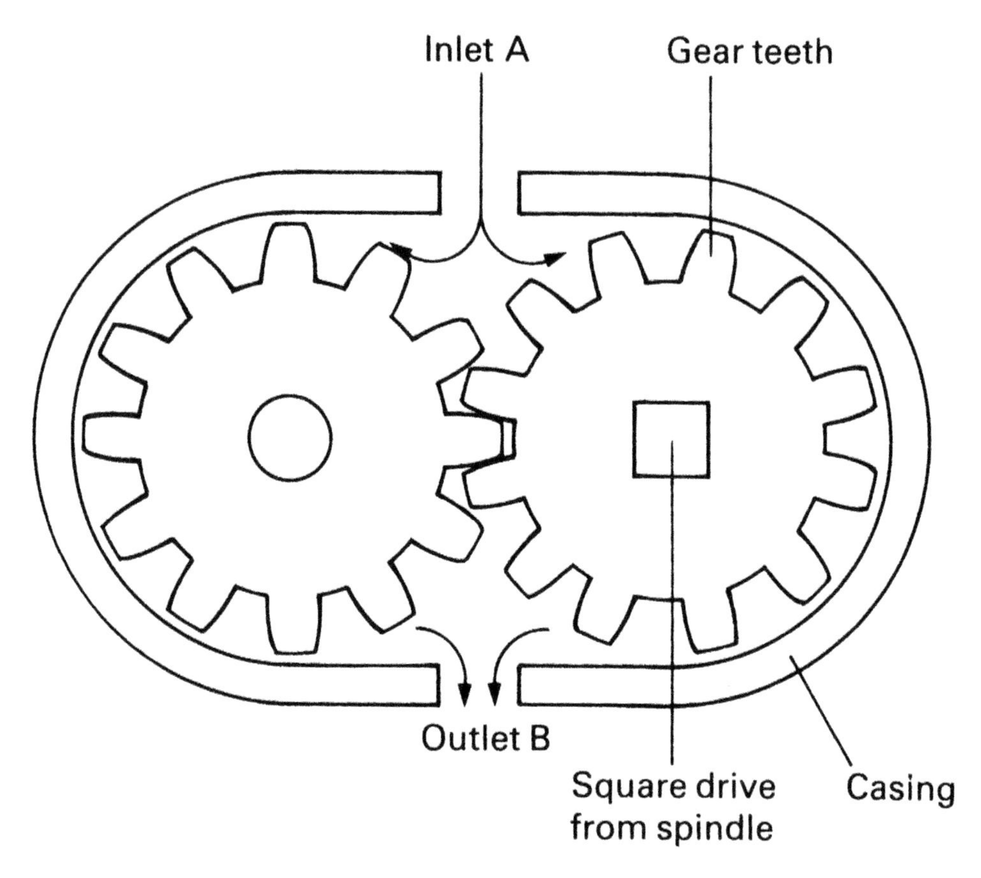

Figure 3.3 Oil pump

Oil pressure relief valve

To prevent the engine oil from exceeding a pre-set figure, an oil pressure relief valve is located in the main oil gallery. Excess pressure may be generated if there is a blockage in the gallery or the engine runs at high speed for a period. The oil pressure relief valve takes the form of a spring-loaded plunger, which is forced away from its seat when the pre-set pressure is reached. Excess oil from the gallery passes through the valve and returns directly to the sump. Removing oil from the main gallery reduces the oil pressure so that the pressure is reduced. When the pressure is lowered the valve is closed by the force of the spring; the oil can no longer return to the sump.

Oil pressure warning light

The purpose of the oil pressure warning light on the dashboard is to warn the driver of low oil pressure. The dashboard light is activated by the pressure-sensitive switch that is screwed into the main oil gallery. The switch makes the light glow when the oil pressure drops below about 30 kPa (5 psi). The light will also come on when the engine is switched

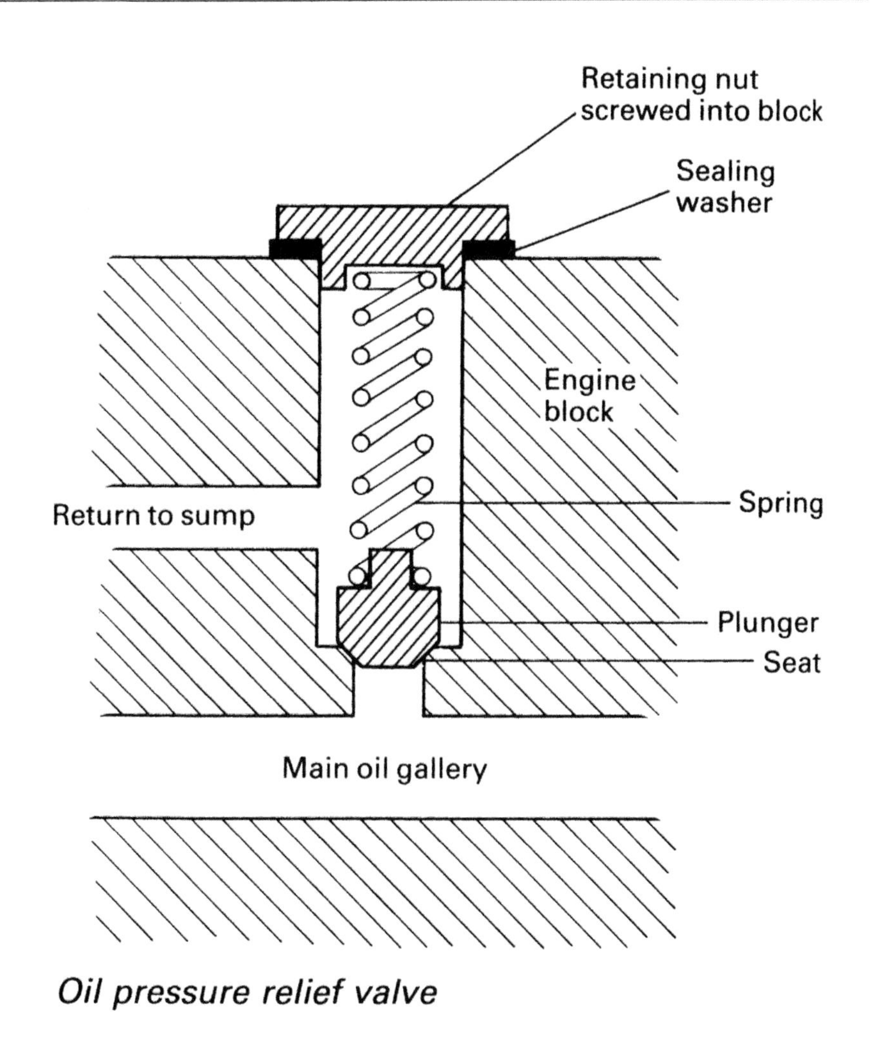

Figure 3.4 Oil pressure relief valve

on but is not running. Higher rated switches are available for tuned engines – typically 100 kPa (15 psi)

Oil filter

It is necessary to keep the engine oil clean and free from particles of metal and carbon that could damage the inside faces of the oil pump or bearing surfaces. This is the job that the oil filter performs. All the oil passes through the filter before it goes to the bearings. The oil filter element is made from a special porous paper; this allows the oil to pass through it but hold back the foreign bodies. The oil filter element is housed in a metal canister that works a sediment trap for the particles that the filter element has prevented from flowing into the main gallery.

Nomenclature

When the moving parts of the engine rub against each other they wear. That is, tiny particles of metal are scraped off or dislodged into the sump or other parts of the engine. Also, the heat and combustion inside the engine create particles of carbon or soot. These particles are referred to as foreign bodies. In filters the heavier foreign bodies drop off the filter walls and fall to the bottom of the canister, or filter housing. This layer of trapping is referred to as sediment.

Servicing the engine lubrication system

Servicing the engine lubrication system entails changing the engine oil and replacing the oil filter. This is done at the intervals set down by the vehicle manufacturer. Typically, this service interval is 16,000 km (10,000 miles). When carrying out lubrication system servicing the following points should be noted:

- The rubber seal on the oil filter must always be replaced
- The sump plug's sealing washer must always be checked and replaced if necessary
- Always check the oil level with the vehicle on a level surface
- After the engine has run, wait for about two minutes before rechecking the oil level and adding more oil
- Never overfill nor under fill the engine; keep the oil level between the minimum (MIN) and the maximum (MAX) marks on the dip-stick

Safety note

Beware of hot engine oil; never change the oil straight after the vehicle has completed a long or fast journey as the oil temperature can be over 110 °C and therefore there is a risk of scalding when removing the sump plug or filter. Always use barrier cream and protective rubber gloves when dealing with dirty engine oil to reduce the risk of contracting dermatitis.

Oil seals

To prevent oil leaking from rotating components an oil seal is needed. There are three different types of oil seals, namely:

- Felt packing
- Scroll seals
- Lip seals

Felt packing, as its name implies, is soft felt packed into a cavity. The rotating shaft turns in gentle contact with the felt. The felt becomes soaked in oil; this reduces the friction against the shaft. The oil also expands the felt so that the seal remains tight on the shaft and an effective oil seal.

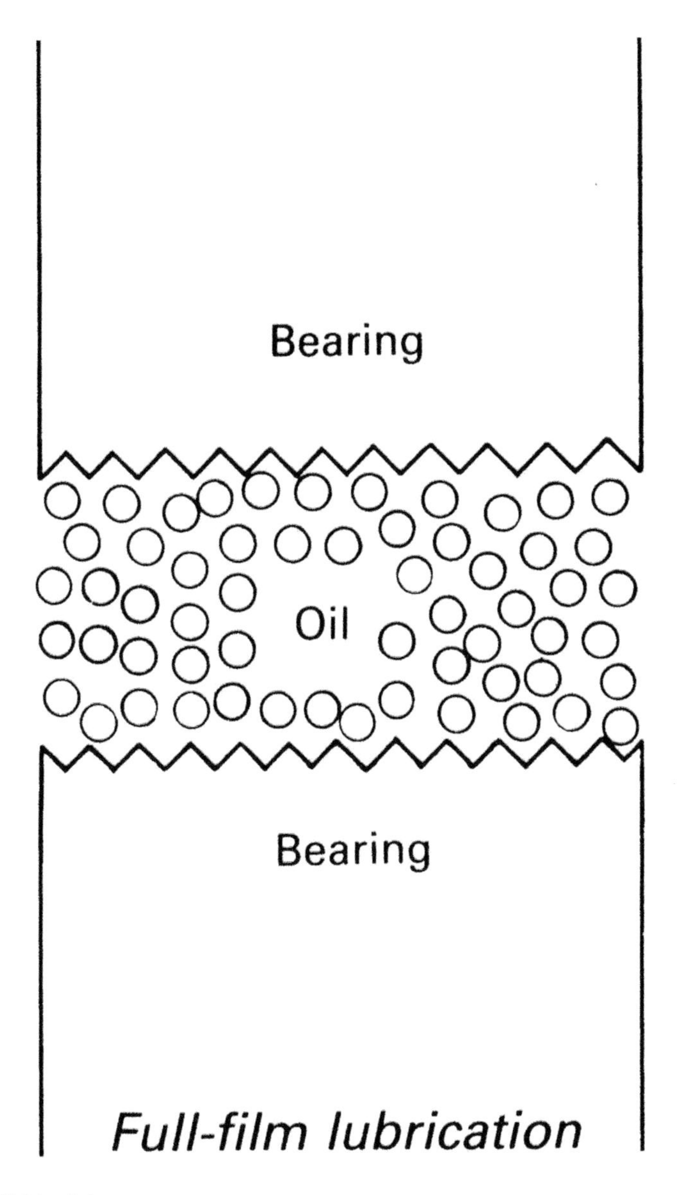

Figure 3.5 Full-film lubrication

The scroll is a groove cut in the rotating shaft that acts like an Archimedean screw. That is, as the shaft rotates the screw draws the oil away from the open end of the shaft and returns it to the sump.

The lip seal forms a hard, spring-loaded knife-edge against the rotating shaft. The pressure exerted by this knife-edge is very high and prevents any oil at all from flowing past the seal under even the most arduous conditions. The lip seal is retained in the oil seal housing by the interference fit of its outer rubber layer.

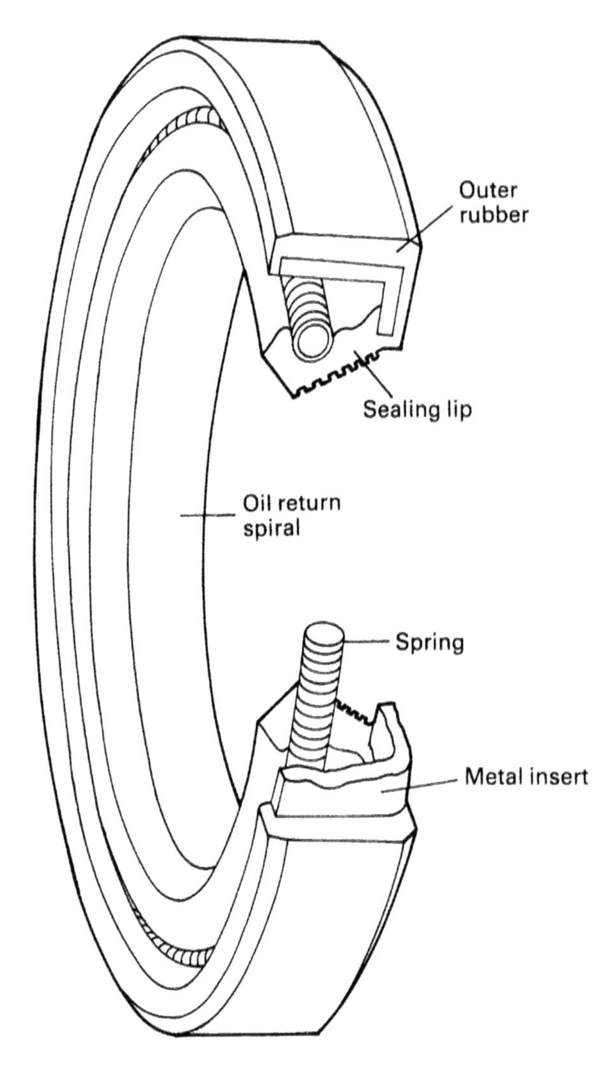

Figure 3.6 Lip-type oil seal

Two-stroke petrol engine lubrication

Lubrication of the two-stroke engine is by a petrol-oil mixture. The lubricating oil is added to the mixture of petrol and air so that it lubricates the crankshaft and big-end bearings when it is in the crankcase, while on its way to the combustion chamber.

The burned lubricating oil is emitted from the exhaust in the form of blue smoke. Because of the effect that this has on the environment, two-stroke engines are only allowed in small motorcycles.

Petrol–oil mixture – low performance type two-stroke engines have their supply of lubricating oil added to the petrol when the tank is topped up. The oil lubricates the main and big-end bearings when the mixture is passing through the crankcase. The rider must add a quantity of oil to the petrol each time the motorcycle is refuelled. The amount of oil will depend on the ratio of petrol to oil, which is required by the engine. A typical ratio is 40:1 (40 parts of petrol to 1 part of oil). In Imperial units this works out at one gallon of petrol being mixed with one-fifth pint of oil. Often the petrol filler cap is designed to incorporate a measure of the correct amount of oil to be added to one gallon of petrol. Or, in SI units this will be one-eighth of a litre of oil to five litres of petrol. The oil used is special two-stroke oil that is made to mix easily with the petrol. However, it is normal practice to rock the motorcycle from side to side after the petrol and oil have been added to ensure that it is a good mixture.

Nomenclature

The term motorcycle is used in this book. Unless they are otherwise especially distinguished, motorcycles includes mopeds, scooters, go-peds and other two- or three-wheeled vehicles.

Total loss system

High-performance two-stroke motorcycles have a separate oil tank, usually located under the rider's seat. Oil is pumped from this by a crankshaft driven pump so that it is mixed with the petrol at the carburetter. This ensures a constant mixture of petrol and oil in the correct proportions. It saves measuring out quantities of oil on the service station forecourt. Sometime an adjustable regulator is fitted so that the petrol–oil mixture can be adjusted to suit different engine running conditions.

The cooling system

Key points

- The purpose of the cooling system is to keep the engine at a constant temperature whilst preventing the overheating of any of the individual components.
- Typically, car petrol engines run at between 80 and 90 °C (180 and 190 °F). This is the best temperature at which to get the best fuel consumption and the least pollution. Diesel engines run about 5 °C (10 °F) cooler.
- There are two main types of cooling system. These are: liquid (water) cooling and air cooling.

Liquid (water) cooling system

FAQs Is water cooling the same as liquid cooling?

Yes, the liquid in the cooling system is water mixed with a number of chemicals and is sometimes referred to by mechanics simply as water. Because it is not just water, its proper name is coolant.

> **Safety note**
>
> Coolant in an engine is likely to be scolding hot and under high pressure. Do not touch cooling system components or remove filler caps when hot.

The liquid cooling system works by using **coolant**, the name for water mixed with other chemicals, to take the heat from the cylinder block and pass it to the radiator so that it is cooled down. That is, the coolant circulates through the engine, where it gets hot, then through hoses to the radiator where it cools down again, finally, back through another hose

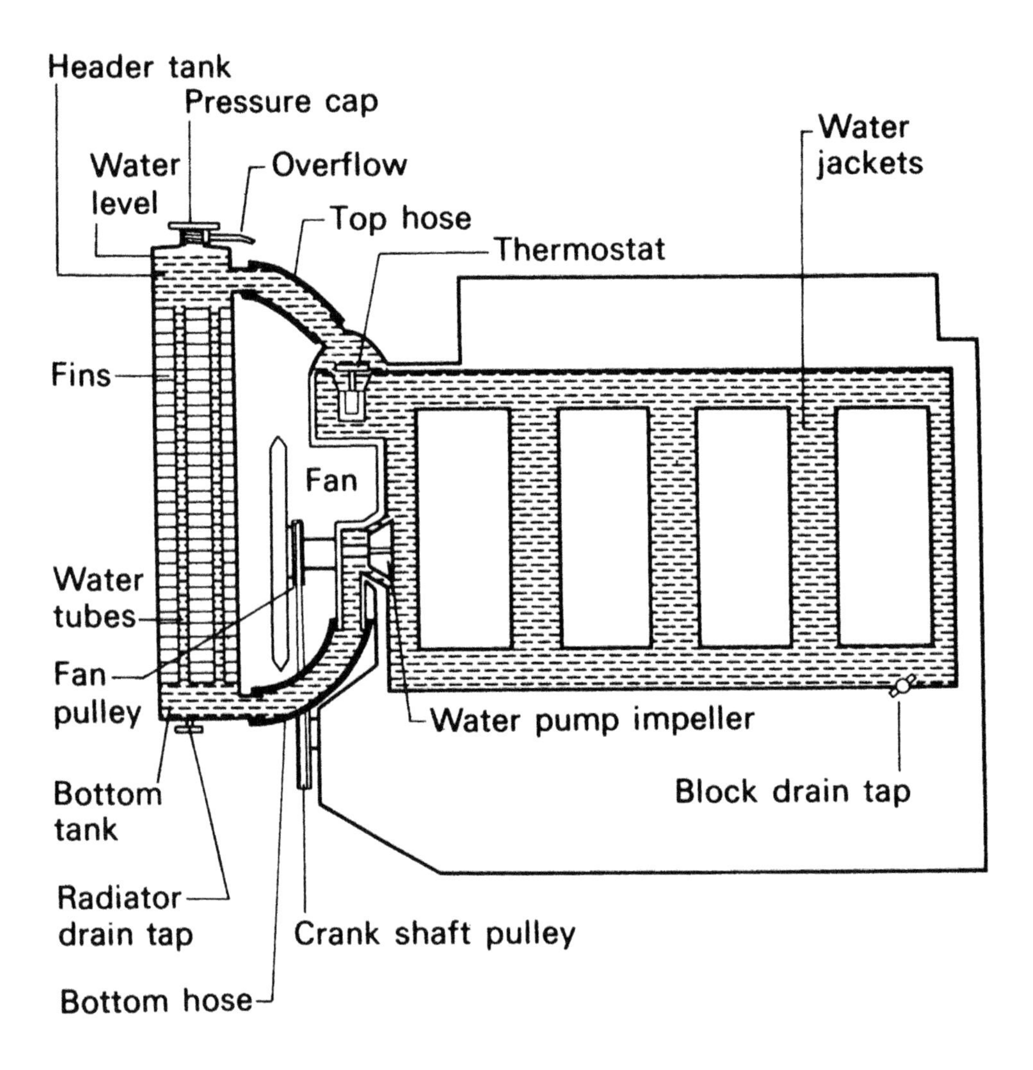

Figure 3.7 Water cooling system

to the engine to go through the process again. When the petrol or diesel fuel is burning in the combustion chamber the temperature gets very hot, about 2,000 °C (3,500 °F). The cooling system therefore needs to work very hard to keep the temperature of the components at about **85 °C (185 °F)**. Most engines run at between about 80 and 90 °C (180 and 190 °F); diesel engines tend to run about 5 °C (10 °F) cooler than petrol engines. The temperature of the engine is kept between 80 and 90 °C(180 and 190 °F) because this is its most efficient temperature, that is to say it will use the least fuel and produce the least pollution.

Coolant has a natural tendency to circulate when heated up in the cooling system. The hot liquid rises, the cooling liquid falls. This is called **thermo-siphon**. As the coolant is heated up in the cylinder block water jackets it rises up; it then passes through the **thermostat** into the **top hose** to the radiator **header tank**. The water then falls through the **radiator core** into the **bottom tank**. As the coolant falls inside the radiator core it is cooled by the incoming air, which passes around the outside of the **radiator fins**. The incoming air is from the front of the vehicle and may be forced along by the fan. The weight of the water in the radiator forces the coolant through the **bottom hose** back into the engine. The **coolant (water) pump** helps the coolant circulate more quickly into the water jackets. For the coolant to be able to circulate, the water level must be kept above the top hose connection so that it can maintain a flow into the radiator.

The coolant

The liquid used in a cooling system is usually called the coolant, because it cools the engine. The coolant is a mixture of **water**, **anti-freeze** and a chemical that **inhibits corrosion** to the metal parts inside the engine. As you probably know, water boils at 100 °C (212 °F) and freezes at 0 °C (32 °F). This means that in winter water could freeze and damage the engine. Typically the coolant mixture boils at about 110 °C (230 °F) and freezes at about –18 °C (0 °F).

FAQs What is anti-freeze made from?

It is a chemical called ethylene glycol.

Coolant can be bought ready-mixed; this is usually advised for specialist engines. For most vehicles anti-freeze, which contains the anti-corrosion chemicals, can be mixed with water. For British winters it is normal to mix a 33% anti-freeze solution, that is 1/3 anti-freeze and 2/3 water, the solution must be measured accurately.

The strength of the coolant can be checked with a special hydrometer called an anti-freeze tester; usually this test must be carried out when the coolant is cold.

Key points

When mixing anti-freeze:

- Always use clean water to prevent damage to the engine
- Check the instructions on the label
- Mix the anti-freeze and water before putting it in the engine

Racer note

Coolant is one of the life bloods of an engine (the other is the oil). So always use the most appropriate coolant for your engine. Look for coolant which contains a wetting agent as well as a corrosion inhibitor and anti-freeze. Aluminium engines are prone to internal corrosion if left only partially drained.

Key points

Anti-freeze loses its properties after about two years, so change it every two years, or at the major services as recommended by the vehicle manufacturer.

Racer note

Coolant temperatures will vary with race conditions; however, whether you win or not you should keep the Champagne in the fridge between 4 and 6 °C (39–42 °F).

Coolant (water) pump

To circulate the coolant quickly a pump is fitted; this is called a coolant or water pump. The coolant pump is fitted to the front of the engine. The bottom hose from the radiator is fitted to the coolant pump inlet so that it is supplied with cooled coolant; the pump outlet is connected directly to the water jacket of the cylinder block.

The coolant pump is driven either by the **fan belt**, or the **cam belt**, depending on the engine design.

Fan

The fan is used to draw in air through the grille or ducting and pass it over the outside of the **radiator fins** so that it cools the coolant inside the radiator core.

The fan may be of the mechanical or the electrical type. The mechanical type is usually attached to the water pump spindle so that it is also driven by the fan belt. The fan will be running all the time that the engine is running. The electrical type is operated by an electric motor which is only switched on when the engine gets hot. There is a temperature-sensitive switch mounted on the radiator, so that when the radiator gets to a pre-set temperature the fan is switched on; when it cools down it is switched off.

Radiator

The radiator is made up of the **header tank**, the **bottom tank** and the **core** between them. The flattened tube type of construction is used on most vehicles; heavy goods vehicles tend to use the round tube variety; the honey comb design is only used on a few luxury or high-performance cars. The radiator fins are to increase the surface area of the cooling zone. That

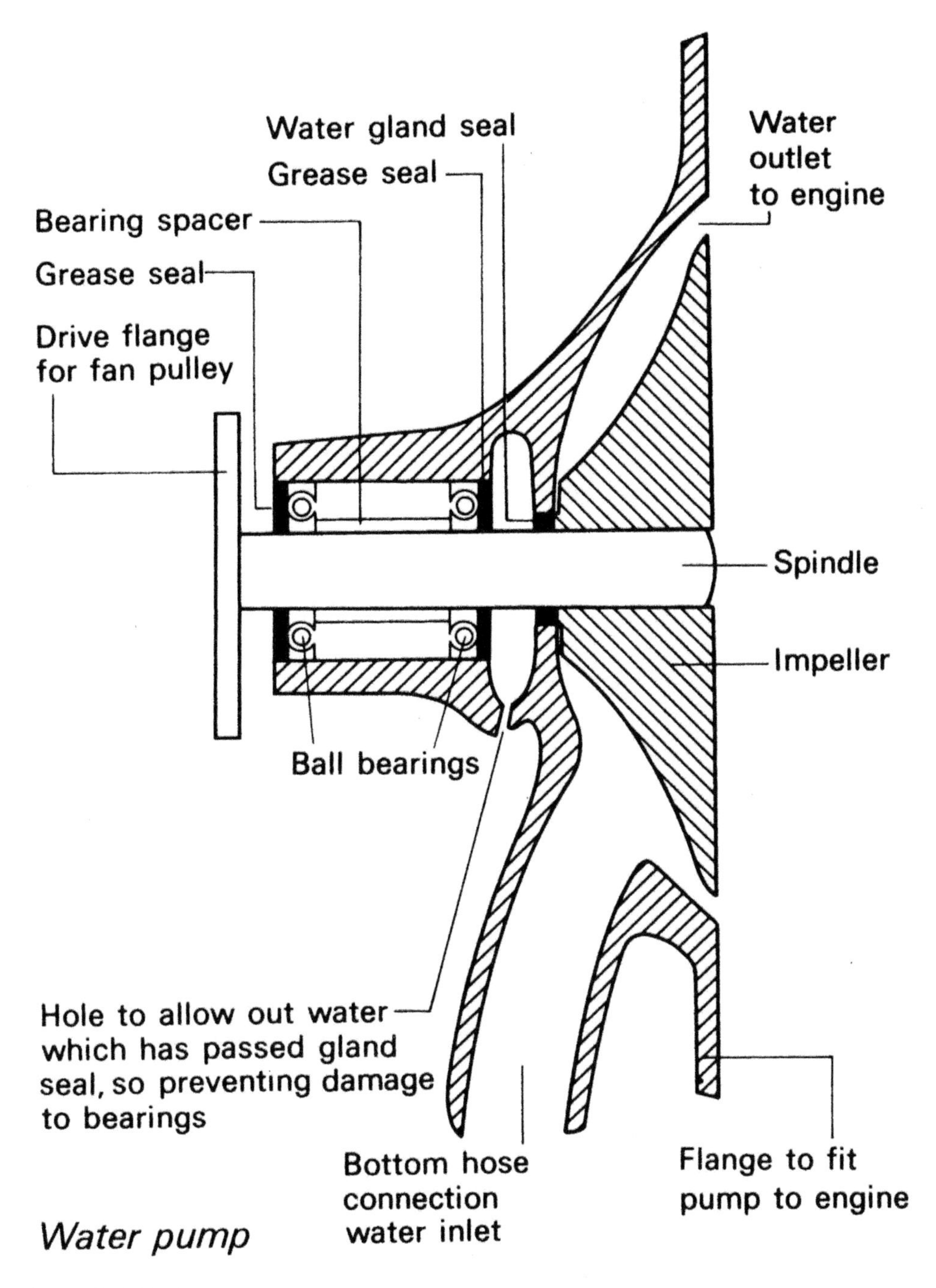

Figure 3.8 Coolant pump

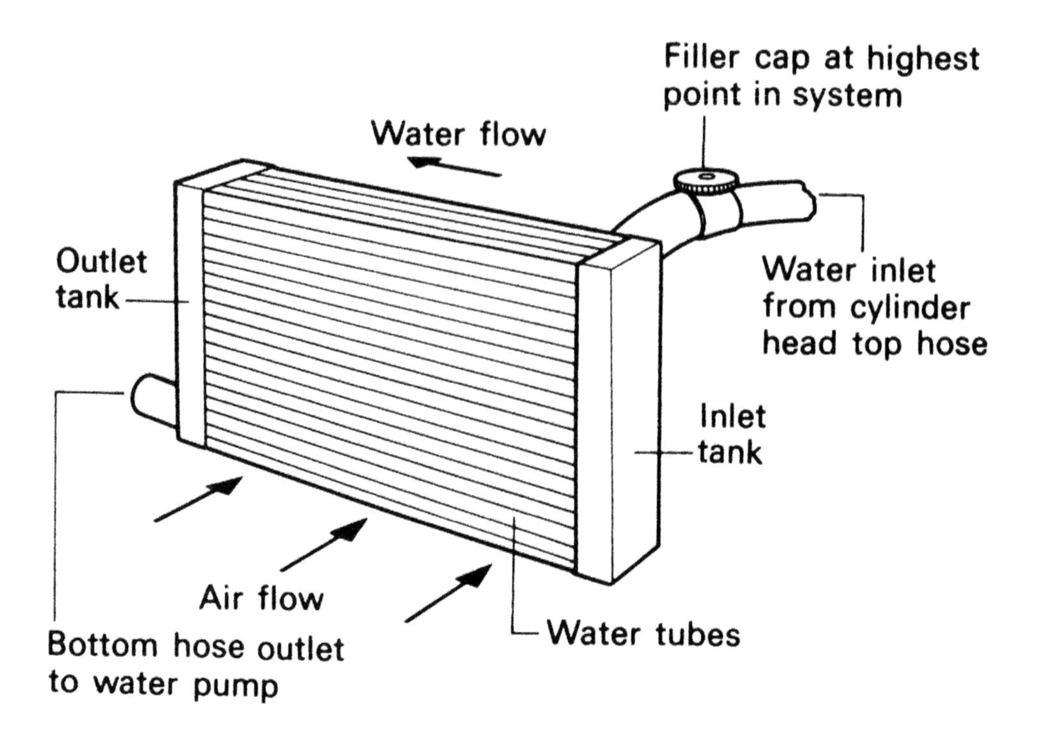

Figure 3.9 Cross-flow radiator

is, the fins dissipate, or spread out, the heat more efficiently so that it is transferred to the air quickly.

The radiator may be either **vertical** in design – with top and bottom tanks stacked with the core, or of **cross-flow** design. The cross-flow radiator has its tanks mounted at each end; this means that the coolant flows from one side of the radiator to the other, rather than from the top to the bottom. The cross-flow radiator is used to give low bonnet height.

Thermostat

The thermostat is fitted in the cylinder head; it is a sort of tap to control the coolant flow between the engine and the radiator. The thermostat is usually in a special connector, one end of the connector holds the thermostat against the inside of the cylinder head, and the other end is attached to the top hose.

When the thermostat is closed the coolant cannot flow; when it is open it can flow freely. The thermostat allows a quick warm-up period by remaining closed until the engine has reached its required temperature. It keeps the engine at a constant temperature by opening and closing as the engine becomes hot or cools down.

The thermostat shown in Figure 3.11 is a wax-stat.

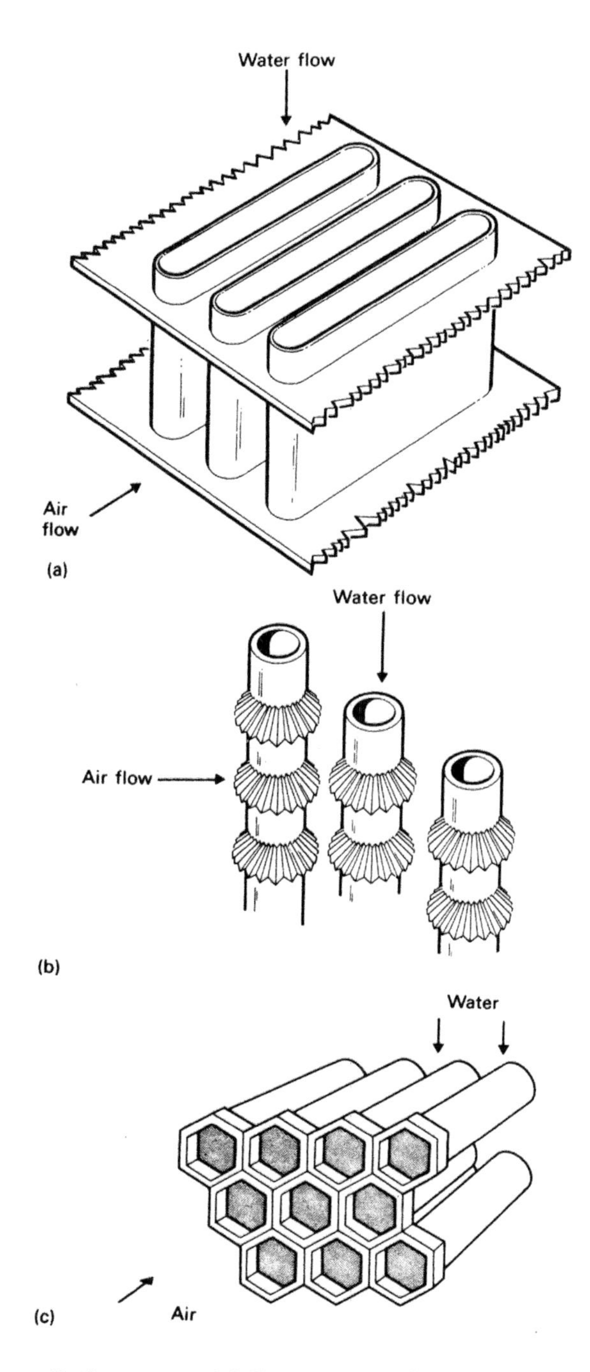

Radiator cores. (a) Flattened tube; (b) round tube; (c) honeycomb

Figure 3.10 Types of radiator fins

FAQs Why is it called a wax-stat?

Because it is a **wax**-operated thermo**stat**.

The metal body or capsule is filled with wax, which expands its volume very rapidly at the temperature at which its designers want it to open, that is usually around 80 °C (180 °F). So

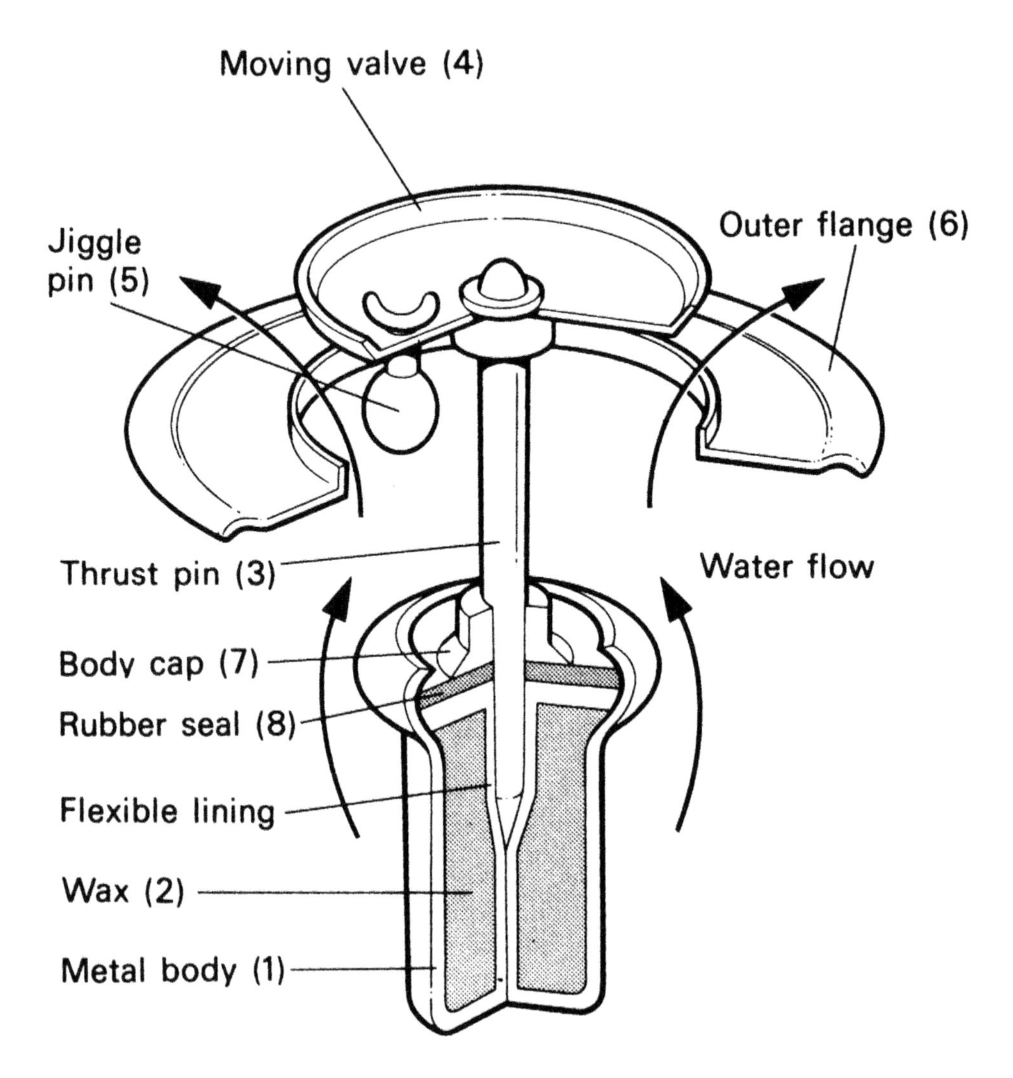

Wax thermostat (open). The valve is closed by an external spring (not shown)

Figure 3.11 Thermostat

when the wax reaches the design temperature it rapidly expands forcing the thrust pin out of the capsule against the pressure of the return spring (not shown). The thrust pin lifts the moving valve to allow coolant to flow from the engine to the radiator through the top hose. When the wax cools it contracts, the return spring closes the valve and returns the thrust pin into the capsule. The jiggle pin is fitted to prevent the formation of a vacuum in the engine water jacket by allowing small amounts of coolant to flow even when the thermostat is closed.

> **Key points**
>
> You'll find the opening temperature of the thermostat stamped on either the rim or the capsule as a two-digit number.

Radiator pressure cap

> **Safety note**
>
> Never remove the radiator cap when the engine is HOT or RUNNING.

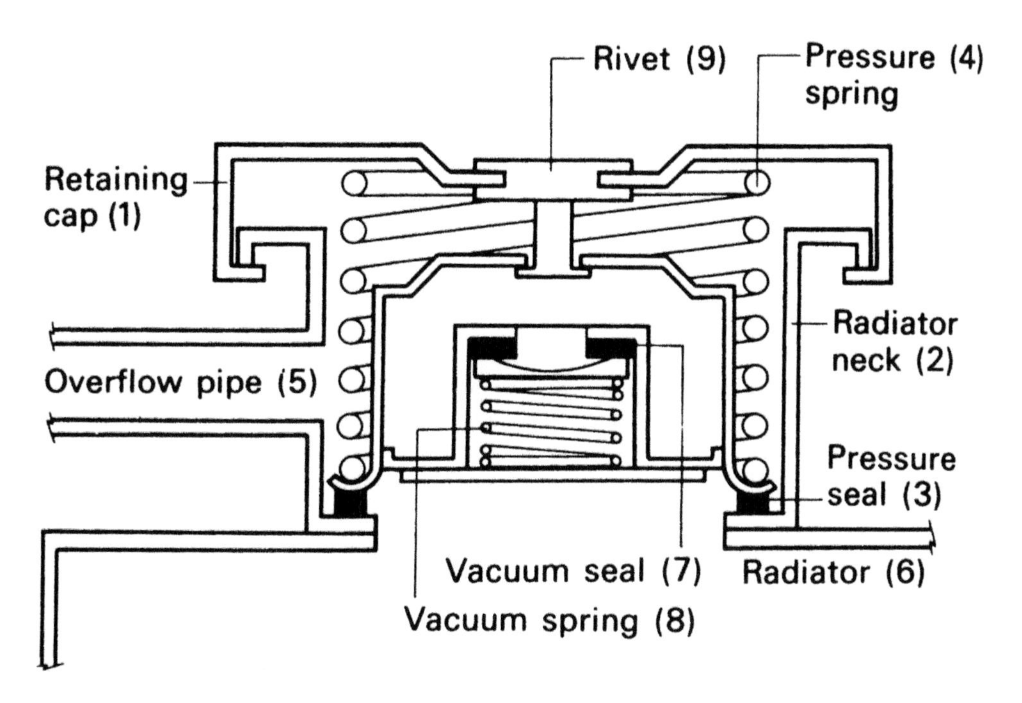

Figure 3.12 Radiator pressure cap

At normal atmospheric pressure, water boils at 100 °C (212 °F). At high altitudes the boiling temperature is reduced. This also applies to the coolant mixture, though the temperatures may be slightly different. To prevent the engine from boiling and overheating, a radiator pressure cap is fitted. The force of the spring in the cap ensures that the coolant is kept under pressure. The higher the pressure, the higher will be the boiling point of the coolant in the system. To prevent the build-up of a vacuum inside the radiator, a vacuum valve and spring are fitted. The vacuum valve prevents the radiator from imploding, or collapsing inwards, when the coolant temperature is reduced and the coolant pressure therefore decreases.

Key points

Radiator caps are made to operate at different pressures – the design pressure is usually written on the outside of the cap – this may be in pounds per square inch (psi) or bars.

The actual retaining cap is secured to the radiator neck with two curved sections. The radiator cap is fitted to the radiator in a similar way to the lid on a jam jar. The pressure seal is held in place against the top the top of the radiator by the pressure spring. Only when the coolant pressure exceeds the pre-set figure, which is stamped on the radiator cap, is the seal lifted against spring pressure. The coolant is then released through the overflow pipe. When the pressure of the coolant in the radiator is released by the coolant running off, the spring will press the seal back onto the top of the radiator. When the radiator cools its coolant pressure decreases. Air pressure from outside the radiator forces the vacuum seal against the force of the vacuum spring to allow air to pass over the seal into the radiator.

Failure of the radiator pressure cap causes overheating and boiling over. The operating pressure of the cap can be tested with a special pressure gauge and the condition of the seals can be inspected visually.

Sealed cooling system

To prevent the loss of coolant, and hence the need for topping up the radiator, most vehicles use sealed cooling systems. An **overflow tank** is fitted to the side of the radiator – sometimes this can be a little way from the radiator on the inner wing panel. A tube from the radiator neck is connected to the overflow tank. Therefore, any coolant that is allowed past the radiator cap will go into the overflow tank. The tube is arranged so that its end is always below the coolant level, therefore, when the radiator cools down and the coolant contracts, the coolant in the overflow tank is drawn back into the radiator to fill the available space. The overflow tank must be kept partially full of coolant to ensure that coolant covers the bottom of the tube.

Hoses

Rubber hoses are used to connect the engine to the radiator so that there is some flexibility between the two components. The engine is free to move slightly on its rubber mountings but the radiator is rigidly mounted to the vehicle's body/chassis. The flexing of the hoses causes them to deteriorate, they usually crack and, if not replaced, they will break or puncture causing the total loss of coolant. The hoses are held in place with clips – the clips usually have a screw mechanism to tighten them up.

> **Key points**
>
> Hose clip drivers are available with hexagon ends; these are much safer to use and less likely to slip and cause damage to the radiator than a screwdriver.
>
> When you are fitting a new hose always check that the surfaces are clean and if possible use a new clip.

Air cooling

Air cooling is used on most motorcycles, a few specialist cars – like old Volkswagens and Porches – and some specialist agricultural machinery.

Air cooling has the advantages of not using a liquid coolant (water) and using fewer moving parts. Having no water, it cannot freeze or leak. However, air-cooled engines tend to be more noisy than liquid-cooled ones. The air-cooling system operates by air entering through the flap valve. The fan, which is driven by the crankshaft pulley, forces the air over the fins on the cylinders. The air is then discharged back in to the atmosphere. The flap valve is controlled by the thermostat, this opens when the engine is hot so allowing air to enter. The flap valve is closed when the engine is cold; this restricts the airflow to give the engine a quick warm up to operating temperature.

Fan belt

A V-shaped fan belt, sometimes called a 'V' belt, is used to drive the fan, the water pump and the alternator. It is important that the belt is free from cracks and shredding. It must be adjusted to give between 10 and 20 mm (½–¾ inch) of free play on its longest side. The fan belt adjustment is usually carried out by moving the alternator on its slotted elongated mounting bracket. The alternator bracket is usually attached to the cylinder block at one end, and given an elongated slot at the other where the alternator is attached. To check the adjustment you should pull and push the fan belt with your finger and thumb on the longest section. Ensure when you are doing this that the engine is switched off and the key is removed.

> **Safety note**
>
> Do not work on a running engine; keep your tie and jewellery from the fan belt area.

Some water pumps are driven by the cam belt; and some fans by electric motors.

Modifications

Lubrication

The first modification to the lubrication system is to use good quality oil and oil filter. There are lots of alternatives for each type of engine, and lots of different makes of each type. The supply of oil to the motorsport industry is both high value and high profit. So

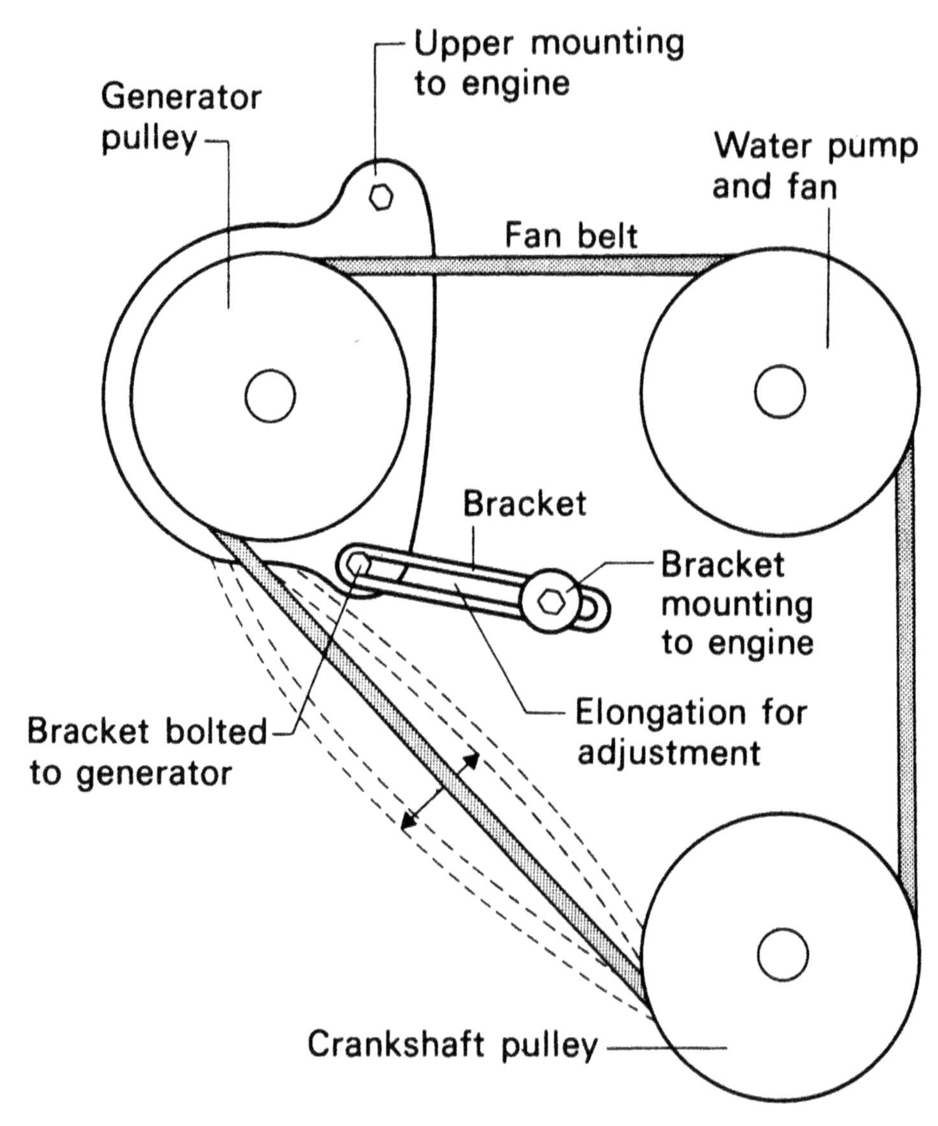

Figure 3.13 Auxiliary drive for water pump and fan

some detailed care in the choice of oil is needed. Many manufacturers have a preferred brand, which is often one of their sponsors. For example, Ferrari recommend Shell and Shell sponsor Ferrari.

Oils are classified by the Society of Automotive Engineers (SAE an institution based in America) for viscosity (how they flow); the common grades are 10W – 40 and 15W – 40. The

Table 3.1 Examples of API oil classification

No	*API classification*	*Application*	*Comment*
1	SJ	Cars made in 2001 or before	
2	SL	Cars made in 2004 or before	
3	SM	Cars made in 2010 or before	
4	SN	Current – introduced in 2010	Check for updates

capability of a particular make of oil to do its job in an engine is classified by the American Petroleum Institute (API). Table 3.1 gives some example; see their website (www.api.org) for more details (useful websites are listed separately in this book).

Actual mechanical modifications that are typically made to the lubrication system are:

Relief valve – the operating pressure of the relief valve may be changed; this may be by changing the spring, adjusting the spring setting or replacing the complete assembly. Usually increasing the oil pressure by about 10% will cope with greater bearing loads caused by engine power increases. Typically oil pressure is 60–100 psi (4–7 bar).

Oil cooler – this is to control the oil temperature and prevent engine damage. Typical oil temperatures in degrees Celsius are: road car 100–120; high-performance car 120–150; race car 150 plus. The oil cooler must be fitted in a place where cold air flows freely – often this is just in front of the engine coolant (water) radiator. The flow of oil through the cooler may be controlled by a thermostat. The oil flow is usually in series with the oil filter.

Sump baffles – these prevent oil surge on corners, so preventing oil starvation because the oil has moved away from the oil pickup; or oil burning by it climbing the cylinder walls. Baffle plates may be welded inside an ordinary sump.

Oil pickup – the modification to this part include fitting a broader mesh filter to allow greater oil flow and moving the open end to the centre of the sump to avoid oil starvation (see sump baffles).

Oil filter – a high flow filter may be fitted to improve oil flow or it may be remotely mounted for ease of changing and incorporation with the oil cooler.

Dry sump – instead of the lubricating oil being in the sump underneath the crankcase a separate oil tank is used. The oil from the engine components fall into the sump; but it is drawn out of the sump to the tank by a separate – scavenge – pump. The pressure pump draws oil from the tank to pressure feed the engine's bearings. This gives the advantages of: lower centre of gravity as the engine can be lowered in the chassis as the sump is very shallow; extra cooling of the oil; reduction of surge and foaming problems.

Questions and skills

1 Using workshop manuals and manufacturer's information sheets, compile a chart to show the recommended lubricants for a few cars of your choice.

2. Dismantle an engine and follow the oil flow path from the sump, through the engine and back to the sump again.
3. Some components are lubricated by pressurised grease from a grease gun through a grease nipple. Describe how a grease gun works.
4. Using the workshop manual for a car of your choice, find out the operating temperature and pressure of the cooling system.
5. On a vehicle of your choice, describe, step by step, how to check the coolant level.

Chapter 4

Transmission

This chapter is about the getting the power and torque transmitted from the engine to the road wheels. If you increase the power output of the engine or fitted a different engine/power unit you will need to ensure that the transmission can cope.

Clutch

The purpose of the clutch is to **transmit the torque**, or turning force, from the engine to the transmission. It is designed so that the drive can be engaged and disengaged smoothly and easily. By disengaging the drive, the clutch allows the gears to be changed smoothly and it provides a temporary neutral position. This allows the transmission gears to be engaged or disengaged whilst the engine is running.

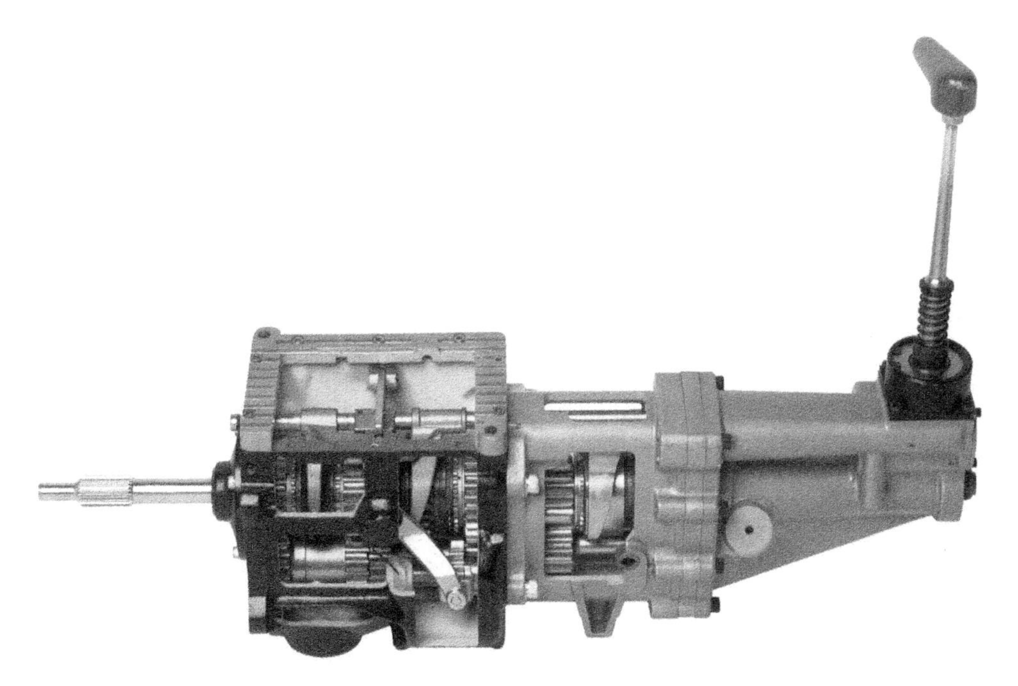

Figure 4.1 Sectioned gearbox

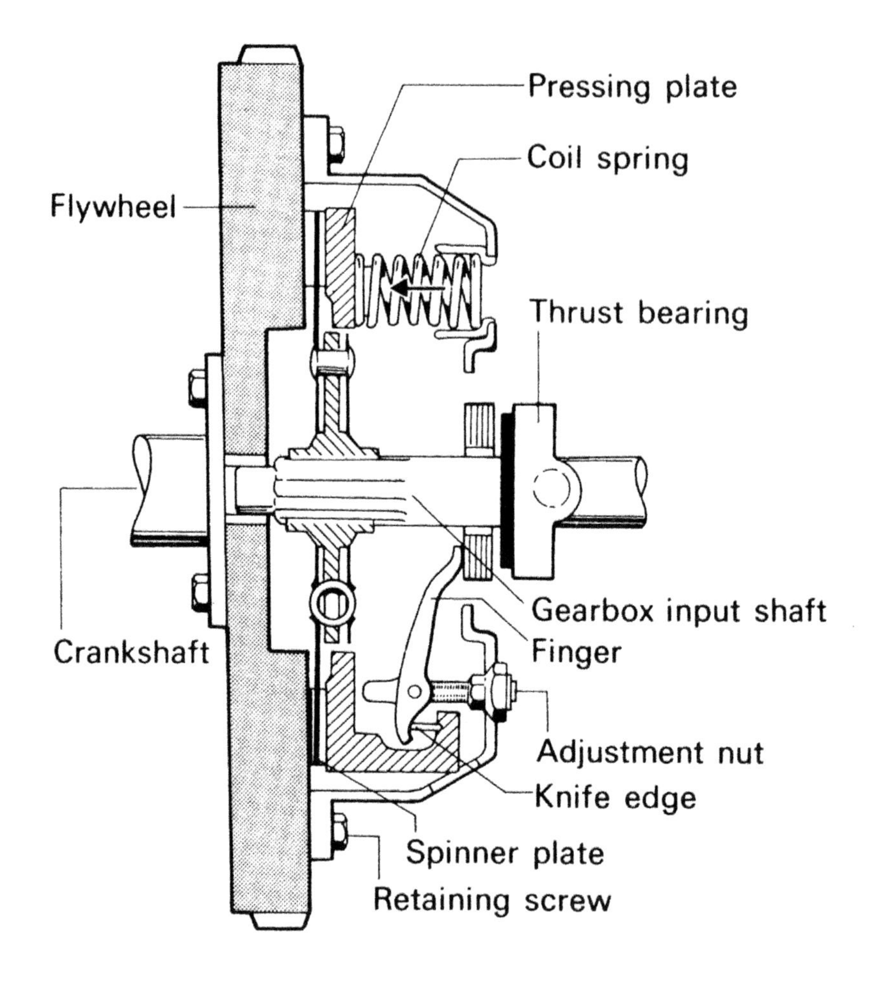

Figure 4.2 Clutch engaged

The clutch assembly is contained in the housing between the engine and the gearbox. On front-engined rear-wheel drive cars this is called the **bell housing**; on other cars it is just called the clutch housing, or clutch casing. The main components of the clutch are the pressure plate, the spinner plate and the thrust bearing. The flywheel also has the job of being part of the clutch as well as its other two functions. The engagement and the disengagement of the clutch are carried out by a foot pedal-operated mechanism on most popular vehicles.

Key points

- The main components of the clutch are the pressure plate, the spinner plate and the thrust bearing

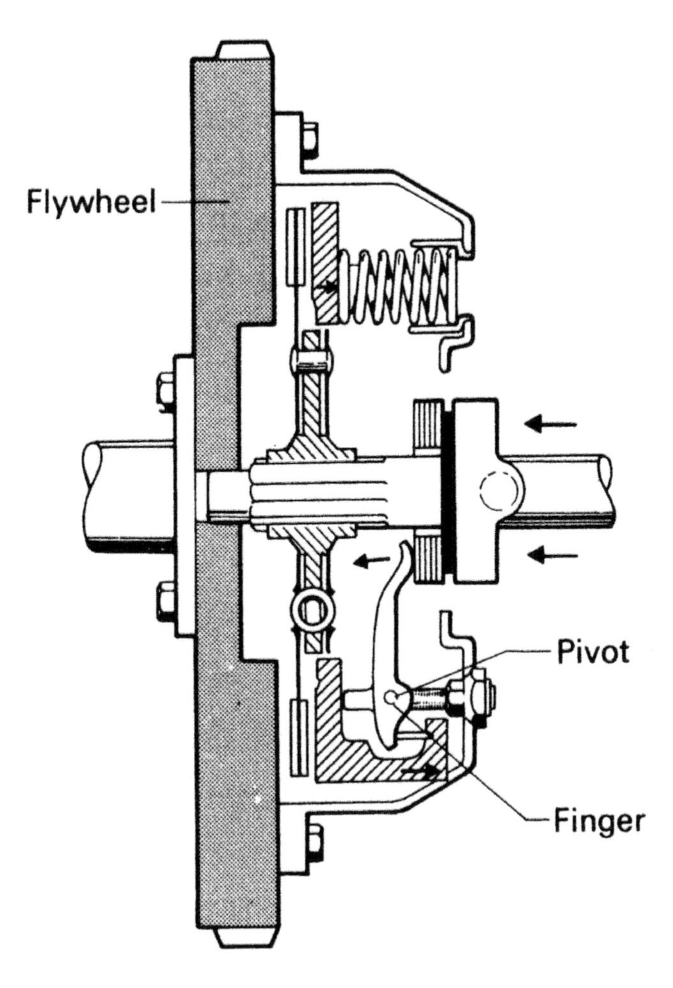

Figure 4.3 Clutch disengaged

- Most clutches have a diaphragm-type spring
- The spinner plate may have a solid or a sprung centre
- Clutch dust is a potential health hazard
- Clutch adjustment is important to ensure that it does not slip

Safety notes

The dust from the clutch plate is a health hazard and breathing it in may cause respiratory problems; use a breathing mask if appropriate. Also clutch components may be hot to touch if the vehicle has just been running; use of mechanics gloves is advised.

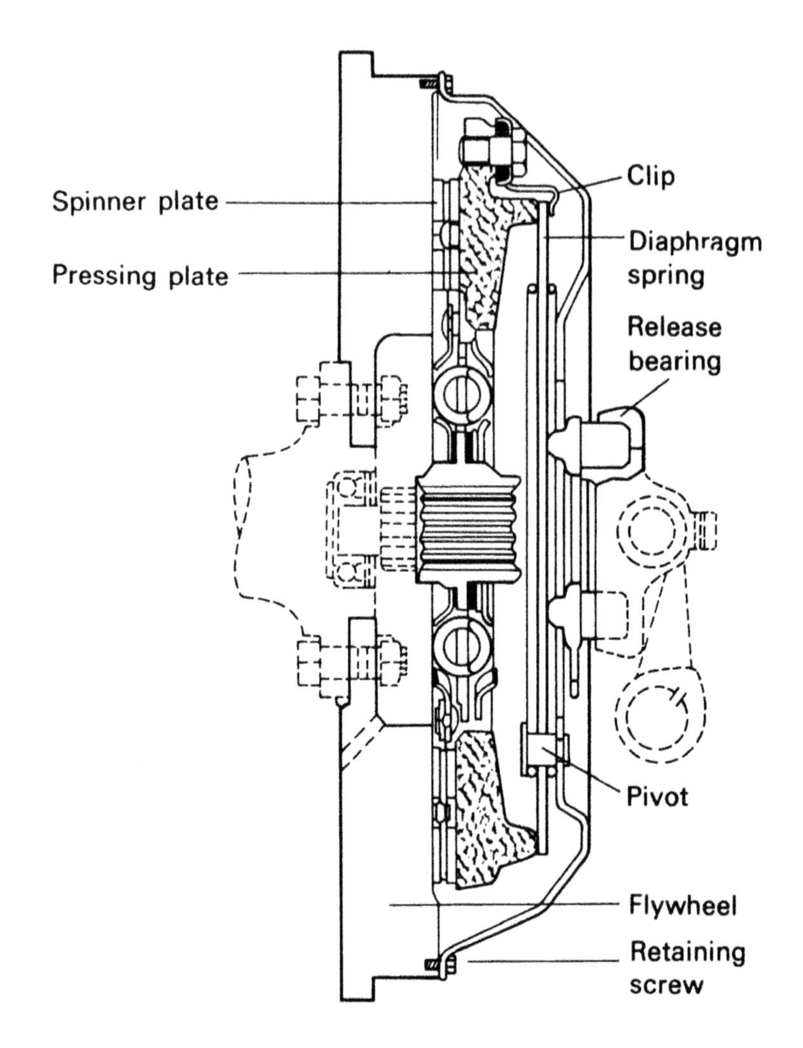

Figure 4.4 Diaphragm-spring clutch engaged

Transmission of torque

The transmission of torque from the engine to the gearbox depends on the strength of the diaphragm spring in the pressure plate, the diameter of the spinner plate, the number of friction surfaces and the co-efficient of friction of the clutch materials. The stronger the spring and the larger the diameter of the spinner plate, the greater the torque that can be transmitted. The clutches of trucks are up to three times the diameter of those on cars, and they can weigh about ten times more. On motorsport vehicles the clutches are of small diameter but have several friction surfaces – typically six plates which means 12 friction surfaces.

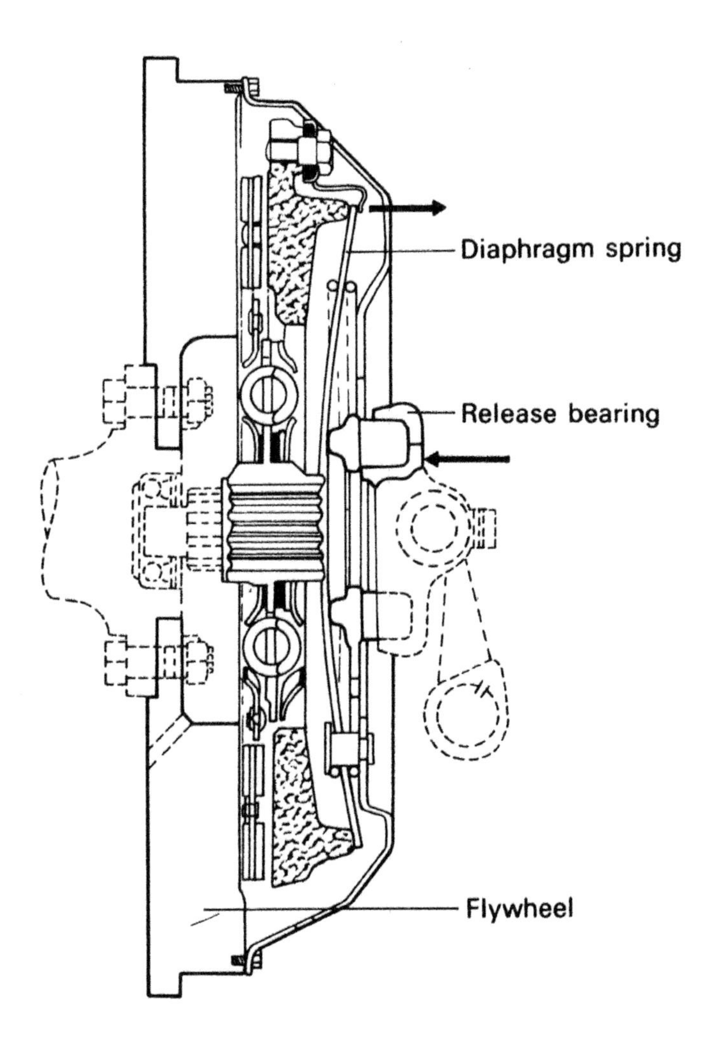

Figure 4.5 Diaphragm-spring clutch disengaged

Diaphragm-spring clutch

The diaphragm-spring clutch is used on most cars. The diaphragm spring is shaped like a saucer, or a deep dinner plate, with a series of radially cut grooves.

In the engaged position the diaphragm spring is shaped like a saucer. The force of the outer rim forces the pressure plate against the spinner plate. The pressure plate cover is bolted to the flywheel and the pressure plate is attached to the cover with flexible metal straps. So that when the flywheel rotates the cover rotates, the cover turns the pressure plate by means of the **metal straps**. Most engines rotate in a clockwise direction when viewed from the front; as the straps must pull, not shove, the clutch must be assembled so that the straps pull in an anti-clockwise direction when seen from the rear.

The diaphragm-spring pivots on the cover using rivets with shoulders as **fulcrum points**. A fulcrum is another name for a pivot, something to swing on. It is against these rivets that the spring forces itself to transmit force to the pressure plate and the flywheel to transmit the drive.

To disengage the clutch the thrust race presses in the middle of the diaphragm spring. This causes the spring to pivot on the shoulders of the rivets so that it lifts its outer rim. This is similar to the action of a jam jar lid, or a CD in its case. The lifting of the outer rim pulls the pressure plate away from the spinner plate so that the spinner plate can rotate freely. The clutch is now disengaged so that the drive is not transmitted to the gearbox.

Racer note

The diaphragm-spring clutch is used on most modern vehicles; but old race cars and other vehicles sometimes use coil spring clutches. It is useful to compare the two as an example of the development of automotive technology and sound engineering principles. The diaphragm-spring clutch has the following advantages: it is smaller – having only one flat spring, it is lighter – as it uses less metal, it has fewer moving parts to break or wear, as it has only one spring it is consistently smooth.

Spinner plate

The spinner plate consists of a **steel hub**, which fits on to the splines of the gearbox input shaft. Attached to the hub is a disc that carries the friction material. The friction material is riveted to the disc. The friction material on old vehicles is asbestos. Asbestos dust is a very serious health hazard; breathing asbestos dust can lead to the disease of asbestosis and subsequent death.

Modern vehicles use an asbestos substitute; however, you should remember that breathing in any dust is to be avoided as it can be harmful to your throat and lungs. Most workshops have a special vacuum cleaner to suck up clutch and brake dust. The clutch parts, which are being reused, should be cleaned down with a special cleaner; usually this is called *brake cleaner* because it is used for cleaning brake dust (this is the same as clutch dust). The rags and other materials, along with the dust, must be disposed of safely in special plastic sacks.

There are two types of spinner plate in common use. These are: the **solid centre** type and the **sprung centre type**. With the solid centre type the drive is directly from the friction lining to the hub. The sprung centre type has a series of coil springs to transmit the drive from the disc, which hold the friction disc to the hub. The springs serve two purposes, they:

- absorb the shock loads when the clutch is suddenly engaged
- absorb the small fluctuations in engine speed and vibrations, giving a smooth transmission of power to the gearbox.

Racer note

Motorsport vehicles usually have solid centre clutch plates. Often these are of the paddle type.

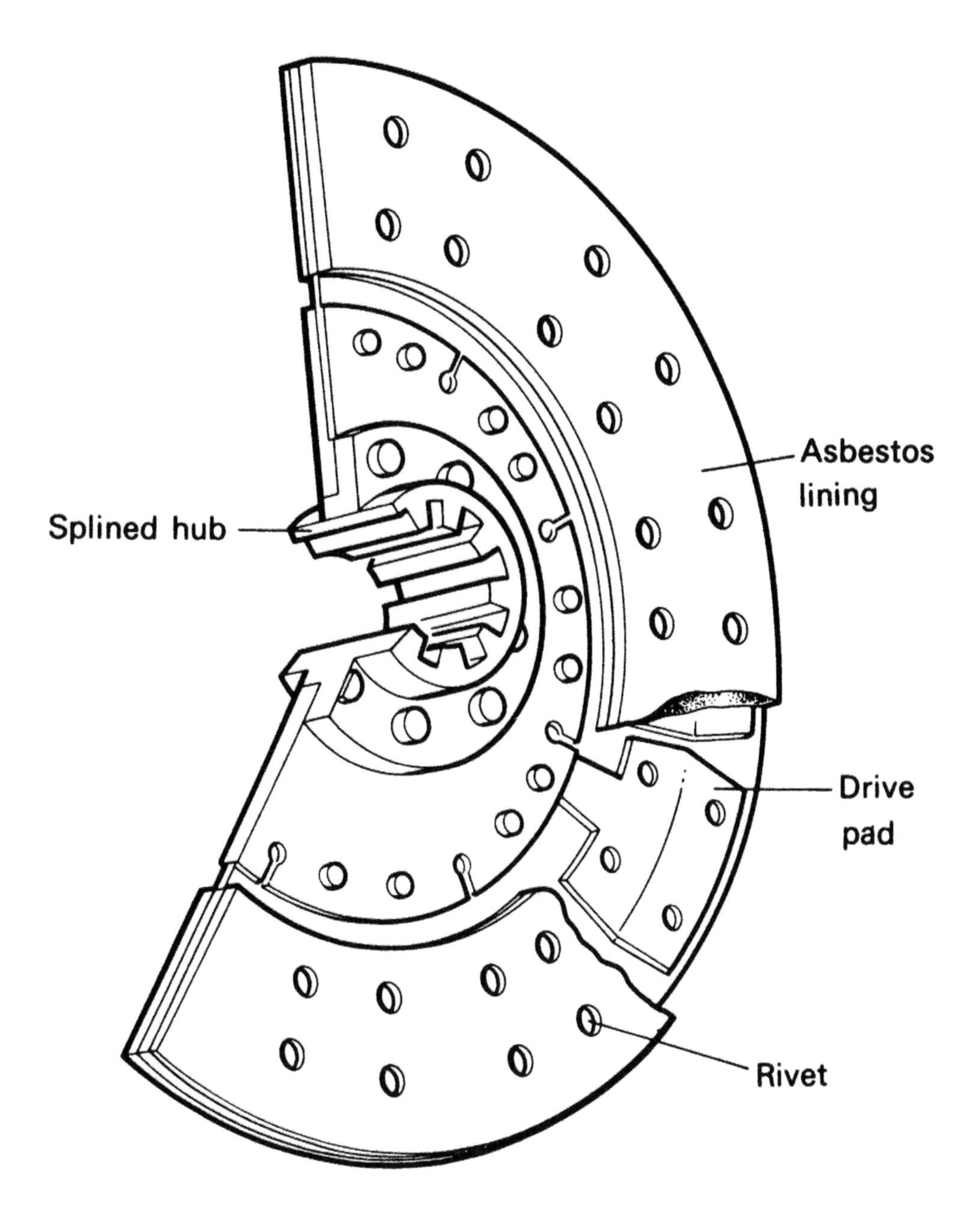

Figure 4.6 Solid centre clutch plate

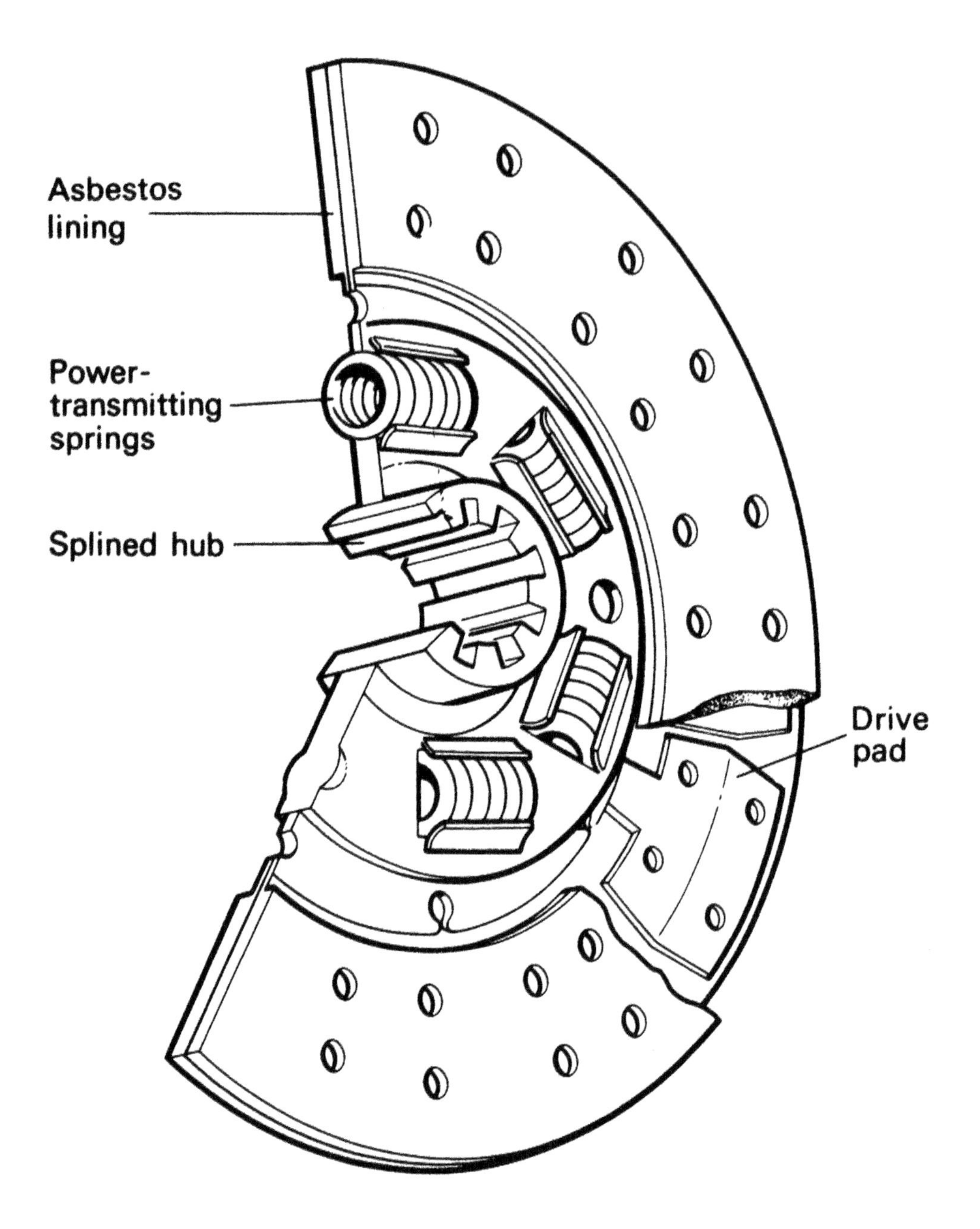

Clutch (friction disc); sprung-centre spinner plate

Figure 4.7 Sprung centre clutch plate

FAQ

What is meant by the clutch is engaged or disengaged? The clutch is engaged when it is in a position to transmit the drive from the engine to the gearbox. That is the clutch spring is holding the spinner plate tightly between the pressure plate and the flywheel, the pedal is in the up position. When the driver presses the clutch pedal to the floor the clutch is disengaged. That is, the pressure is taken off the spinner plate so that is can rotate freely between the pressure plate and the flywheel. When the clutch is disengaged the drive from the engine is not transmitted to the gearbox.

Operating mechanisms

The clutch is disengaged by the driver pressing down the pedal on the left; it is engaged by releasing the pedal. When not operating the pedal, the driver must rest his or her foot away from the pedal. Some cars have footrests for this purpose. The operating mechanism must have a small amount of free play to ensure that the clutch is fully engaged. The pedal action operates the thrust race that moves the clutch diaphragm. The connection between the pedal and the thrust race may be by one of several different linkages. You'll find the popular ones are rod, cable and hydraulic; we'll look at each different type in detail.

Rod clutch

The rod clutch method of operation is the simplest. A steel rod is connected between the pedal and the clutch cross-shaft. When the driver depresses the **pedal,** the **rod** is pulled and the **cross-shaft** is rotated. The clutch cross-shaft transmits the movement from the rod to the clutch thrust bearing. You'll find these on old vehicles and some industrial plant such as forklift trucks. The problem with this mechanism is that is can transmit engine vibrations to the clutch pedal.

Cable clutch

The cable clutch is used on a large proportion of cars and vans. The inner twisted wire **Bowden cable** moves inside a steel outer guide cable (sometimes called a sheaf). The pedal pulls the inner cable, which in turn moves the clutch cross-shaft. The outer cable acts to provide a guide and controls the length for adjustment purposes. This system is similar to a bicycle brake or gear cable; it is adjusted in a similar way – that is by using a screwed nipple to change the length, taking out the slack. The flexibility of the cable means that vibrations from the engine are unlikely to be transmitted to the clutch pedal.

Hydraulic clutch

Both the rod and the cable mechanisms rely on mechanical linkages, the lengths of the levers effect the mechanical advantage. A more sophisticated system is the hydraulic clutch. This system uses **hydraulic fluid**, like that used in hydraulic brakes, to transmit the movement from the clutch pedal to the cross-shaft in the bell housing. This system has its mechanical advantage; the difference between the force applied by the driver's foot on the pedal and the actual force that the clutch thrust race applies to the pressure plate diaphragm is built into the hydraulic system.

The clutch pedal moves a pushrod, which in turn pushes a **hydraulic piston** into the clutch **master cylinder**. The hydraulic fluid above the piston is forced along the connecting tube. In turn the fluid forces the **clutch cylinder** piston against a short operating rod, which transmits the force to the clutch cross-shaft. These systems are very smooth in operation, being used on many cars and trucks. It is essential to ensure that the clutch master cylinder reservoir is kept topped up to the correct level with the correct type and grade of hydraulic fluid. Most cars and trucks use the same fluid for both the brakes and the clutch.

Nomenclature

Mechanical advantage is the leverage gain, or torque multiplication, given by a mechanical mechanism. For instance, you might need to apply a force of 100 N to the clutch pedal and move the pedal 40 mm to disengage the clutch. The linkages in the clutch mechanism may have increased this force at the thrust race to 1,000 N and reduced the distance travelled to 4 mm. The mechanical advantage would be 10 to 1. It is the same principle as when you jack up a car weighing over a tonne using only one hand. The hydraulic jack uses the different sizes of piston to achieve the same results. The hydraulic clutch works in a similar way to the hydraulic jack.

Clutch adjustment

It is essential that there is enough **free play** in the clutch mechanism for the thrust bearing to be clear of the pressure plate. This is needed to ensure that the clutch does not slip. It is also important that there is not too much free play, otherwise the clutch may not be able to be disengaged completely.

The normal amount is 2–4 mm (1/8th inch) of free play at the thrust race, or cross-shaft lever. This will be the equivalent of about 20 or 25 mm (1 inch) of free play at the pedal.

Adjustment is by screw threads on the cable of operating arm. However, many clutches are self-adjusting by means of a ratchet type of device.

Clutch faults

The main faults likely to occur on a clutch are slipping and grabbing. Slipping is when the clutch is not transmitting the drive; it might be brought about by:

- the spring being fatigued or tired
- oil on the friction lining
- the friction lining being worn down to the rivets
- lack of free play

Grabbing is when the clutch cannot be engaged smoothly. That is the clutch suddenly grabs and takes the drive up with a thud. Grabbing may be caused by:

- spring damage or uneven wear
- wear in the mechanism or friction lining
- a broken spinner plate hub

Transmission

Key points

- The transmission includes the gearbox, the final drive gears and the propeller shaft and/or the drive shafts
- The gearbox allows the engine speed to be varied to suit the road conditions
- The final drive gears give the vehicle its overall gear ratio
- Universal joints and constant velocity joints are used in the transmission
- To allow the driven wheels to turn at different speeds when cornering a differential gear is used
- Straight cut, helical, double helical and epicyclic gears are used

Component layout

The gearbox is fitted behind the engine and the clutch on **conventional layout** vehicles – that is front-mounted engine driving the rear wheels. The **bell housing** that covers the clutch is usually part of the **gearbox casing** and connects the gearbox to the engine. On conventional cars a ring of bolts around the bell housing secures the gearbox to the engine. The weight of the gearbox is supported at the rear by a cross-member that attaches the gearbox to the chassis. The gearbox is held on the cross-member using a rubber mounting; this gives flexibility and prevents vibrations from being passes to the chassis and body.

A similar layout of components is used on both **front wheel drive** (FWD) and **rear wheel drive** (RWD) vehicles, excepting that the relative position of the gearbox in the vehicle is different. That is, FWD engines are usually mounted across the engine compartment so that the gearbox is to one side. On some FWD cars the gearbox is below the engine. On RWD cars the gearbox may be to one side of the engine, or it may be mounted like a conventional layout but in reverse order. You should look at some typical cars and trucks to be able to visualise the location of the gearbox and the other transmission components.

On **mid-engined** cars the engine is in front of the gearbox; they may be mounted **longitudinally** as on Formula Ford for example or **transversely** as MGTF.

Propeller shaft

On conventional layout vehicles, this includes many vans and trucks, the gearbox is mounted on the rear of the engine and supported by the chassis; the rear axle is mounted on the road springs. The function of the propeller shaft is to transmit the drive from the rear of the gearbox to the rear axle, so propelling the car. The rear axle moves up and down with the road springs as the vehicle travels over bumps. This movement means that the **angle** of the propeller shaft between the gearbox and the rear axle changes as the car moves along the road. To accommodate changes in this angle, a moving joint is fitted at the ends of the propeller shaft. As the rear axle moves up and down it also tends to rotate if it is mounted on leaf springs. The rear axle moves in an arc; this is called the **nose arc**. As the axle rotates this arc causes the distance between the gearbox and the axle to change. To allow the propeller shaft to increase and decrease in length to accommodate movement of the axle a **sliding joint** is fitted to the propeller shaft. The sliding joint allows up to 75 mm (3 inch) of variation in length of the propeller shaft.

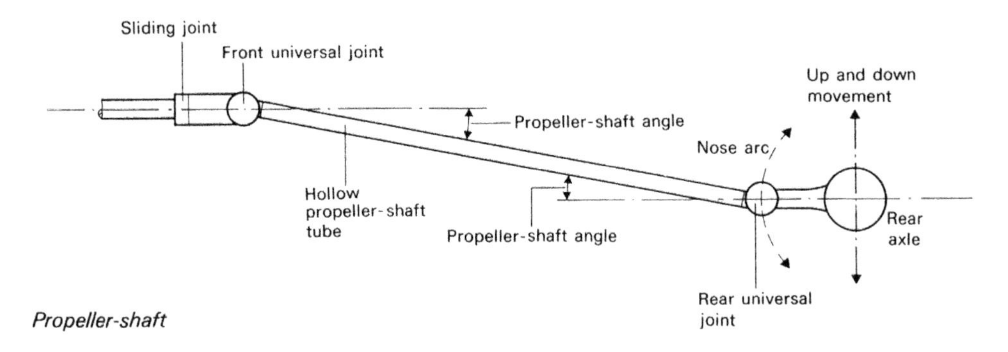

Figure 4.8 Propeller shaft layout

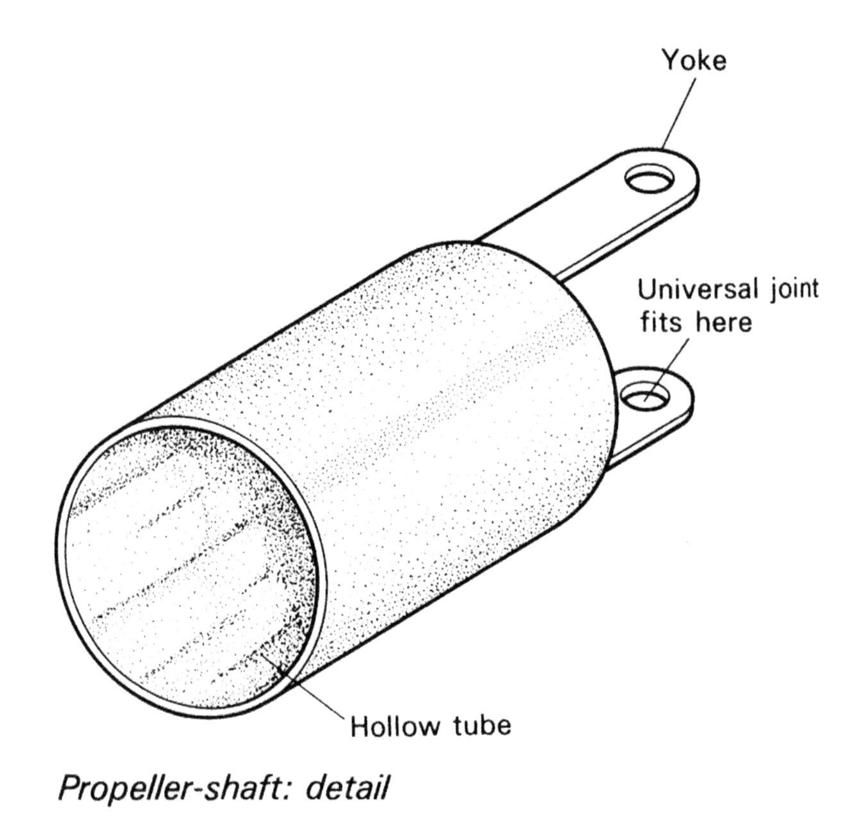

Figure 4.9 Propeller shaft detail

The propeller shaft is a hollow metal tube; each end is welded to a flange to accommodate the universal joints and the sliding joint. The hollow tube is used because it is both light and strong.

> **Racer note**
>
> If you change a gearbox and need a shorted propeller shaft there are a number of engineering firms who will make up a custom propeller shaft and ensure that it is balanced correctly.

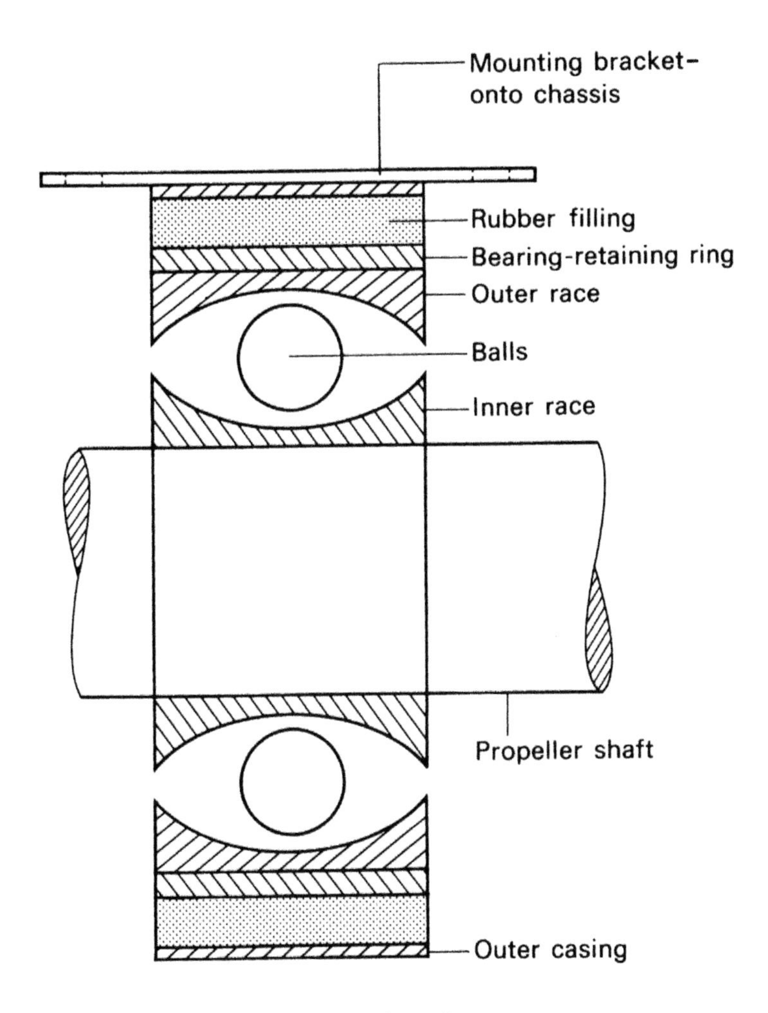

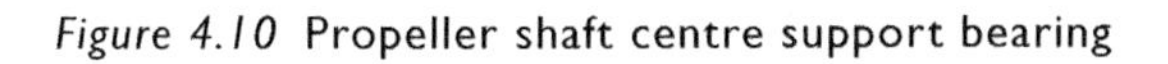

Figure 4.10 Propeller shaft centre support bearing

Support bearing

On long vehicles it is necessary to use two short propeller shafts instead of one long one. This ensures that the power is transmitted from the gearbox to the rear axle smoothly without propeller **shaft whip** or **wind-up**. The support bearing is a rubber-mounted bracket with a ball type bearing in which the inner ends of the front and the rear propeller shafts rotates. The bracket is attached to the chassis between the gearbox and the rear axle. The support bearing is especially important at high speeds.

FAQs What is the difference between whip and wind-up of the propeller shaft?

Whip is when the propeller shaft bends out in the middle, it forms a sort of bow shape like a skipping rope as it spins round. Wind-up is when the propeller shaft twists in torsion and store energy, which can make the transmission jerk. The shape, if exaggerated, would look like a telephone handset cord, or a piece of twine.

Drive shafts

On rear wheel drive (RWD) and front wheel drive (FWD) vehicles, and cars fitted with independent rear suspension (IRS), the drive shafts transmit the power from the differential to the rear driving wheels. The differential divides the drive between the two rear wheels. The differential divides the drive between the two wheels, allowing the outer wheel to turn faster than the inner when cornering. Drive shafts are like short propeller shafts. The drive shafts may contain a universal joint, a **constant velocity joint** (CV joint) and a sliding joint. The constant velocity joint is needed to ensure that the drive is transmitted evenly at all times.

Gearbox

Function

The job of the gearbox is to allow the car to accelerate and climb hills easily, and to provide a means of reversing. This is done by using a selection of gear trains that enable changes to be made in the ratio of engine speed to wheel speed and, in the case of reverse gear, the direction of rotation.

Principles of gearing

The reason for needing a gearbox is that the engine only develops usable power over a limited range of speeds – called the **power band**. The speed at which power is developed

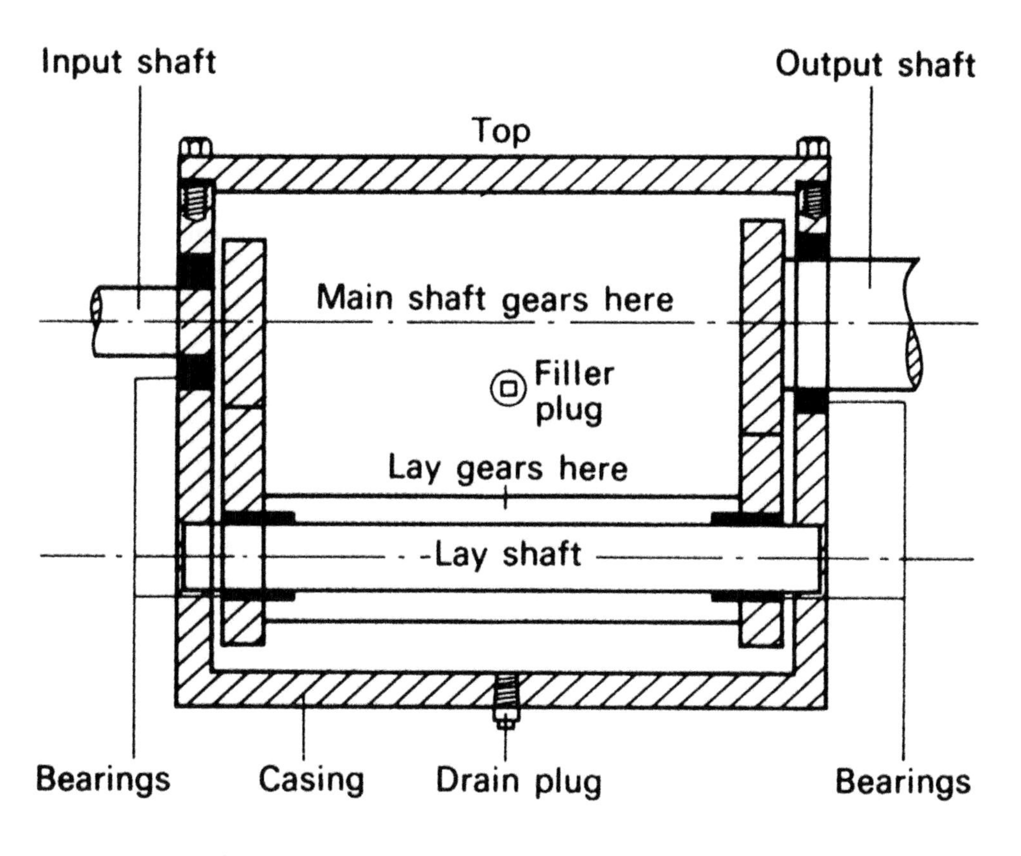

Figure 4.11 Typical gearbox layout

depends on the type of engine. For example, trucks develop their power at low speeds, typically 3,000 rpm. Racing cars and motorcycles are the opposite in that they develop their power at very high speeds; this can be up to 20,000 rpm. Most cars and vans develop their usable power between about 2,000 and 5,000 rpm. This means that if only one gear were fitted, to be able to set off from rest the car would need a gear ratio to give about 10 mph at 2,000 rpm. At 5,000 rpm the top speed would be 25 mph. At the other end of the scale, to have a top speed of 90 mph at 5,000 rpm 2,000 rpm would give 36 mph. As you can see these are like first and top gears on a typical small car.

The gearbox acts like a lever, enabling a small engine to move a very heavy object. This is like how a tyre lever enables the tyre fitter to apply great force to the tyre bead.

Also remember that the gearbox provides a means of reversing the car and a neutral gear position.

Racer note

The gearbox has four main functions. These are that it gives:

1 Low gears for acceleration, moving heavy loads and climbing steep gradients
2 High gears to enable high-speed cruising
3 A neutral gear, so that the engine can be running whilst the car is stationary
4 A reverse gear so that the car can be manoeuvred into parking spaces and garages

Gear ratio

The gear ratio of any two meshing gears is found by the formula:

Gear ratio = Number of teeth on driven gear / Number of teeth on driver gear
This is usually written = Driven / Driver

Where two gears mesh together, the gear ratio is:

Gear ratio = B / A
= 50 / 25
= 2 / 1

This is written 2:1 (say 2 to 1).

This means that for each two turns of A, B will rotate one turn; hence two (turns) to one (turn). That is the gear B will rotate at half the speed of gear A. In other words, B will rotate at half the number of revolutions per minute compared with gear A

If equal size pulleys and ropes were attached to the shafts to which gears A and B are fixed, as in the diagram, it would be possible to use the 10 kg (22 lb) weight to balance the 20 kg (44 lb) weight. This is because the turning effort, or torque, is increased proportionally to the gear ratio. Although the speed is halved, the turning effort is doubled. This effect of the gear ratio is used when climbing steep hills or pulling heavy loads, such as a trailer.

Gearbox ratios

The gears used in a typical gearbox are in **compound** sets of gears. For example, in first gear four gear wheels are in mesh and transfer the power from the clutch to the propeller shaft. An example can be seen in Figure 4.15 – with the input gear A and the two lay shaft gears B and C, and the output gear D. By using more gear wheels, in what is called a compound train, smaller gear wheels can give a bigger gear ratio whilst taking up less space.

To calculate the gear ratio in the diagram we have to decide which gears are driven, and which gears are drivers. As A is the gear that gives the input, this is a driver. Gear B is driven by gear A. Gear C is attached to the lay shaft like gear B and therefore is turning at the same speed. So we can take C as a driver to D; D is therefore a driven gear.

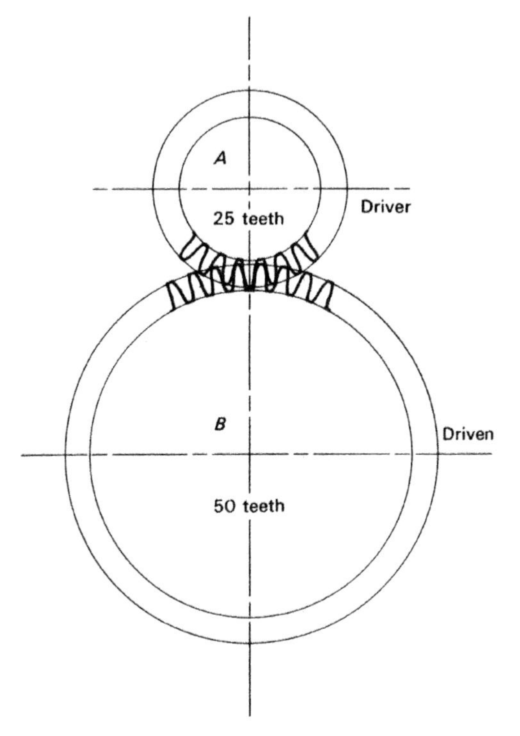

Gear ratio

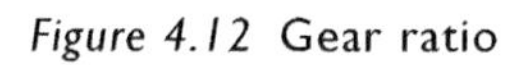

Figure 4.12 Gear ratio

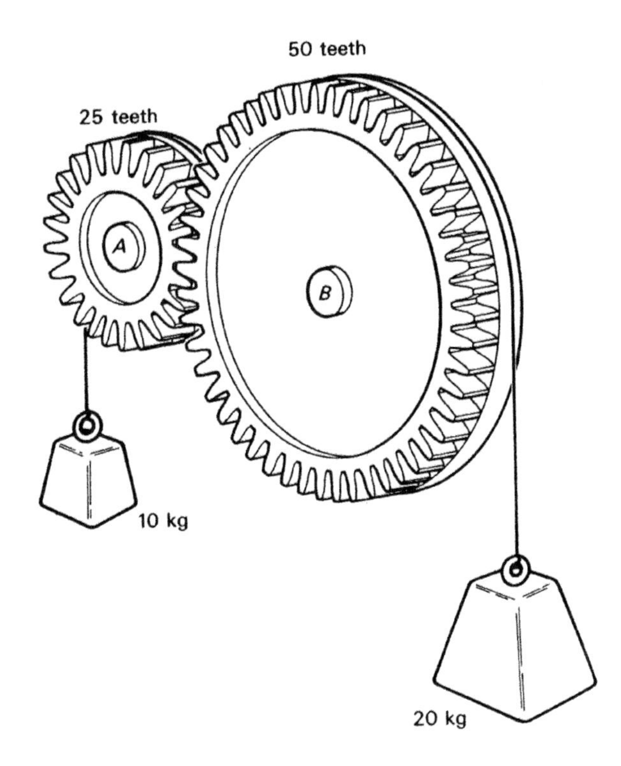

Gear ratio: turning effort

Figure 4.13 Turning effort

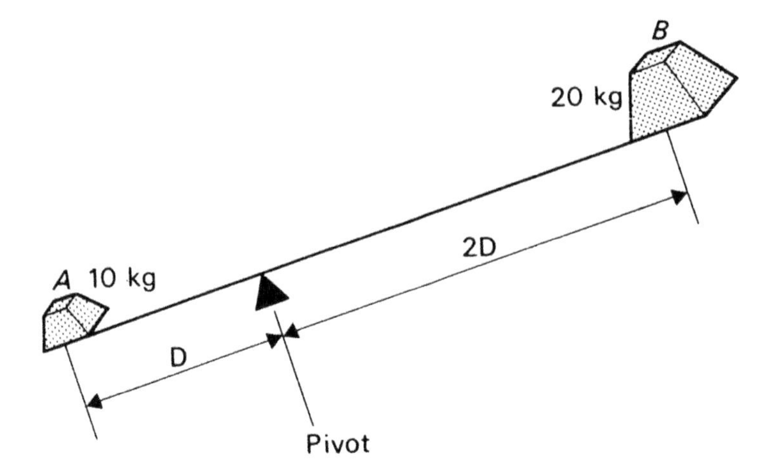

Figure 4.14 Gear leverage

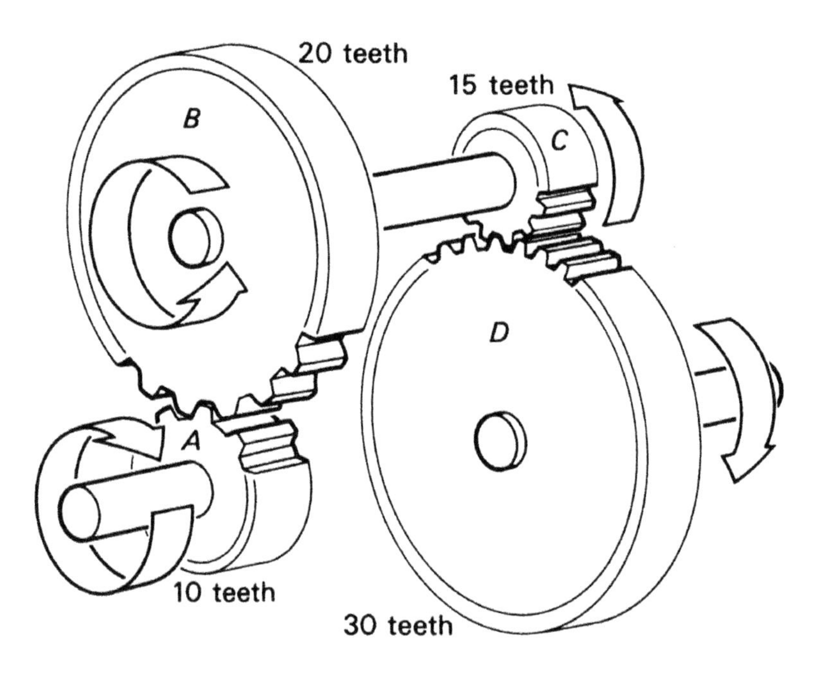

Figure 4.15 Compound gears

> **Racer note**
>
> Gear ratios are a bit confusing at first; it is a good idea to look a sectioned gearbox to understand what is happening in the metal as you might say. If you cannot get to see a sectioned gearbox, try to find an old gearbox that you can take apart. Also, most workshop manuals have lots of pictures of gearboxes, which might help you to understand how the gears run together.

Compound gear ratio = Driven / Driver * Driven / Driver

In our example:

= B / A*D / C

If the number of teeth on each wheel is:

A = 10
B = 20
C = 15
D = 30

The formula would become:

Gear ratio = 20 / 15*30 / 15
= 600 / 150
= 4 / 1
= 4 : 1

Final drive gear ratio

The final drive ratio is the ratio of the speed of the gearbox output to that of the road wheels:

The final gear ratio = Number of teeth on crown wheel/Number of teeth on pinion.

For example, with a 50-tooth crown wheel and a 10-tooth pinion:

Final drive ratio = 50 / 10
= 5 : 1

Overall gear ratio

The overall gear ratio is the ratio of the speed of the engine to the speed of the road wheels. This is found by multiplying the gearbox ratio by the final drive ratio:

Overall gear ratio (OGR) = Gearbox ratio * Final drive ratio

If the gearbox ratio is 2.5:1 and the final drive ratio is 3:1, then:

$$\begin{aligned} \text{OGR} &= \text{Gearbox ratio} * \text{Final drive ratio} \\ &= 2.5 * 3 \\ &= 7.5{:}\ 1 \end{aligned}$$

Layout

The power enters the gearbox through the input shaft that is splined at its outer end into the clutch spinner plate. The power is passed through the constant-mesh gears to the lay shaft, then through the selected gear sets to the output shaft. The output shaft is connected to the propeller shaft that transmits the power, or turning force, on to the rear axle.

The gearbox casing is usually made from cast iron. It holds the gears firmly in place in relation to each other and provides a reservoir for the lubricating oil.

Nomenclature

Many of the gearbox parts have alternative names, so if you are using a workshop manual you should always check out exactly which part is being referred to. Here are a few alternatives.

Input shaft – first motion shaft, primary shaft, clutch shaft, spigot shaft or jackshaft.
Lay shaft – second motion shaft or counter shaft.
Output shaft – main shaft, third motion shaft.

On FWD and RWD gearboxes the layout is the same as on conventional ones, excepting that the output shaft may directly connect to the final drive and differential gears instead of the propeller shaft.

Gear teeth

There are three main types of gear teeth in use in gearboxes; these are spur gear, helical gear and double helical gear. Each kind of gear can be identified by the shape of its teeth.

Spur gear – this has straight teeth, like a cowboy's spur. This type of gear is also called straight cut. Straight cut gears can only carry a limited load and they are noisy in operation. You can hear straight cut gears wine and occasionally rattle.

Helical gear – so called because if the shape of these teeth were projected, as around a long tube, the shape formed would be a helix. Another example of a helix is the screw thread on a bolt. Because the tooth is longer than the gear is wide, it is stronger than the equivalent straight cut gear. Helical gears are used in the gearboxes of most cars as they are quiet in operation. The disadvantage of helical gears is the side thrust. The two meshing gears will have a side thrust that increases with the applied load and the angle of the teeth. The lay shaft gears are machined as a unit, so the side thrust is passed through the metal. The main shaft gears are usually free to slide on the main shaft, so thrust washers are needed to hold them in place.

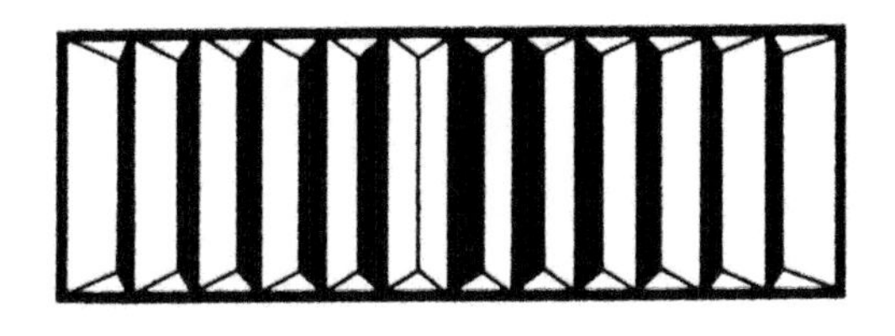

Spur gear

Figure 4.16 Straight cut gear teeth

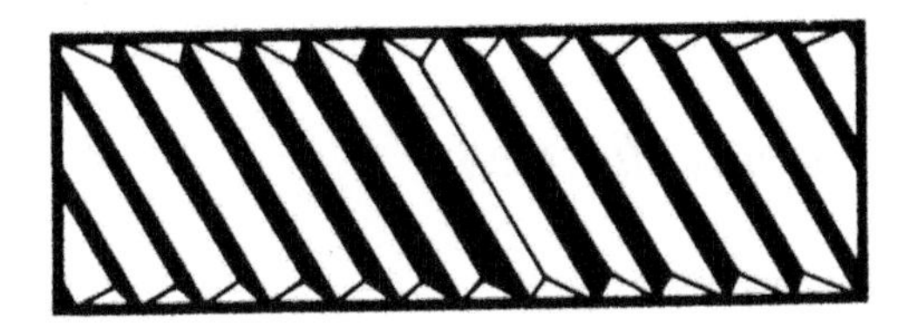

Helical gear

Figure 4.17 Helical gear teeth

Double helical gear

Figure 4.18 Double helical gear teeth

Double helical gears – are made like two rows of opposing helical gears. This is machined from one piece of metal so that the side thrust on one half of the gear balances the side thrust on the other half. Hence, there is no tendency of the gear to move sideways on the main shaft. Double helical gears are used in the gearboxes of trucks and buses where it is important to be able to transmit high loads.

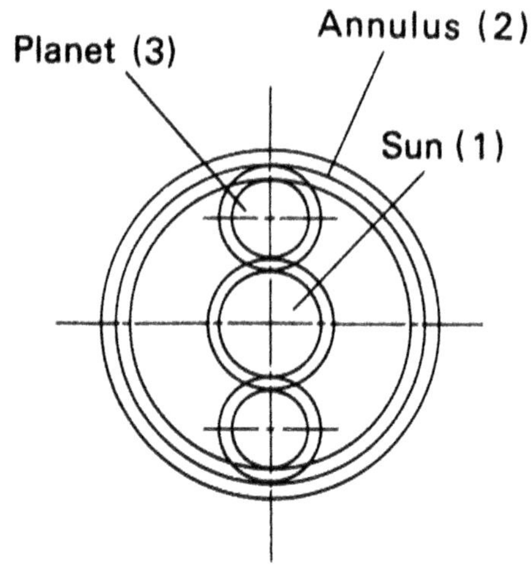

Figure 4.19 Epicyclic gears

Epicyclic gears

Epicyclic gears are used in automatic gearboxes and Sturmey Archer bicycle hub gears. This is a type of gear arrangement where sun and planet gears run inside an annulus. This arrangement allows a range of ratios to be obtained from one gear set. In the diagram the inner gear is called the sun gear (1), the outer toothed part is the annulus gear (2), and the small gears between the sun gear and the annulus are called planet gears (3). Various ratios can be obtained from one set of epicyclic gears by locking each of the different sections in turn to the gearbox casing. For example, in an automatic gearbox the annulus can be held by a brake band so that the power is transmitted from the sun gear to the planet gears. In this case, the planet gears are turning, that is running around the inside of the annulus. If the carrier of the planet gears are held, the sun gear will rotate the planet gears on their spindles, which will turn the annulus. The latter gear ratio would be the lower.

This system of gearing is very compact when compared to conventional gearbox arrangements; this is the reason for its use in automatic gearboxes, cycle gear hubs and overdrive units. However, epicyclic gears are expensive to manufacture and they require great skill to assemble.

Gearbox operation

There are two types of manual gearboxes in use, the sliding-mesh gearbox and the constant-mesh gearbox. The sliding-mesh gearbox is so called because the gears slide into mesh with each other. The constant-mesh gears get their name from being constantly in mesh with each other.

Sliding-mesh gearbox

The general layout of the sliding-mesh gearbox is shown in Figure 4.20. The power enters the input shaft from the clutch. This turns the input gear, which turns the lay shaft. The input gear and the lay shaft gear that turns with it are called constant-mesh gears. The power is transmitted to the main shaft from the lay shaft by whichever main shaft gear is slid into contact. The engagement of each gear is detailed individually; the numbers are those of the gears in the sliding-mesh gearbox diagram.

First gear

First gear on the main shaft (8) is slid into mesh with the first gear on the lay shaft (14). So the power from the input gear (2) goes to the lay shaft (11), then from the first gear on the lay shaft (14) to first gear on the main shaft (8). The sliding gears are splined on the main shaft (9) so that the shaft turns when the gears are turned.

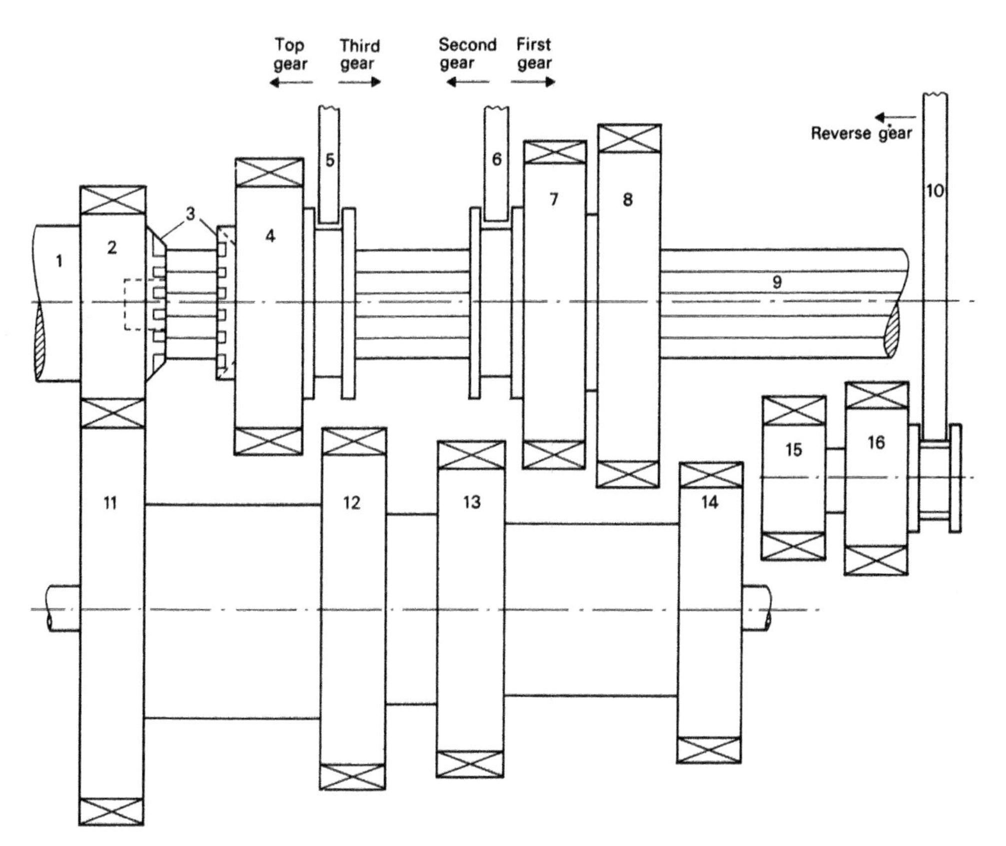

Four-speed sliding-mesh gearbox

Figure 4.20 Sliding-mesh gearbox

Second gear

Second gear on the main shaft (7) is usually connected to first gear (8) so that only one selector fork (6) is needed to engage first gear or second gear. The fork moves backwards (right in diagram) for first gear and forwards for second gear. The second gear on the main shaft (7) is slid along into mesh with second gear on the lay shaft (13) so that the power path is input gear (2) to lay shaft (11), second gear lay shaft (13) to second gear main shaft (7), through the splines to the main shaft (9).

Third gear

Third gear is engaged by meshing the third gear on the main shaft (4) with the third gear on the lay shaft (12). So that the power path is input gear (2) to lay shaft (11), third gear (12) to main shaft third gear (4) and that turns the main shaft (9).

Fourth gear

Fourth gear is engaged by meshing the dogteeth (3) on the front of the third gear main shaft (4) with those on the input gear (2). The top gear is therefore direct drive. The input shaft turns the main shaft (9) at the same speed, the drive being through the dogteeth (3). The constant-mesh gears (2 and 11) turn the lay shaft, but it is only idling, it is not transmitting any power.

Reverse gear

Reverse gear is engaged by moving the reverse idler into mesh (15 and 16). The reverse idler is a shaft with two gears attached. One gear (16) meshes with the first gear on the main shaft, the other (15) with the first gear on the main shaft. The power path is thus input gear to lay gear to idler gear to first gear on the main shaft. Looking at the front of the gearbox the gears and shafts will rotate in the following directions:

Input gear: clockwise
Lay shaft gears: anti-clockwise
Reverse idler gear: clockwise
First gear main shaft: anti-clockwise

Neutral

When no gears are in mesh this is neutral, and in this position no drive is transmitted.

FAQs Does the lay shaft turn when the gearbox is in neutral?

Yes, the constant-mesh gears are turning all the time, so the lay shaft must be turning too.

Constant-mesh gearbox

All the gears, except reverse, are in constant mesh, not transmitting power, but idling, except when they are engaged. The main shaft gears are not splined to the main shaft, but run on bushes on the main shaft. This means that the gears and the shaft can turn independently of each other. Figure 4.21 shows the general layout. To engage gear, the synchromesh units that are splined onto the main shaft are slid into mesh with the dogteeth on the gears, so that the gear turns the synchromesh unit that transmits the power to the main shaft. The power paths of each gear are detailed in the following few paragraphs. The numbers refer to those on the constant-mesh gearbox diagram in Figure 4.21.

Neutral

In neutral gear no power is transmitted as none of the synchromesh hubs are engaged with the gears. However, all the gears are turning when the engine is running and the clutch is engaged. The input gear turns the lay shaft through the constant-mesh gears.

First gear

The synchromesh hub (9) is slid on the shaft (13) so that it engages with the dogteeth on the main shaft first gear (11). The power path is then input gear (2) turns the lay shaft (15),

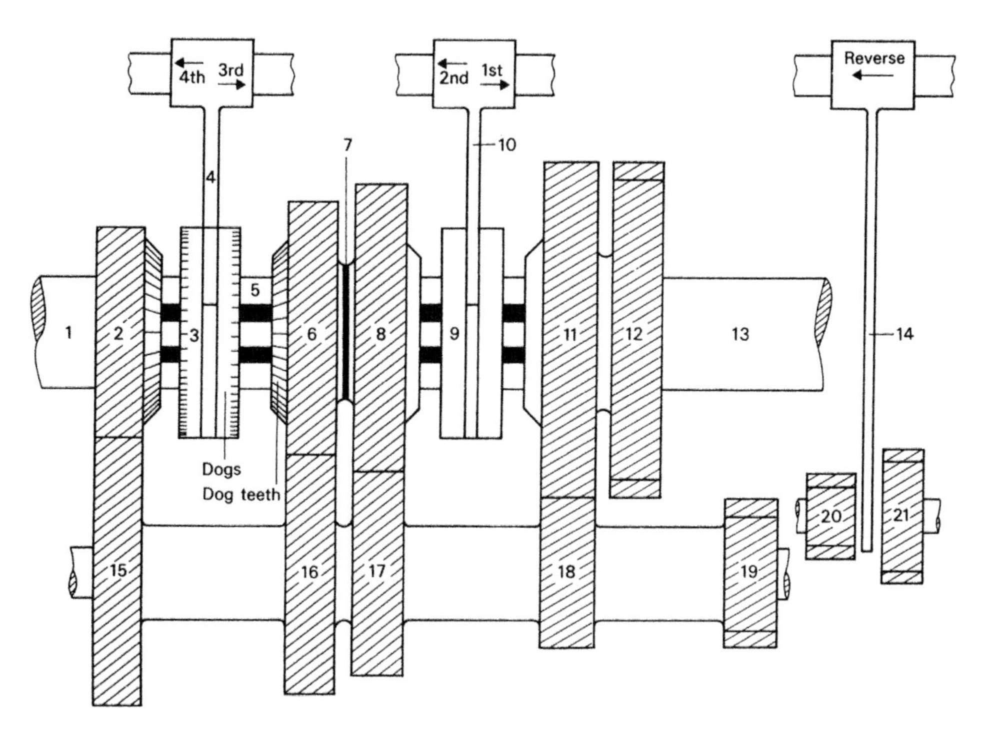

Four-speed constant-mesh gearbox. Gears 12, 19, 20 and 21 are spur gears, the others are helical gears

Figure 4.21 Constant-mesh gearbox

which turns the first gear on the main shaft (11) through (18). The power then goes through the dogteeth to the synchromesh hub (9) and through its splines to turn the main shaft (13).

Second gear

The synchromesh unit is slid in the opposite direction to engage with the dogteeth on the side of the second gear on the main shaft (8). The power path is thus input gear (2) lay shaft (15 and 17), second gear main shaft (8), synchromesh unit (9), and main shaft (13).

Third gear

The other synchromesh unit (3) is slid into mesh with the third gear on the main shaft (6). The power path is similar to first and second gear, being input gear (2), lay shaft (15 and 18), third gear main shaft (6), dog clutch (3) and the main shaft (13)

Fourth gear

Fourth gear is engaged by sliding the synchromesh unit (3) into mesh with the dogteeth on the side of the input gear (2). This gives direct drive in the same way as the sliding-mesh gearbox. That is through the dogteeth, the synchromesh hub and the splines to the main shaft.

Reverse gear

The reverse gear is usually engaged in the same way as in the sliding-mesh gearbox. That is the idler gears (20 and 21) are slid into mesh with a reverse gear on the lay shaft (19) and one on the main shaft (12).

Nomenclature

The sliding-mesh gearbox is also referred to as a crash gearbox because of the noise made by inexperienced drivers using them. The constant-mesh gearbox is often called the synchromesh gearbox because of its use of synchromesh hub units. Porsche invented the synchromesh hub unit when he worked on VW cars in the 1930s. You will not find a fully sliding-mesh gearbox on a modern ordinary car; but you will find cars that have one or two gears of the sliding type and the rest synchromesh. You will also find sliding-mesh gears on some racing cars and some motorcycles.

You should also note that the first motion shaft gear and the lay shaft gear that it meshes with are called constant-mesh gears too.

Synchromesh hub

A synchromesh hub is shown in Figure 4.23. It consists of a hub, which is splined to the main shaft and an outer sleeve, which is splined to the hub. Spring-loaded balls hold the outer sleeve in the neutral, or the engaged, positions.

When initial pressure is applied by the selector to the outer sleeve the pressure of the ball also moves the hub section. This causes the conical surface on the hub to give the initial interference to the cone on the gear involved. This interference adjusts the speeds of the mating shafts to the same speed. That is it synchronises them. Further pressure by the selector pushes the outer sleeve over the balls against the spring pressure until they engage in the next row of grooves. At this point the outer hub has engaged its inner splines with the dogteeth on the gear. Therefore the gear is fully engaged. Drive from the gear is passed from the dogteeth to the outer sleeve, to the hub and to the shaft.

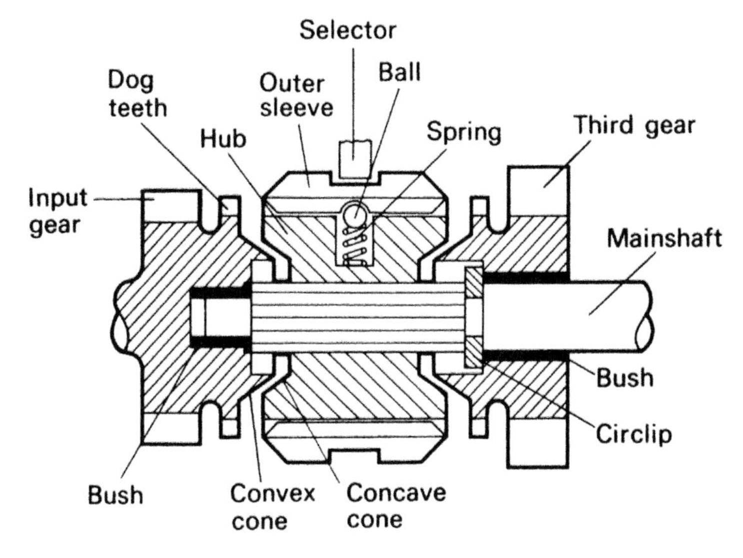

Figure 4.22 Synchromesh hub

Synchromesh hub

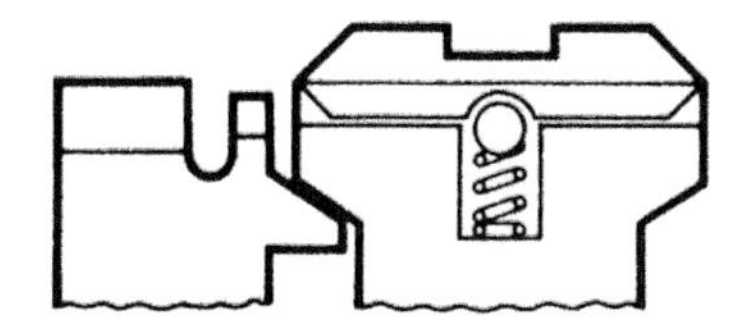

Synchromesh unit operation: initial interference of cones' synchronising speeds

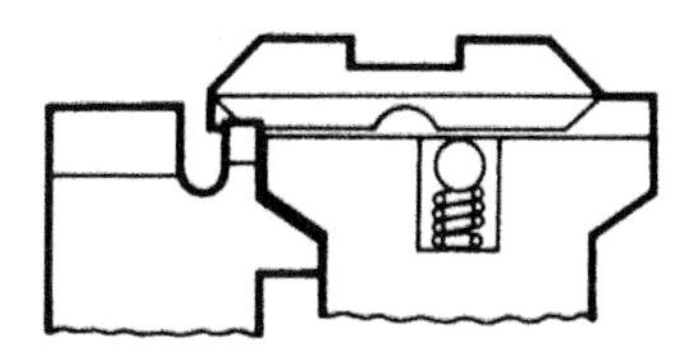

Synchromesh unit operation: drive gear engaged; outer sleeve over dog teeth

Figure 4.23 Synchromesh operation

Selector

The diagram shows a typical selector, or selector fork to give it its full name. The fork is moved by a selector rod through a mechanism attached to the gear lever.

Detents

To ensure that the gears are held in the selected position, a system of detents is used. Examples of detents are shown in the diagrams. One uses a spring-loaded ball; another uses a plunger. The C-shaped detent is a mechanical locking device.

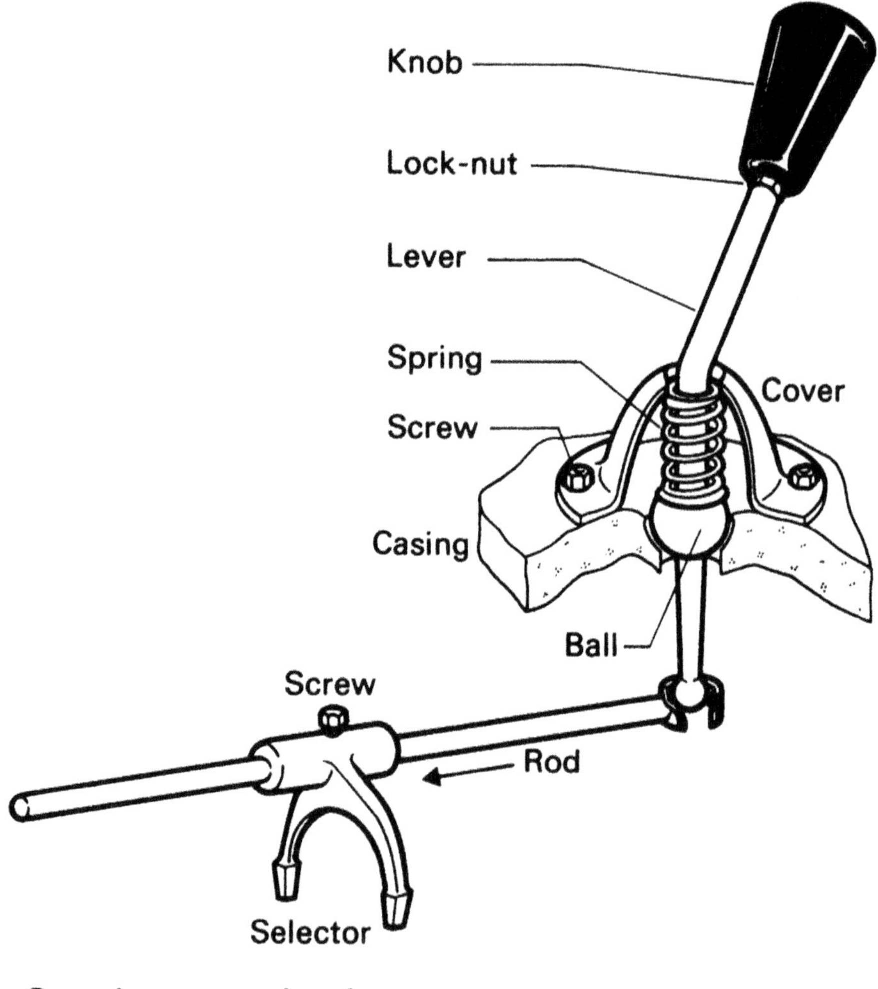

Figure 4.24 Gear lever set-up

Gearbox removal

You will usually need to remove the gearbox to replace the clutch as well as to repair the gearbox itself. Most gearboxes are located underneath the vehicle; this means that you will need a hoist, or a pit, to gain sufficient access. On most cars the gearbox is connected to the engine by a ring of bolts around the bell housing; some form of rubber mounting to the body or chassis will also be used. The exact procedure for removing a gearbox will be given in the workshop manual appropriate to the vehicle, but the general procedure is as follows:

- Disconnect the battery earth terminal for safety
- If possible drain the gearbox oil
- Remove the speedometer cable
- Disconnect the gear lever mechanism
- Disconnect the clutch mechanism
- Remove the drive shafts or the propeller shaft
- Support the engine
- Use a cradle to support the gearbox
- Remove the bell housing bolts and the mounting bracket
- Withdraw the gearbox on the cradle

Racer note

Gearboxes are very heavy; you should always work with a colleague to get a team lift when removing them.

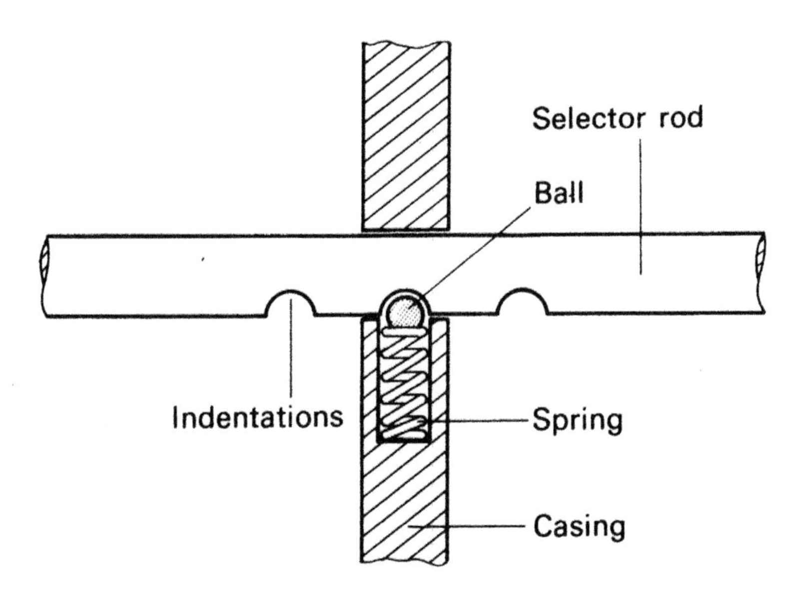

Figure 4.25 Gear selector detent

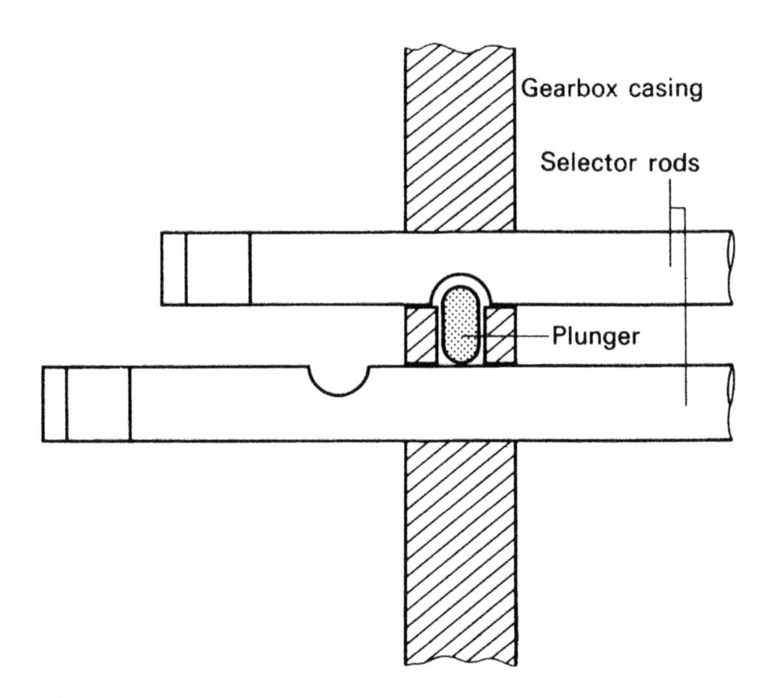

Selector detent

Figure 4.26 Gear detent

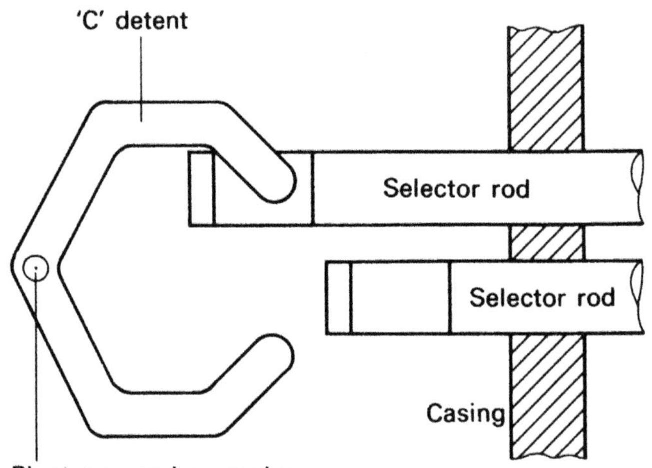

C-shaped detent (allows only one gear to be engaged at a time)

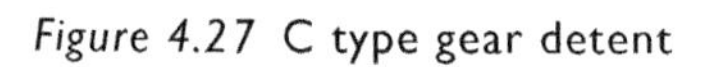

Figure 4.27 C type gear detent

Gearbox stripping

The method of stripping gearboxes varies with the make and model; you should check with the workshop manual before carrying out any stripping to prevent damage to the gearbox or personal injury. Figures 4.28–4.32 show some common procedures.

Gearbox lubrication

The gearbox oil varies in type and viscosity with the vehicle. So you must check the workshop manual before topping up or changing the gearbox oil. It is common to find the same oil, such as a 10/40 SAE, used in the gearbox as in the engine. However, many vehicles, especially trucks, use oil such as EP 90 SAE.

Figure 4.28 Straight cut gear set

Figure 4.29 Sectioned gearbox casing

Figure 4.30 Input side of gearbox

Figure 4.31 Open casing

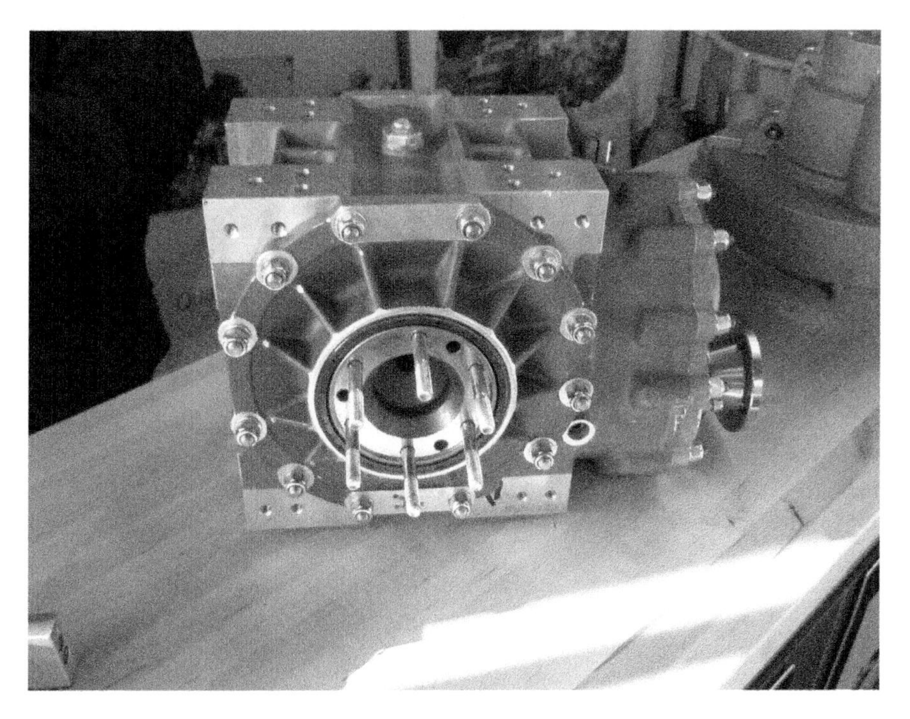

Figure 4.32 Output drive side

Racer note

Some types of oil can be identified by their smell. EP and Hypoy classifications of gear oil contain a large amount of sulphur, which gives the oil a pungent, and easily recognisable, smell.

Final drive arrangements

The rear axle

Conventional layout vehicles have live rear axles. That is the axle transmits the driving force as well as supporting the vehicle. A typical rear axle carries inside it the following components:

- Final drive gears
- Differential gear assembly
- Half-shafts
- Rear hubs and bearings
- Lubricating oil

Final drive gears

The final drive may be in the form of a crown wheel and pinion in a truck, or other conventional layout vehicle rear axle; or an output pinion and differential gear in a FWD vehicle. On conventional layout vehicles the crown wheel and pinion turn the drive through 90 degrees as well as providing an additional gear ratio. On FWD, RWD and mid-engine cars with a transverse layout, there is no need to turn the drive through an angle.

There are a number of different types of bevel gears used for the final drive gears; three of them are straight bevel gears, spiral bevel gears and hypoid bevel gears.

Nomenclature

Bevel gears are ones cut so that they mesh to form a 90 degree angle. That is the turn the drive through a 90 degree angle.

Straight bevel gears – these have straight cut teeth so they are very noisy in operation.

Spiral bevel gears – have the pinion on the same centre line as the crown wheel, like the straight bevel gears, but the teeth are cut at an angle so that they are both strong and quiet in operation.

Hypoid bevel gears – these gears have the centre line of the pinion below that of the crown wheel. This is to allow the propeller shaft to be as low as possible, so the vehicle floor can be low and the handling is more stable. The gear teeth are both cut at an angle and curved. Hypoid bevel gears must be lubricated with special hypoy oil to withstand the very high pressures, which are caused by these two gears rolling into mesh.

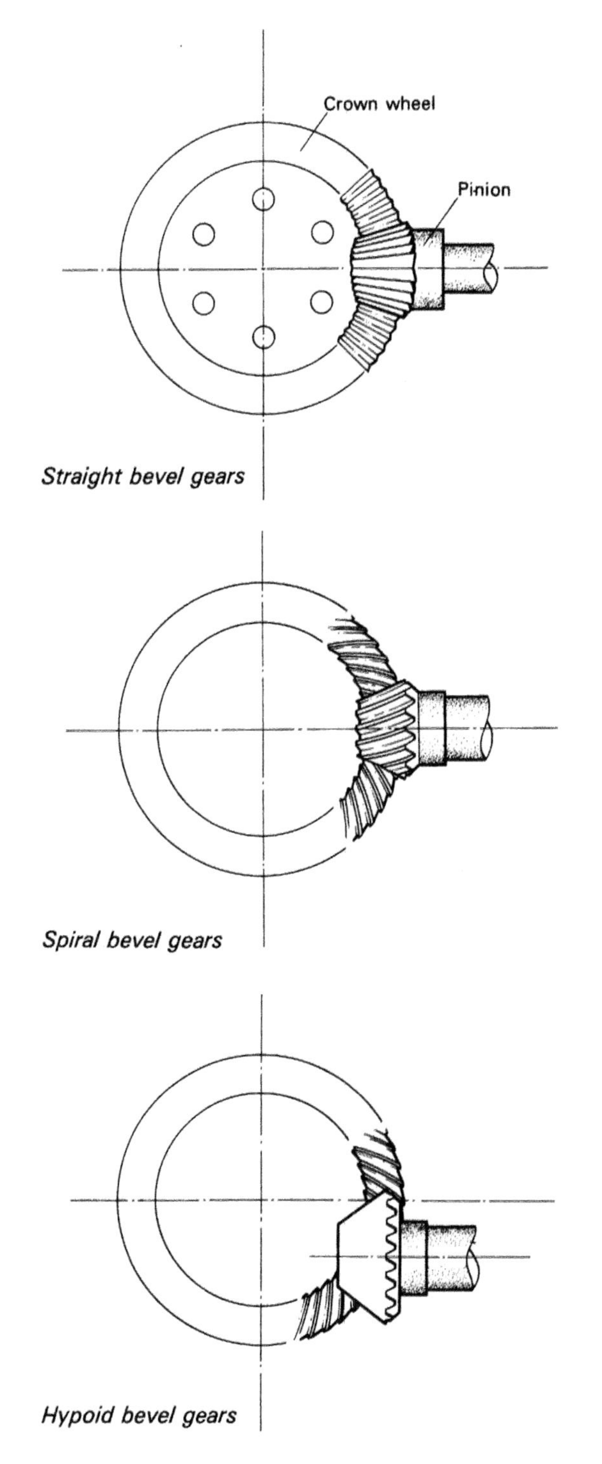

Figure 4.33 Types of bevel gear

Worm and wheel – is an alternative to the more usual crown wheel and pinion. The worm looks like a very large bolt; it turns the gear teeth on the outside of the wheel. The wheel looks similar to a crown wheel, but the teeth are on the outer perimeter, not at an angle. The worm may be located below the wheel to give low chassis height, or above the wheel to give maximum ground clearance. The advantage of the worm and wheel drive is that it can transmit very high loads and give very low gear ratios. The gear ratio is calculated by dividing the number of teeth on the wheel by the number of starts on the worm. The term starts means the number of threads that go to make up the worm; this is usually between three and five:

Gear ratio = Number of teeth on wheel / Number of starts

Racer note

The worm and wheel is a very important engineering principle. You will only find worm and wheel drive on specialised vehicles, such as heavy duty dump trucks and off-road contractors' plant. However, you will find the principle applied in a number of

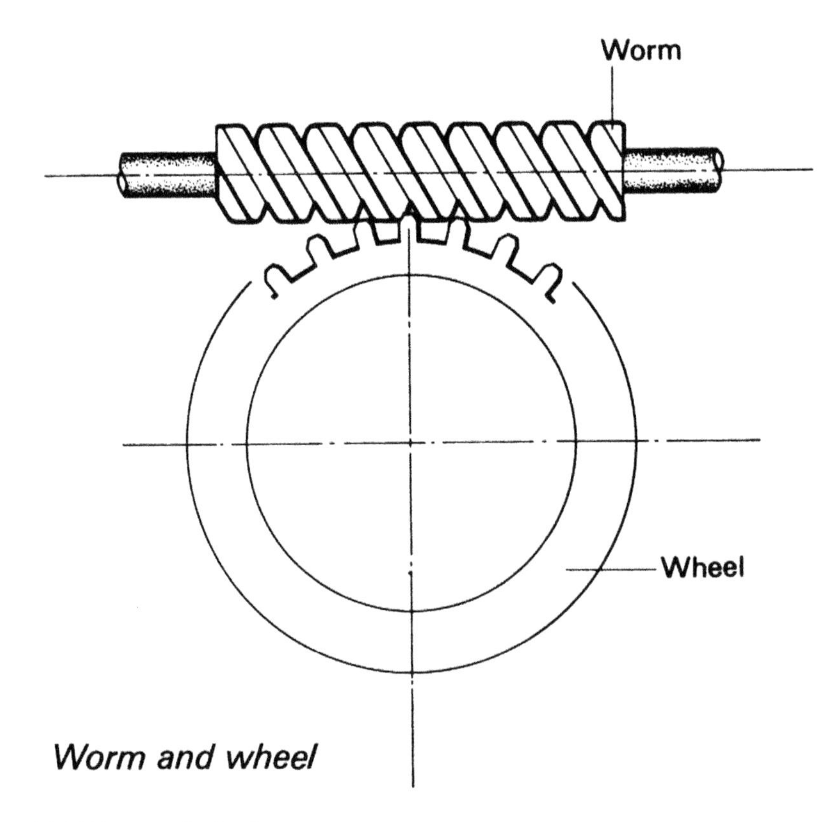

Figure 4.34 Worm and wheel

ways in engineering; a good example is the screw driven hose clip, the most popular one being known by the trade name Jubilee Clip.

Differential

The differential gears are mounted inside a carrier; the **crown wheel** is screwed to the outside of the **carrier**. Having a look at the diagrams might help. The differential carrier is mounted on bearings that are carried by the axle casing. The differential gear components comprise of two sun-gear wheels and two planet-gear wheels. The sun-gear wheels are attached one to each end of the axle shafts; and the planet gears are free to rotate on a cross-shaft, which is located in the differential carrier.

The functions of the differential are to:

- Allow on wheel to turn faster than the other does. For instance, when going round a corner, the outer wheel will travel further and hence faster than the inner wheel
- Divide the driving force, or torque, evenly between both wheels

Operation of the differential

Figure 4.36 shows the essential gears; it does not represent a complete differential. Under normal straight-line conditions the pinion rotates the crown wheel, which turns the differential carrier. The carrier turns bodily taking the planet gears with it. It is important to note that the planet wheels do not spin on the cross-shaft, they pass the driving torque on to the sun wheels, so that the sun wheels turn at the same speed as the crown wheel and both axles are therefore rotated. The speed and the turning force are the same for both axles and therefore both of the road wheels.

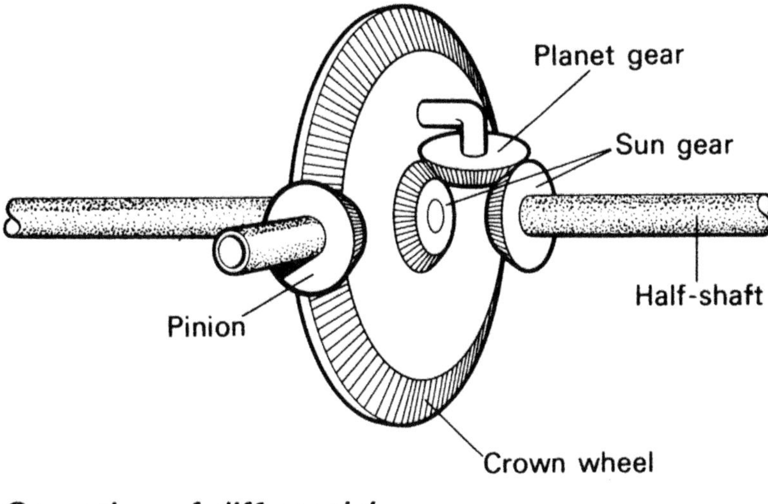

Figure 4.35 Differential operation

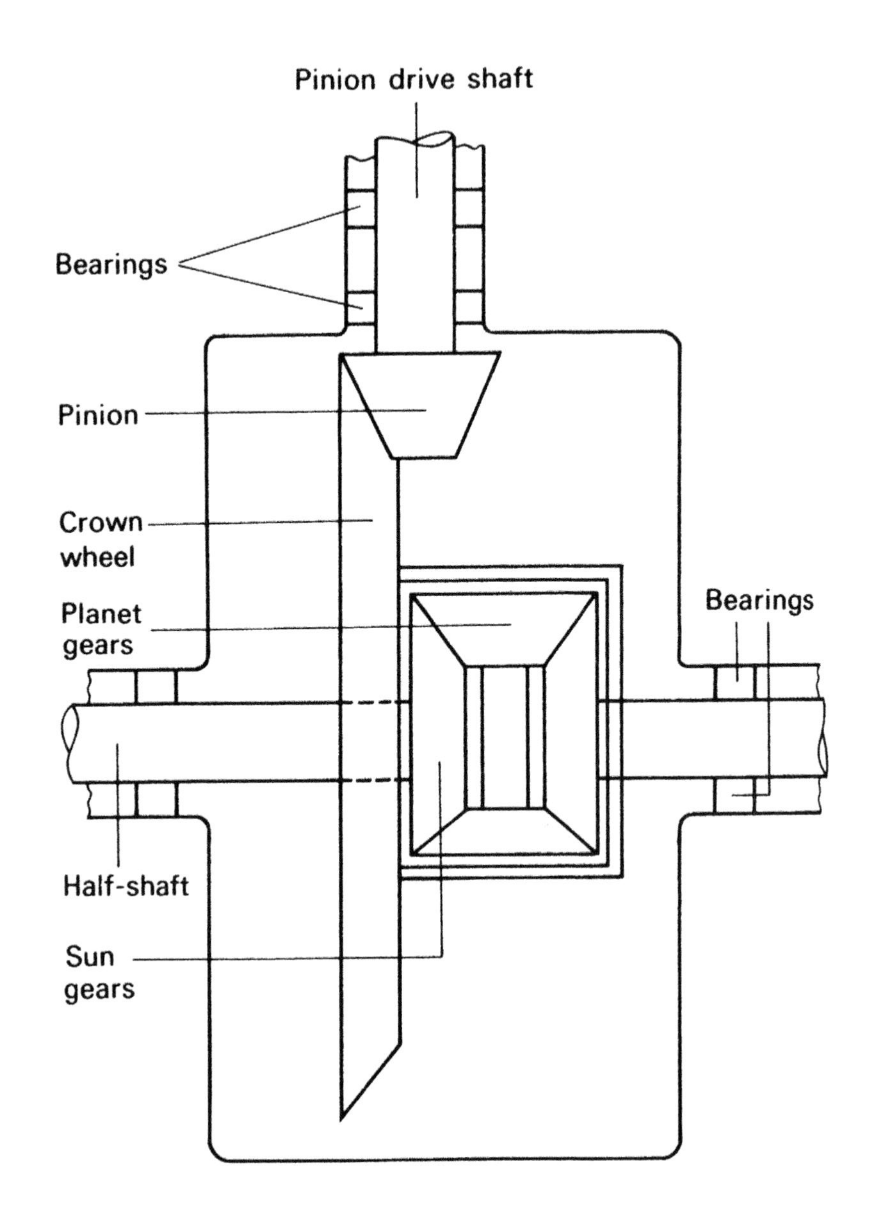

Figure 4.36 Differential assembly

Figure 4.37 Limited slip differential

Figure 4.38 LSD with crown wheel and pinion

On cornering, the inner wheel drive shaft and the sun gear on the end of it are slowed in comparison to the outer one. Therefore there is a speed differential (difference). The planet wheels rotate around the slowed sun gear, spinning on their cross-shaft. This movement is passed onto the outer sun gear, which therefore turns faster. The outer sun gear gains the speed lost by the inner sun gear.

Half-shafts

The half-shaft connects the sun-gear wheel to the hub, so transmitting the drive to the road wheel. The inner end of the half-shaft is splined to the sun-gear wheel in the differential. The outer end of the half-shaft has a taper or a flange assembly to connect to the hub.

Hub/axle assemblies

The rear hub and axle assembly is subjected to considerable forces, namely:

- The driving force to the wheels
- The braking force
- The load of the vehicle

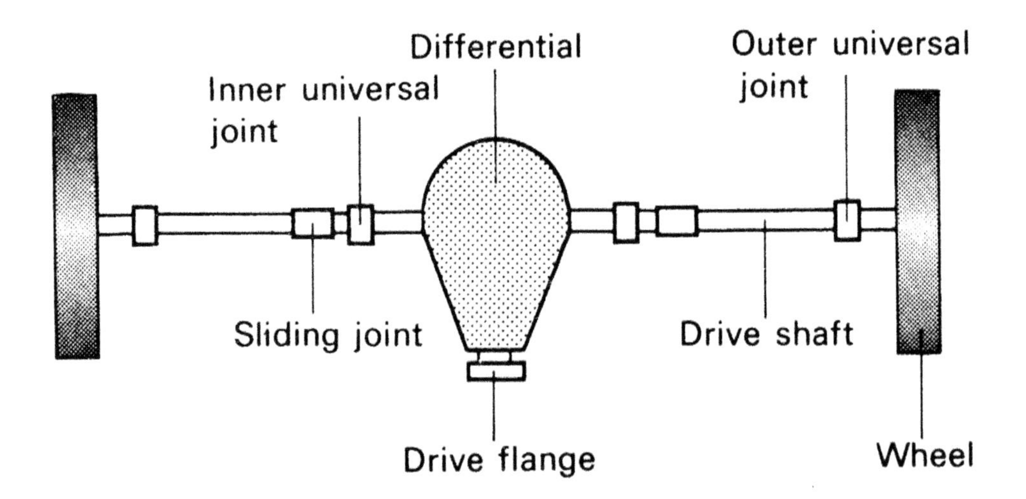

Figure 4.39 IRS layout

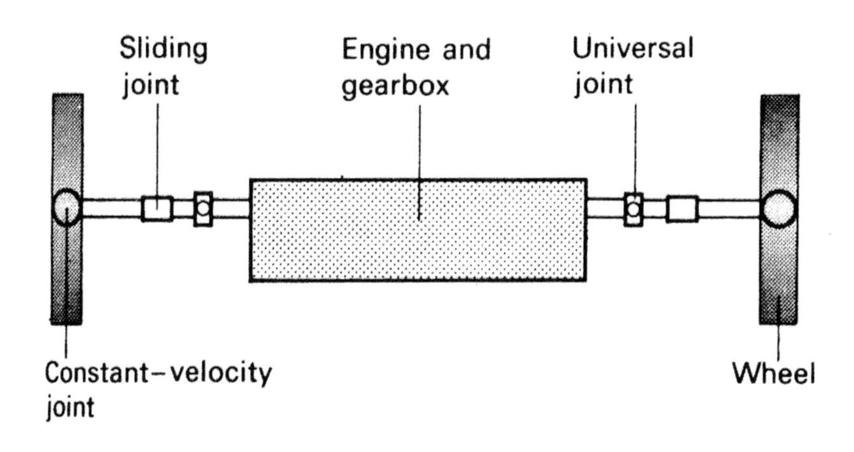

Figure 4.40 FWD layout

These forces put bending and shear stresses on the rear axle. To provide sufficient strength without undue weight, one of three types of hub/axle bearing arrangements may be used. The three arrangements are semi-floating, three-quarter floating and fully floating. Figures 4.41–4.43 illustrate the different arrangements.

Semi-floating

The semi-floating assembly is the least strong arrangement because the bearing is between the inside of the axle casing and the half-shaft. The problem is, if it breaks the wheel will fall off completely.

Three-quarter floating

The three-quarter floating arrangement is much stronger than the semi-floating type. The bearing is located between the hub and the axle casing. The major part of the weight is supported by the bearing – hence its name – the axle is three-quarter floating.

Fully floating

The half-shaft in this arrangement carries none of the vehicle's weight, hence the name fully floating. The hub is supported by twin ball, or roller, bearings. This arrangement is used on the axles of almost all trucks and buses. An added advantage is that both the half-shaft and the differential can be changed without jacking up the vehicle.

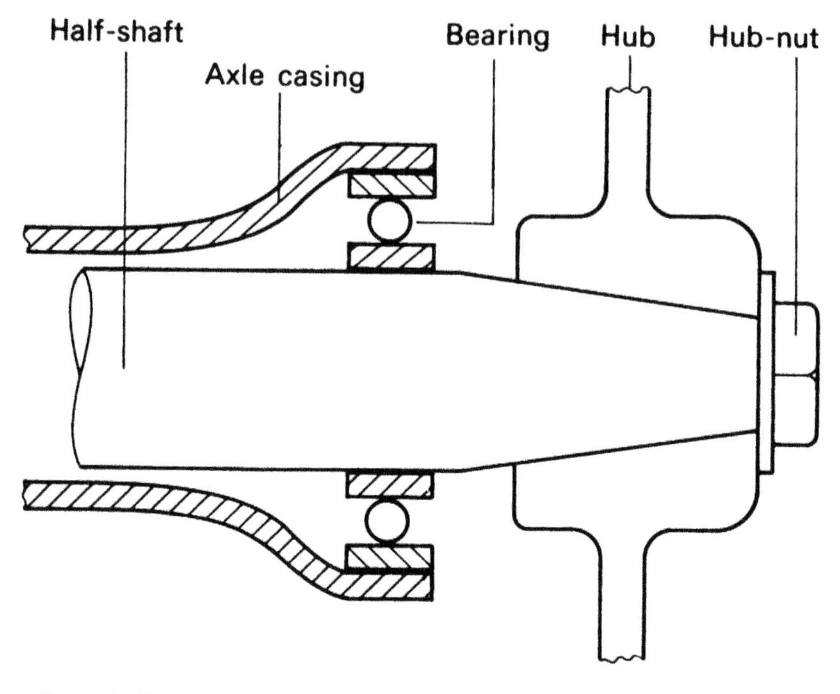

Figure 4.41 Semi-floating hub and axle arrangement

Drive shafts

Drive shafts connect the differential gears to the hubs. To allow for movement they are usually fitted with some form of joint.

Hook joints

The Hook joint was invented by the physicist Robert Hook (1635–1703). Although the Hook joint is its proper name, it is known more commonly as the universal joint.

Nomenclature

The universal joint is another name for the Hook joint. In the motor industry is also known as a 'Hardy Spicer' joint; Hardy Spicer is the trade name of the main manufacturer. This is similar to calling a vacuum cleaner a 'Hoover'.

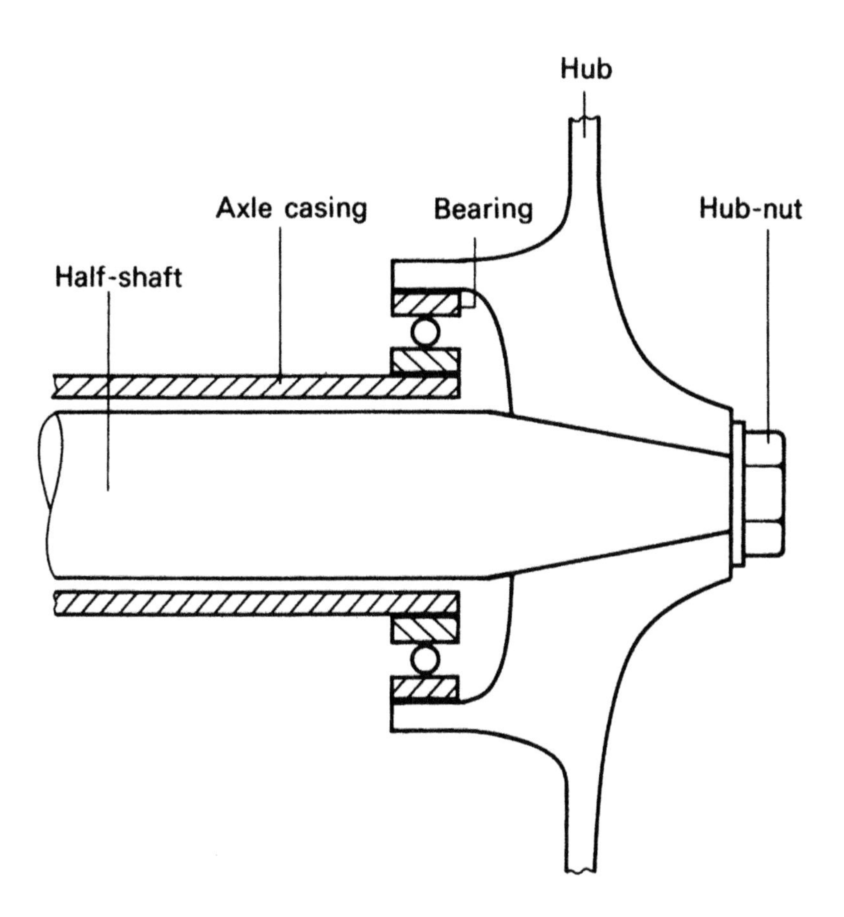

Figure 4.42 Three-quarter floating hub axle arrangement

The Hardy Spicer joint is constructed from a cruciform member that rotates in four cups each containing needle roller bearings. Two of the cups fit into the flange that is attached to the rear of the gearbox, the opposite two are attached to the flange at the end of the propeller shaft. At the other end of the propeller shaft is a similar arrangement for the final drive pinion. The needle roller bearings are usually pre-packed with special grease so that they are sealed for their life. On some vehicles the joints are lubricated by the means of grease nipples. A grease gun must be used to apply grease every 15,000 km (10,000 miles).

Hardy Spicer joints are fitted in pairs to give an even transmission of power. The yokes must be fitted to the propeller shaft in the same plane. Where a detachable sliding joint is

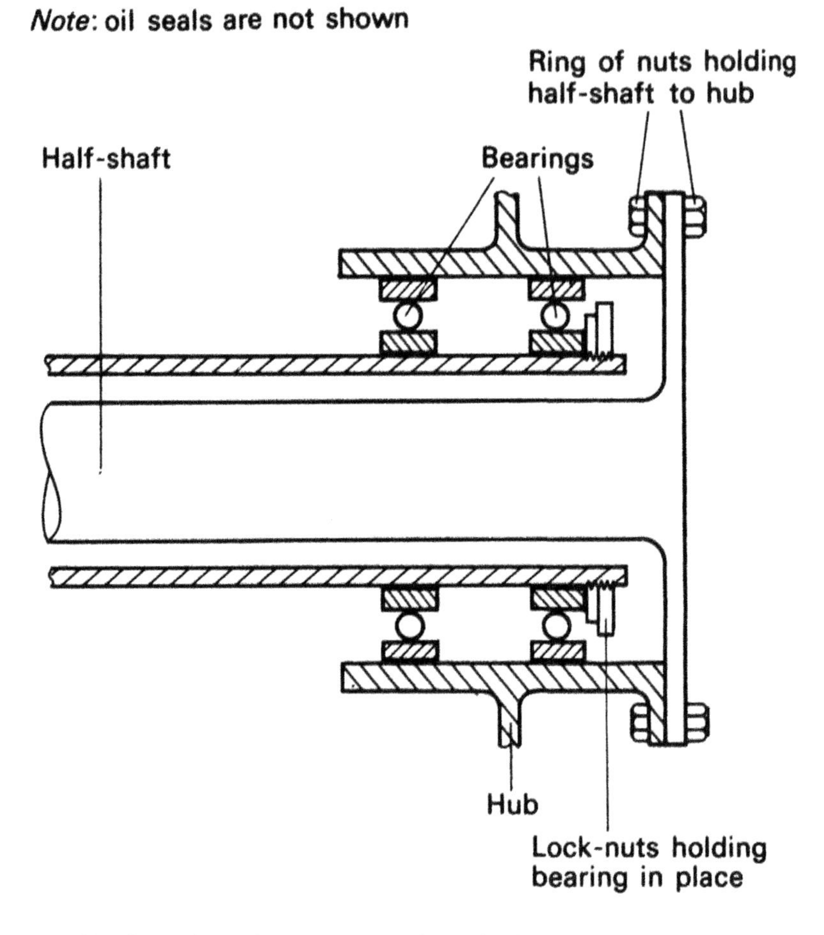

Figure 4.43 Fully floating hub axle arrangement

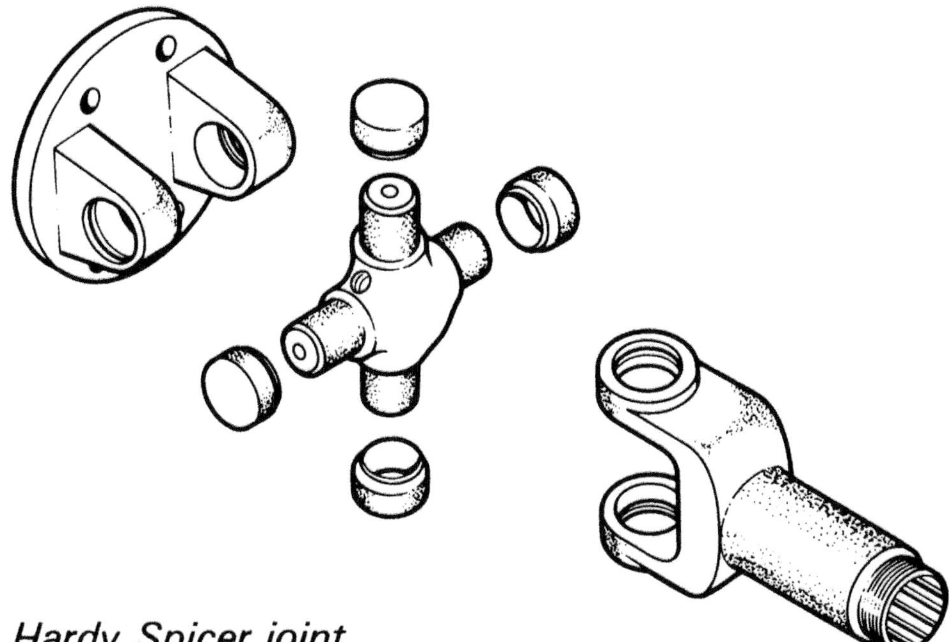

Figure 4.44 Propeller shaft joint – Hardy Spicer type

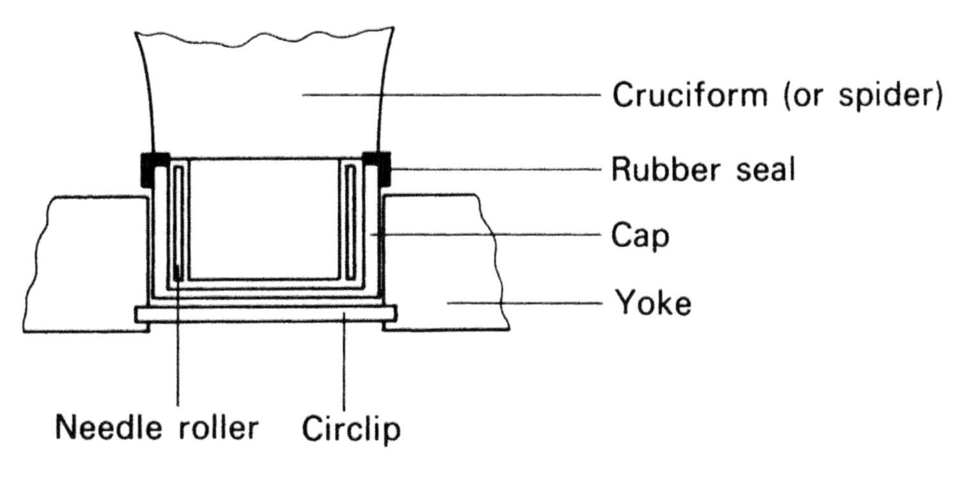

Figure 4.45 Cup in propeller shaft joint

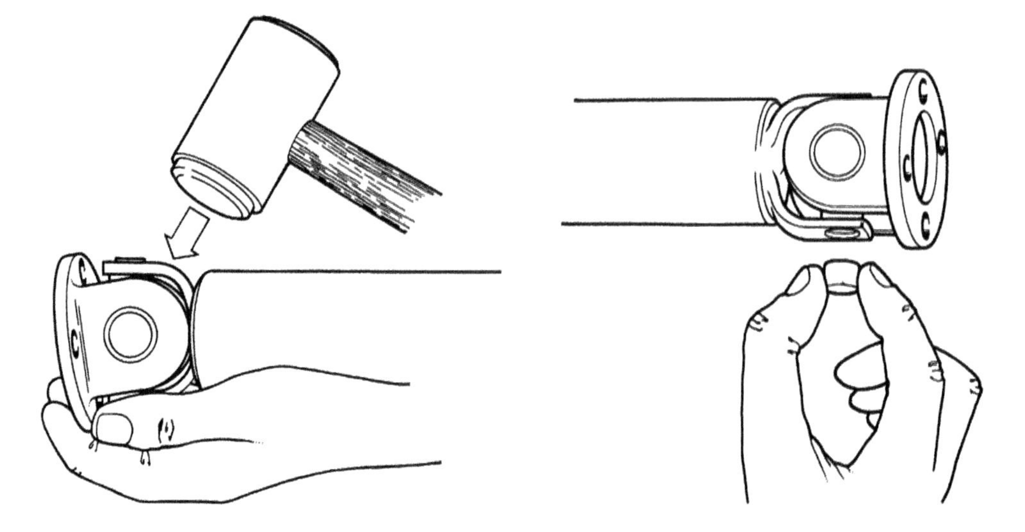

Changing a universal joint

Figure 4.46 Removing propeller shaft joint cups

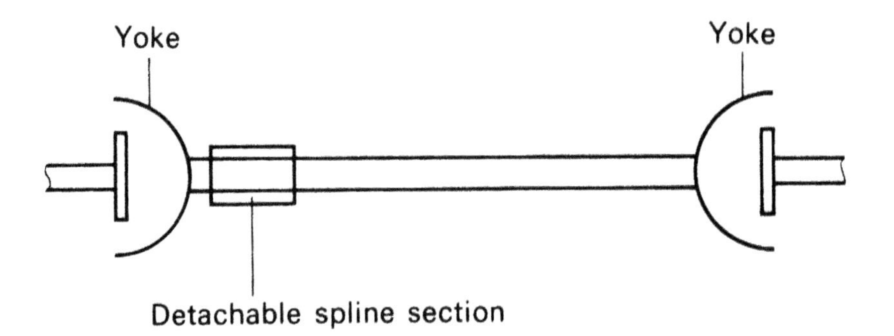

The yokes of the propeller-shaft fitted in the same plane

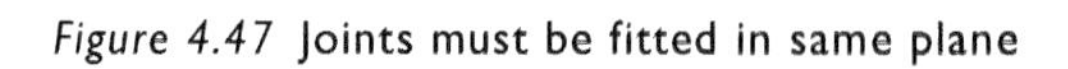

Figure 4.47 Joints must be fitted in same plane

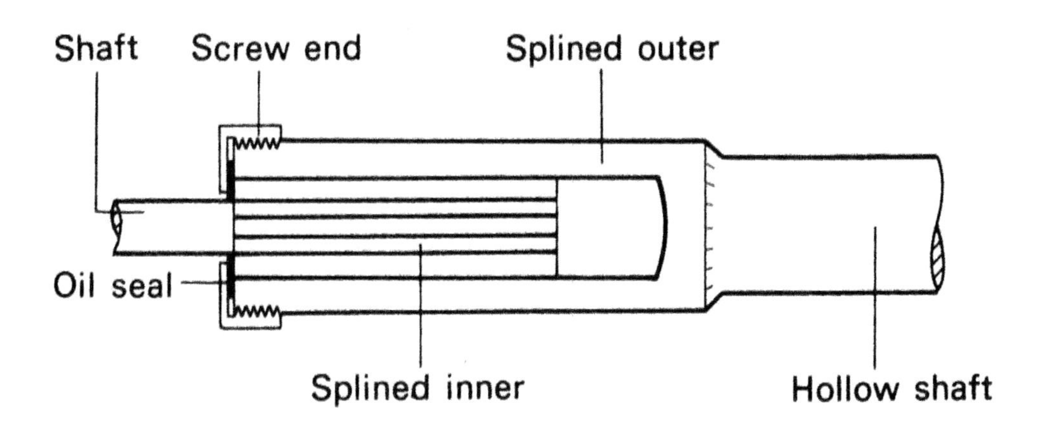

Figure 4.48 Sliding joint

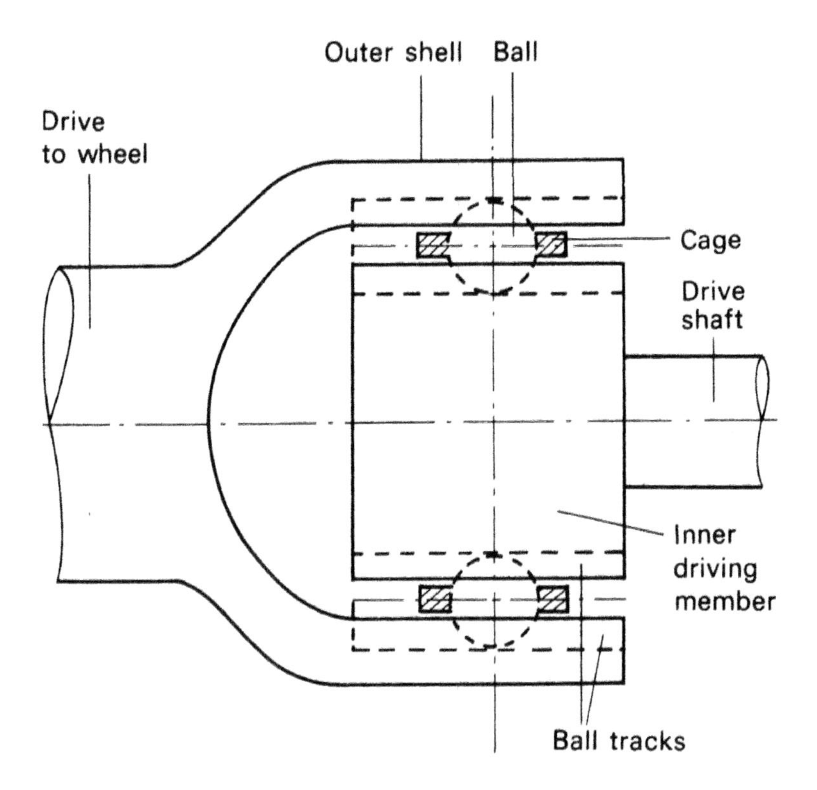

Figure 4.49 Birfield CV joint

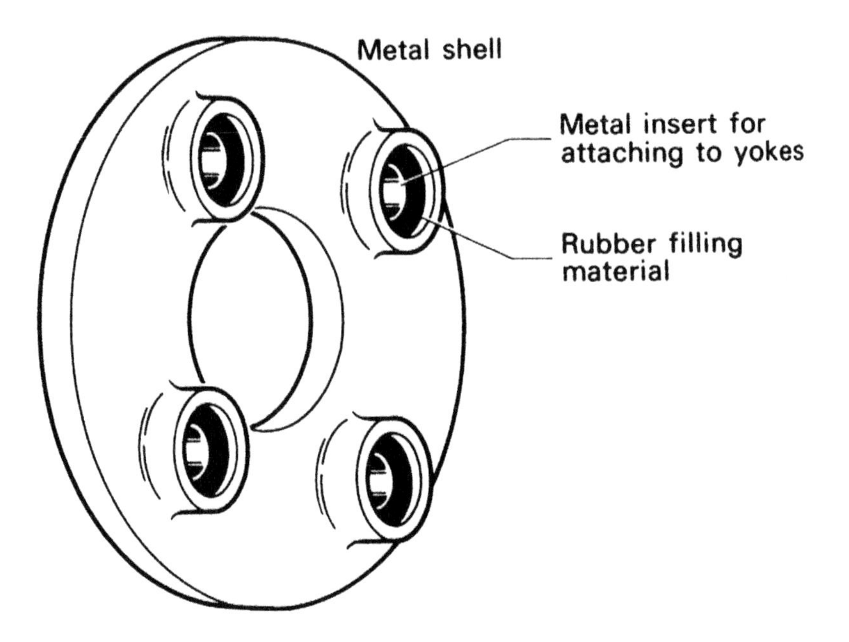

Layrub coupling

Figure 4.50 Layrub coupling

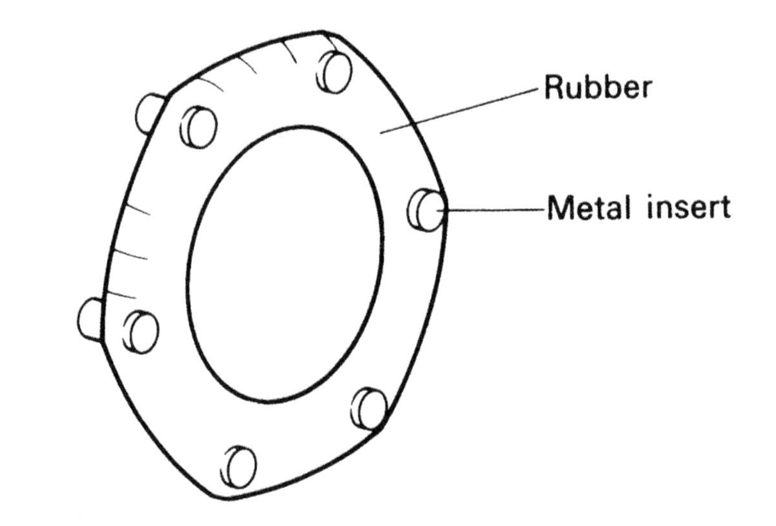

Rotaflex coupling

Figure 4.51 Rotaflex coupling

fitted it may be possible to refit it with the yokes out of alignment; this will lead to an uneven transmission of power and make the vehicle shake.

CV joints

To give a drive without fluctuations in speed; a constant velocity (CV) joint is used. This is important on front wheel drive cars where using a Hook joint would lead to steering vibrations. The joint can transmit drive through an angle of more than 15 degrees without variations in velocity. The CV joint is usually fitted in the centre of the suspension upright. A popular type of joint is the Birfield joint; this uses an inner driving member with grooves and an outer driven member with groove too. There are balls in a cage that run in both grooves, so that the power is transmitted from the inner part of the joint to the outer part through the balls. The balls are simply large ball bearings. The CV joints transmit all the driving forces; to ensure that they are correctly lubricated a special grease is used.

Replacing a drive shaft joint gaiter

The CV joint must be kept correctly lubricated and free from water and dirt. As well as the practical requirements of protecting the CV joint, or the sliding joint, it is a requirement of the MOT test that the drive shaft gaiters are kept in good condition.

To replace a gaiter on a CV joint it is necessary to remove the road wheel, disconnect one of the suspension joints and swing the suspension upright to one side, then pull back the damaged gaiter so that the CV joint can be freed from the drive shaft.

Before you fit a new gaiter you should always ensure that the joint is clean and free from water, and then apply the special CV joint grease before fitting the new gaiter.

Racer note

Transmission components are, because of where they are on the car, likely to be very dirty; and the oil or grease that is used collects dirt easily. Where possible you should always try to clean off the components before you work on them. There are many types of de-greasers; if you use one which does not need water then you will have less risk of the part rusting. However, when you are working with any de-greaser you must use protective gloves and goggles and be careful to keep it off any skin.

Checking transmission oil level

Most gearbox oil levels are checked using a **level plug**. That is, the gearbox has a plug on its side; oil is added through the hole when the plug is removed. To add the oil you need a squeeze bottle and a plastic tube. The oil level is correct when the oil is level with the lower part of the hole.

Some gearboxes have short **dipsticks** attached to a bung on the top of the gearbox. You need to replace the plug to check the oil level.

Figure 4.52 Four-paddle spinner plate

Figure 4.53 Six-paddle spinner plate

In all cases, remember to only check the oil level when the vehicle is on a **level surface**. Also, be careful that gearbox oil can be hot, so let it settle for a short while after driving the car.

Modifications

Clutch

The clutch transmits the power from the engine to the gearbox – it must be capable of doing this. Typical clutch modifications include:

Operating mechanism – the operating mechanism may be changed – a hydraulic linkage can be set up to provide a better mechanical advantage as well as cope with different clutch positions to that of a cable mechanism.

On motorcycle engine systems, such as Suzuki Hybusa and Yamaha R1 engine and gearbox set-ups, the need to disengage the clutch is removed on up changes by having a system that cuts the fuel supply when the gear change is moved. This slows the engine momentarily to allow a smooth clutchless change, saving effort and time.

Clutch springs – to transmit the extra torque of a tuned engine the clutch springs will require a higher force when using the same size and type of clutch plate.

Clutch spinner plate – for competition purposes a solid spinner plate will be used. This may be of the paddle type for reduced rotating mass – hence quicker gear changes as the speed of rotation can be changed more quickly.

Clutch material – a variety of materials are used for clutch friction plates (spinner plates); these include: carbon, ceramic – often sintered to the steel and bronze.

Multi-plate clutch – a multi-plate clutch mechanism allows the transmission of greater torque by the use of more friction surfaces. This also allows the construction to be of smaller diameter – contributing to a reduction of rotating mass and hence better acceleration. Multi-plate clutch kits are available for most engines commonly used in motorsport.

Gearbox

The main change with the gearbox is to get the correct ratios. For a circuit car it is common to have different gear sets available for each different circuit. This is easily changed with a Hewland-type gearbox where the gear cluster is removed in one unit assembly from the rear of the gearbox without removal from the car.

On road cars often the only alternative is an after-market dog box – these are usually simply close ratio gear sets to enable the use of a narrow power band.

Drive shafts

Up-rated drive shafts and joints may be needed if the power output is increased. With popular models of vehicles it is often the case that the drive shafts, joints and the suspension and brake assemblies can be used from the more powerful models on the tuned lower range models.

Questions and skills

1 Look at a cable clutch and describe where the ends of the outer cable are attached.
2 Find out how clutches are operated on motorcycles.
3 Take apart an old gearbox and, using the workshop manual, identify the parts and calculate the gear ratios.
4 Using workshop manuals, or data sheets, compare the gear ratios for a number of different vehicles of your choice. If possible include some trucks.
5 Using an old drive shaft, or propeller shaft, remove and dismantle the CV joint or universal joint.

Chapter 5

Running gear

The body or chassis with running gear attached is sometimes referred to as the **rolling shell** or **rolling chassis**. Often race and rally cars are sold like this so that the new owner can fit an engine and transmission of choice. This chapter looks at the parts which make up the running gear part of the rolling shell or chassis. Chassis and bodies are covered separately.

Suspension and steering

The front suspension and the steering mechanism are combined into one unit that carries out two distinct functions. The rear suspension is usually separate from other functions. On front

Figure 5.1 Typical front suspension

wheel drive cars, the drive shafts with their CV joints pass through, and turn with, the steering and suspension. In this section we are only looking at the suspension and the steering.

Key points

- The suspension and steering are combined at the front of the vehicle; rear suspension is usually much simpler
- Three popular types of suspension systems are beam axle, MacPherson strut and wishbone
- Three popular types of springs are coil springs, leaf springs and torsion bars
- Suspension joints must be checked at major services, before events and, if appropriate, before the MOT test
- Castor, camber and KPI are important angles with regard to handling
- The Ackermann angle affects true rolling motion of the wheels
- Wheel alignment is checked using optical gauges

Function of the suspension mechanism

The function of the suspension is to **connect the road wheels to the body/chassis**; it is designed to prevent the bumps caused by the road-surface irregularities from reaching the occupants of the car. This is to make the car both pleasant to ride in and easy to drive. The suspension also protects the mechanical components from road vibrations, so making them last much longer.

The suspension consists of a system of movable **linkages**, and a **spring** and a **damper** (also called a **shock absorber**) for each wheel. The **tyre** also forms part of the suspension; the flexing of the tyre absorbs small irregularities in the road and helps to keep the noise to a minimum. You might imagine what it was like before cars were fitted with pneumatic rubber tyres.

Function of the steering mechanism

The steering mechanism is needed to guide the car along its chosen path. The principle of steering is that the **front wheels exert a force to guide the car in the chosen direction**; the rear wheels will follow the front ones. The front wheels must exert enough force on the road to ensure that it maintains its set course. Steering the front wheels, that is turning them from side to side, is achieved by a set of **rods** and **levers** that are operated by the steering wheel acting through either a steering box or a rack and pinion unit.

Before we look at the actual components involved in the suspension and the steering mechanism we will define some of the special terms which are used. For example, **castor**, **camber**, **king-pin inclination**, **Ackermann principle** and **wheel alignment**.

Castor

Castor or, to give its full name, the castor angle is the angle that the **swivel pin**, or king-pin, is arranged to lean backwards in the **longitudinal plane**. By giving the swivel pins a castor angle of 1 or 2 degrees, the imaginary centre line meets the road before the centre of the wheel. This distance is called **castor trail**. The castor angle gives the front wheel a self-aligning, or self-centring, action. The wheel follows the pivot, so keeping the car on a straight course and reducing the need to manually straighten the steering wheel after negotiating a corner. This can be likened to the castors on a trolley. The castors swing round so that the trolley can be pushed in a straight line. You can see this in action on any supermarket trolley.

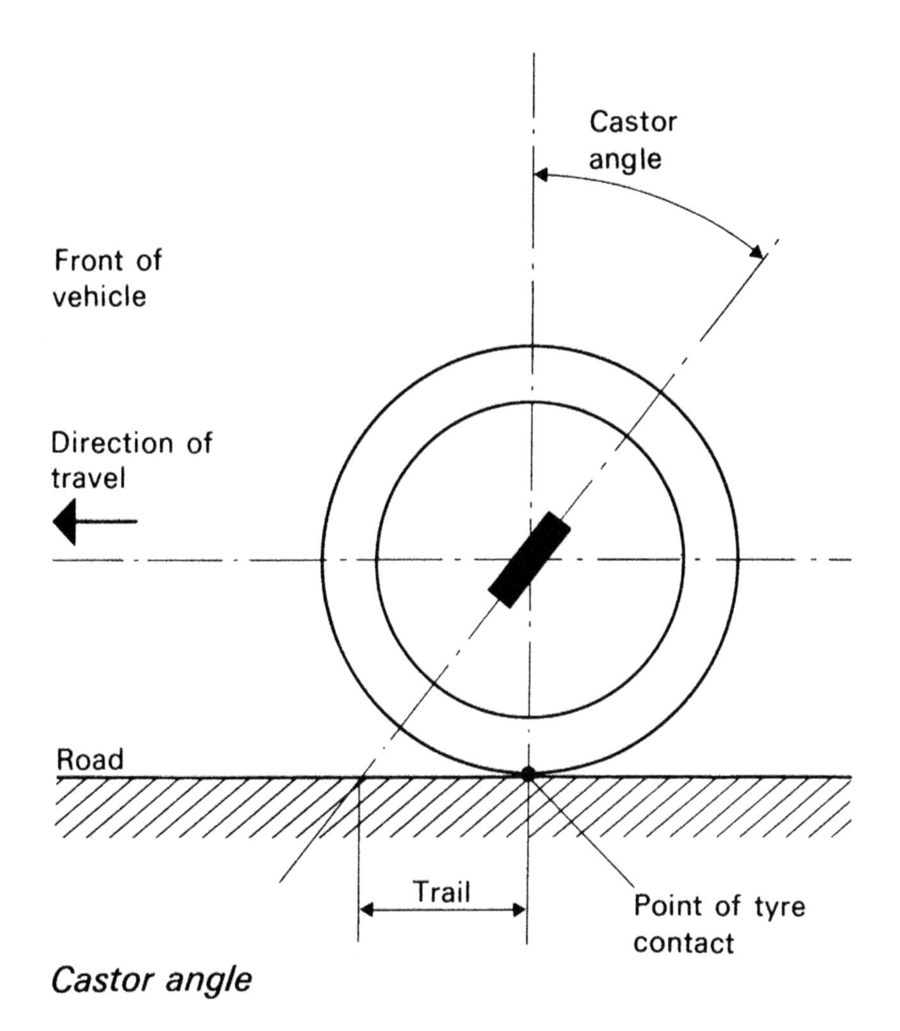

Figure 5.2 Castor angle

> **Racer note**
>
> The angles shown for castor, camber and KPI in the diagrams are greatly exaggerated so that you can see them easily. The actual angles are very small, generally between 0.5 and 2 degrees. To measure them you need specialised equipment.

Camber

Camber is the inclination of the road wheel in the **transverse plane**. The wheel is generally inclined outwards at the top; this is called **positive camber**. If the wheel were inclined inwards at the top, it would be **negative camber**.

> **Nomenclature**
>
> The term kingpin is also used for any kind of suspension pivot mechanism, especially when referring to the suspension angles. If a car does not have a kingpin it can still have a kingpin inclination angle; that is the angle that the suspension has that is the equivalent of a kingpin on an older car. Yes it sounds strange, but lots of the words that we use about cars are strange or are based on old ideas.

Kingpin inclination (KPI)

KPI is the inclination of the kingpin in the **transverse plane**, which is the same plane as the camber. Normally kingpin inclination is inwards at the top; this is **positive** KPI. **Negative** KPI is when the kingpin is inclined outwards at the top.

When the camber and KPI angles meet at the road surface this is called **centre-point steering**. The actual intersection of the lines is in the middle of the patch where the tyre meets the road.

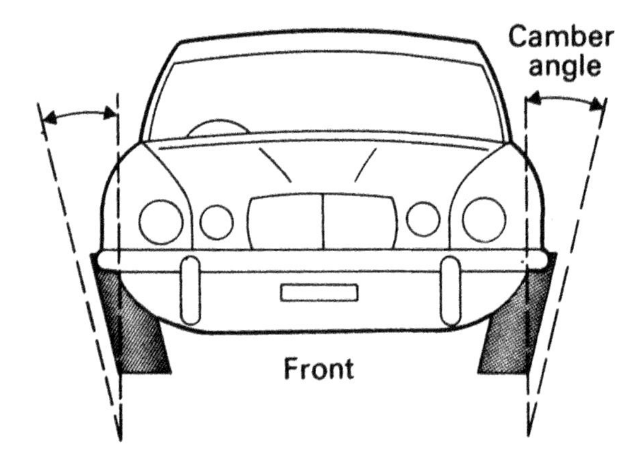

Figure 5.3 Camber angle

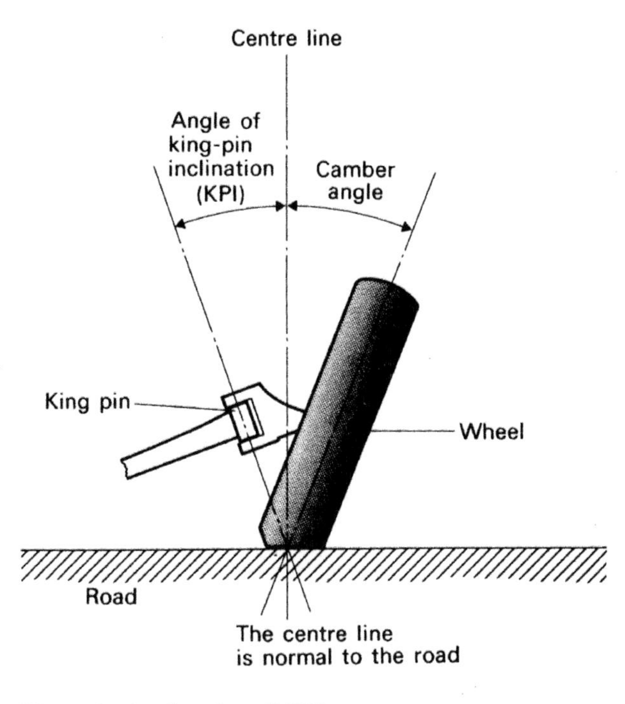

Figure 5.4 Camber angle and KPI

Figure 5.5 Camber gauge

Ackermann angle

When a vehicle turns a corner all its wheels must rotate about a **common point** or the tyre treads will be **scrubbed**. If you think of the vehicle going round a full circle; all the wheels must be turning about the centre point of that circle. This is achieved by taking the radius lines from the centres of the wheels and steering them so that they are tangential to the radius lines. That is the position of the wheels and the radius lines are at 90 degrees. The two rear wheels are on the same radius line; this is because they cannot be steered. The front wheels can be steered; they are on separate axes. The steering mechanism is designed so that the **inner wheel is always turned through a greater angle than the outer.** In the diagram, A and B are the angles through which the wheels have to be turned. The inner angle A is greater than the outer angle B.

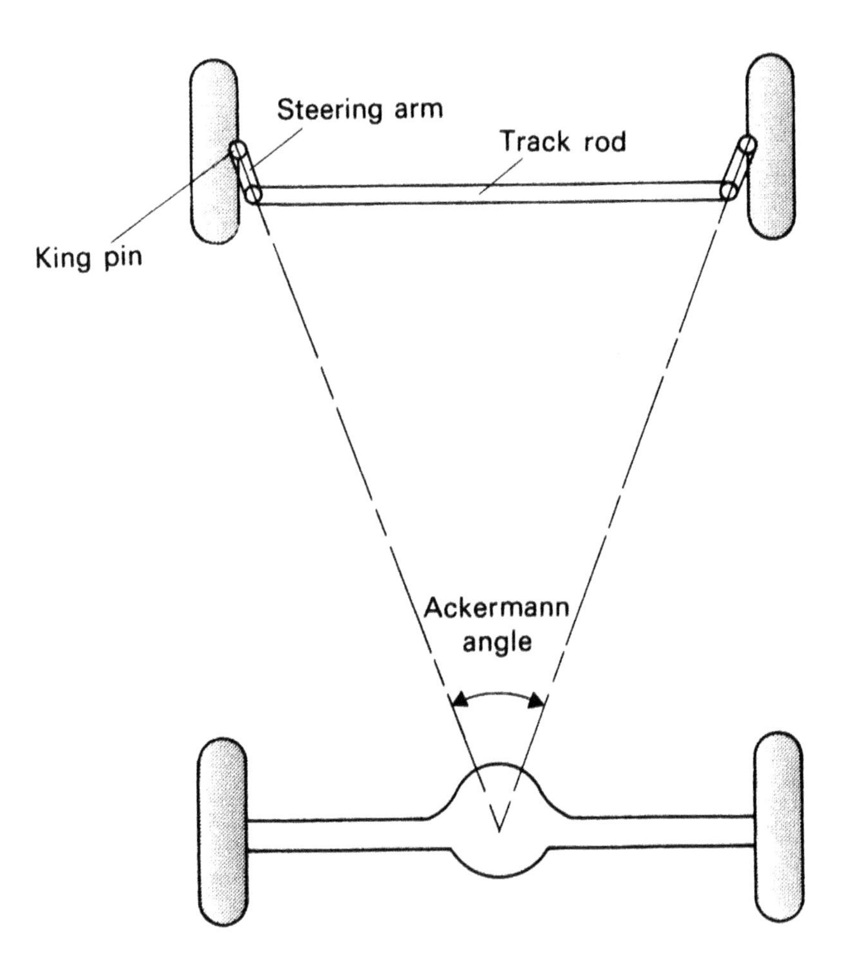

Figure 5.6 Ackermann angle

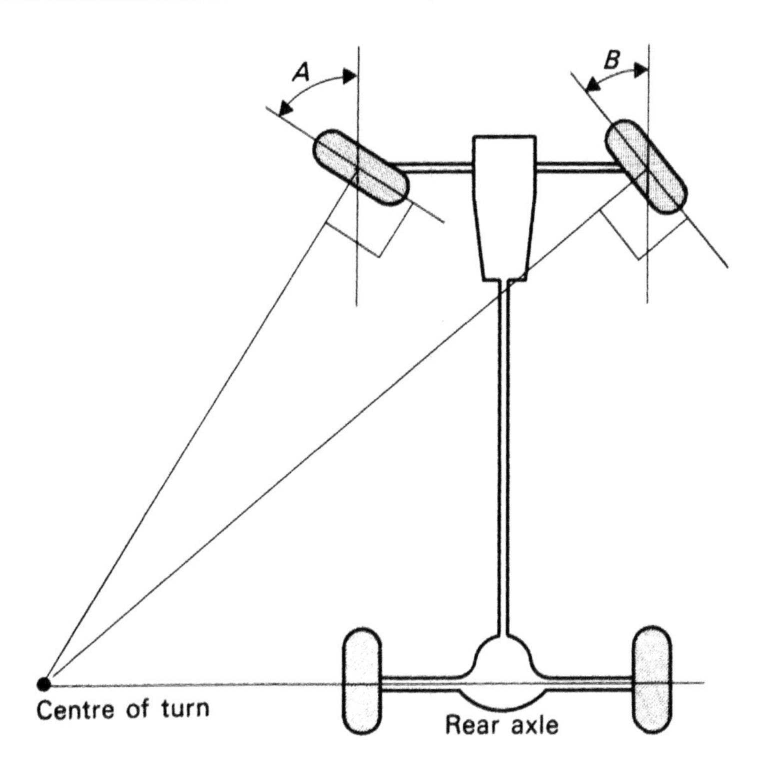

Figure 5.7 Centre-point steering

FAQs What is true rolling motion?

True rolling motion is when the wheels roll without side-slip. If the car is travelling in a straight line, and the wheel alignment is set at zero, that is there is no toe-in or toe-out, then the wheels will be rolling truly. If the car, with Ackermann, is turning a corner slowly, then all the wheels should roll truly

Wheel alignment (toe-in and toe-out)

The front wheels are arranged so that they either **toe-in** or **toe-out**. Toe-in is when the front of the wheels is closer than the rear of the wheels. That is A is less than B in the diagram. Toe-out is when A is a greater distance than B. Typically toe-in or toe-out is 2 mm. You should look in the **workshop manual**, or on the **data chart**, to find out the setting for any particular vehicle. If the wheel alignment is not set correctly the tread will be scrubbed-off the front tyres. That is because the wheels are not rolling truly – the wheel and the tyre move sideways. This will cause the tread to be scrubbed, like using a rubber eraser on a pencil drawing.

Figure 5.8 Turn-table

Figure 5.9 Turn-table scale

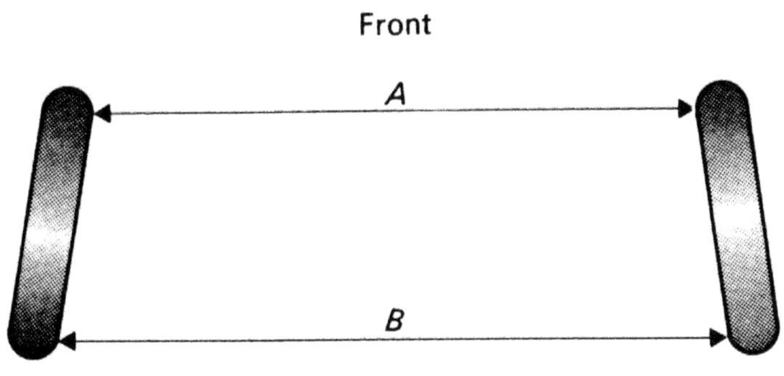

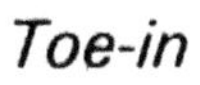

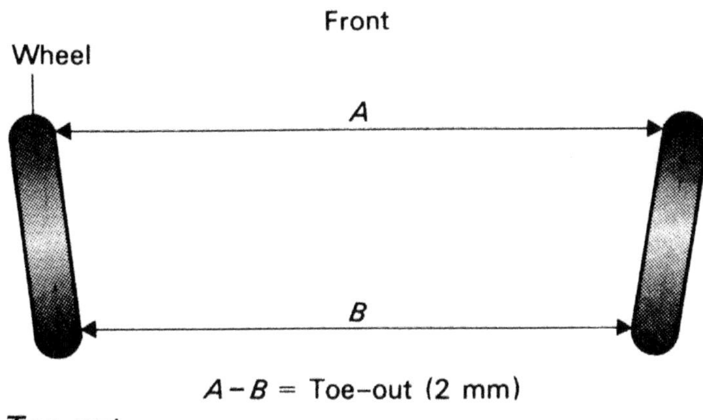

Figure 5.10 Toe-in, toe-out

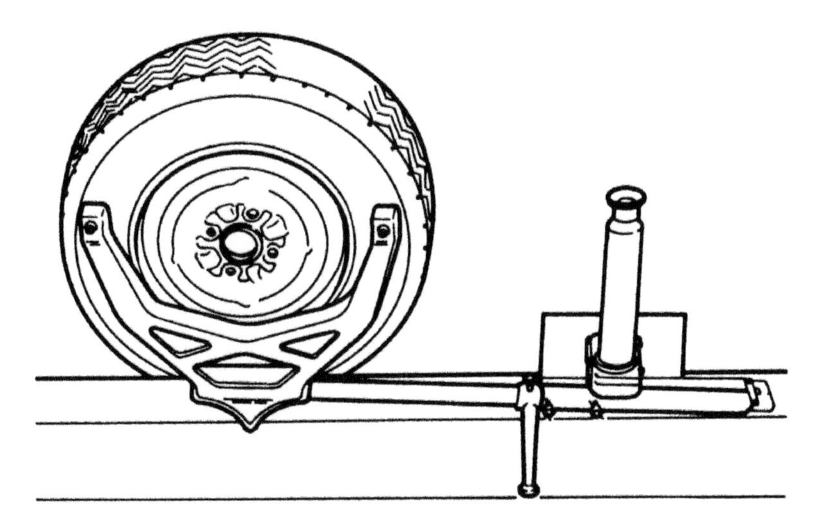

Figure 5.11 Wheel alignment gauge

Springs

The spring absorbs the shock when the wheel hits a bump in the road. The spring must be strong enough to support the vehicle and its load; but be able to be compressed when a bump is hit. There are a number of different springs; the popular ones are coil springs, leaf springs and torsion bar springs. All these different springs are made from specially tempered medium-carbon steel called spring steel.

Coil spring

The coil spring is usually fitted around the shock absorber or McPherson strut. It is made from round section spring steel that is wound to shape. Coil springs offer a large amount of suspension movement and lightness.

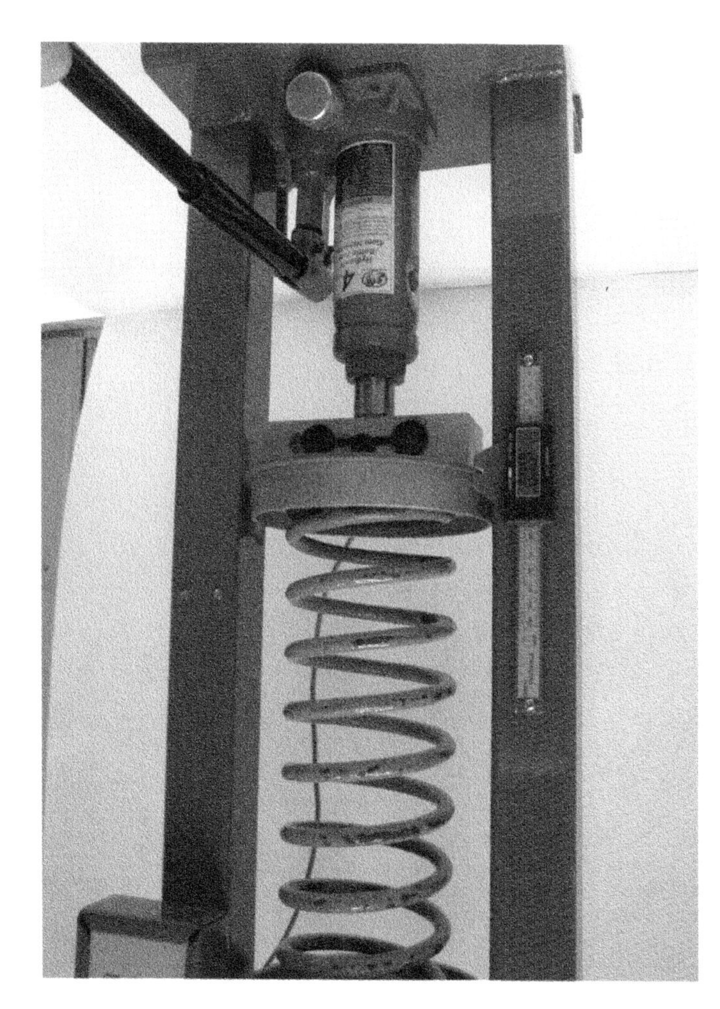

Figure 5.12 Spring stiffness tester

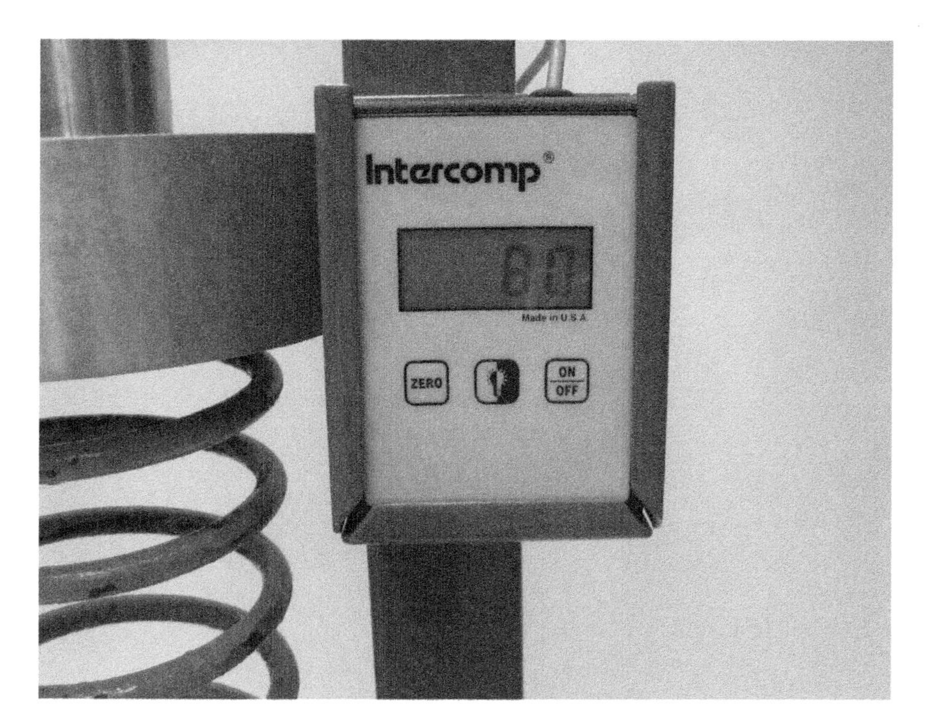

Figure 5.13 Spring test readout

Figure 5.14 Spring compressor tool

Figure 5.15 Ball joint tool

Leaf spring

These are made from a number of flat metal sections; rather like thin leaves. The main leaf has an eye at each end so that it can be attached to the chassis. Leaf springs can carry a lot of weight; but suspension travel is limited.

Torsion bar spring

The torsion bar is a round bar that twists when it is loaded. It has the advantage that its weight is carried fully on the chassis; it can also be fitted in positions where height is limited. The disadvantage is that the amount of suspension travel that is allowed is very small.

Suspension layout

There are lots of different suspension layouts; this section looks at some of the more popular ones.

Beam axle suspension

The axle beam goes transversely across the vehicle; it is attached to the chassis with leaf springs. The leaf spring consists of a main leaf which has an eye at each end and several other supporting leaves. Both the axle beam and the spring are made from medium-carbon steel. The front eye of the spring is attached to the chassis with a fixed shackle; the rear eye has a swinging shackle. The spring is attached to the axle with u-bolts. The steering mechanism is achieved with a kingpin and reverse-Elliot linkage.

The disadvantage of beam axle suspension is that when one of the wheels hits a bump the wheel on the other side is also tipped. This is because both wheels are attached to the same component that is the axle beam. When one wheel hits a bump the whole vehicle is made to tip up sideways.

Independent suspension

With independent suspension each wheel is suspended independently of the others. Some vehicles have independent suspension on the front wheels only – this is called **independent front suspension (IFS).** At the rear it is called **independent rear suspension (IRS).** When one wheel of a car with independent suspension hits a bump only that wheel is deflected upward; the other wheels are not affected and the car remains level.

Wishbone suspension

Wishbone suspension is so called because the arms are shaped like chicken wishbones. The coil spring is usually mounted **concentrically** over the shock absorber. The shock absorber is attached to the lower wishbone and the chassis.

McPherson strut suspension

The McPherson (say mac'fur'son) strut is a combination of a suspension swivel pin and a shock absorber unit in one assembly. The spring is a coil spring, which is mounted concentrically on the McPherson strut. McPherson strut suspension is very easy to remove from the vehicle; the top mounting goes to the inner wing and the lower one to the track control arm. With the strut removed you can then replace the faulty part working on the bench. However, if you are replacing any of the components which require removal of the spring, then you need to compress the spring a little to prevent it flying and possibly causing damage or personal injury. Compressing the spring involves fitting spring compressors – hooks with screw-threaded parts – to two of the coils and screwing up the threaded parts until they take the spring load off the mountings.

Shock absorbers

The purpose of the shock absorber is to dampen the spring action and reaction. The shock absorber stops the vehicle from bouncing each time it hits a bump in the road. There are two types of shock absorbers. These are telescopic and lever arm and they are easily identified by their shape and mountings.

FAQs Is a damper the same as a shock absorber?

Yes.

Telescopic shock absorber/damper

The telescopic damper/shock absorber gets its name from its telescope-like shape and action. The cylinder is filled with a special type of oil, called shock absorber fluid. The lower part of

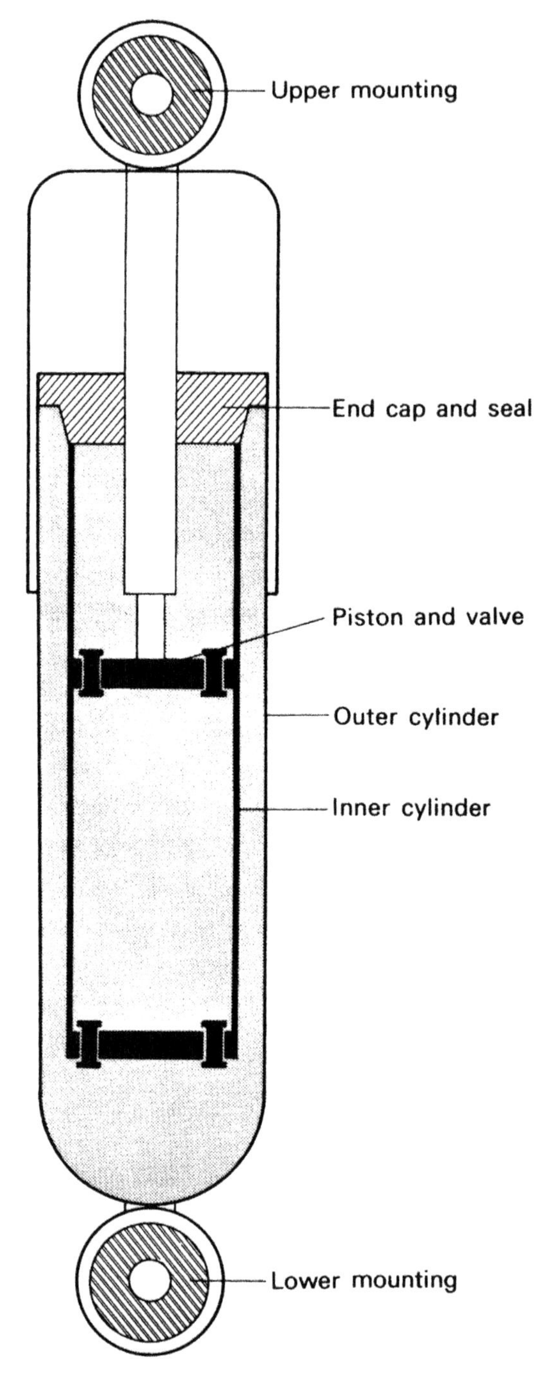

Figure 5.16 Telescopic damper

the cylinder is connected to the suspension with a mounting eye. The upper part of the shock absorber, which comprises the piston and valve assembly, is attached to the vehicle's chassis with the upper mounting eye. The piston and the valve assembly move up and down inside the cylinder with the movement of the suspension.

When the wheel hits a bump the suspension travels upwards, shortening the distance between the mountings. In this situation the piston travels down the cylinder. The resistance of the fluid slows the movement of the piston, so dampening the shock load on the suspension. When the wheel has travelled over the road bump and the suspension rebounds, that is the wheel travels down again and the piston moves down in the cylinder as the distance increases again, the resistance of the fluid dampens the suspension movement and prevents the suspension from travelling too far. The shock absorber therefore dampens the suspension movement and stops the vehicle from bouncing like a rubber ball every time it hits a bump.

Nomenclature

Bump is when the suspension is compressed; that is the wheel goes up into the wheel arch and the body goes down towards the road. Rebound is the opposite of bump; that is the wheel goes down and the body goes up. Commentators on rallies often use the word 'jounce,' this is another word for bump – you'll see it when the car lands back on the road after going over a hump-back bridge. This take-off and jounce is referred to as '**yumping**,' the Scandinavian pronunciation of jumping.

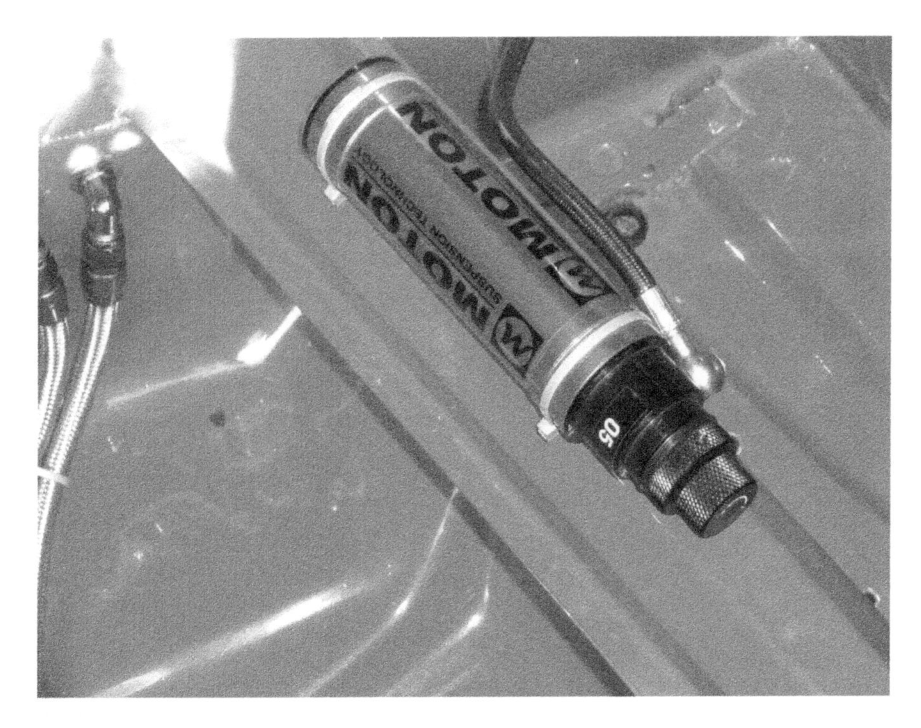

Figure 5.17 Damper reservoir

Lever arm shock absorber

The lever arm shock absorber works in a similar way to the telescopic shock absorber. The body of the shock absorber is bolted to the vehicle's chassis; the lever arm is attached to the suspension and therefore moves up and down with the suspension. The arm is attached to a rocker assembly inside the body. The rocker assembly moves two pistons in parallel bores. One piston goes up as the other piston goes down; this is called double acting. The piston movement is given resistance by shock absorber fluid in the same way as the telescopic one. Lever arm shock absorbers usually have provision for their fluid to be topped up – telescopic ones generally do not.

Bounce test

When you are checking the shock absorbers on a vehicle you should:

- Inspect the mounting for damage or wear – the rubber bushes should be tight
- Inspect the body for damage or dents
- Check the seals for fluid leaks
- Carry out a bump test – that is press down on the wing and let it go quickly. It should not go up and down more than three times

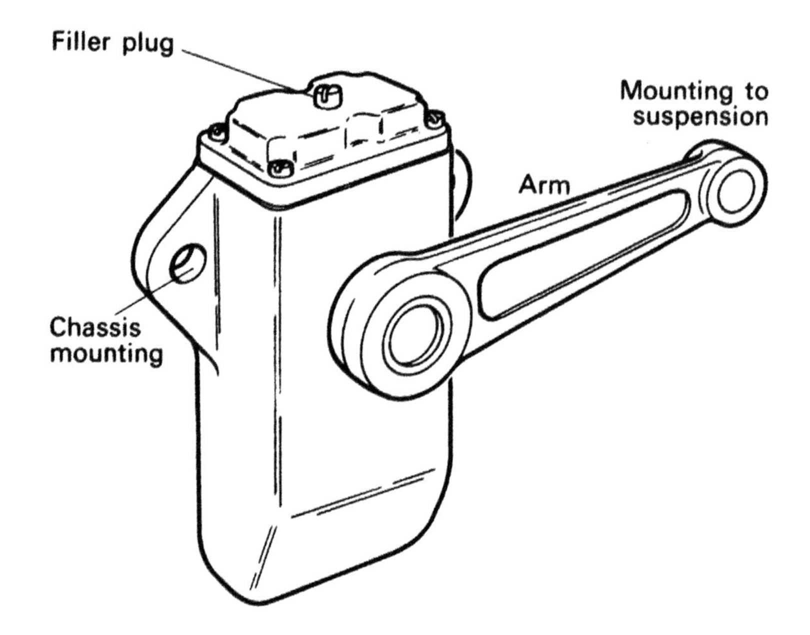

Figure 5.18 Lever arm shock absorber

Steering linkages

Various layouts of steering linkages are used, depending on the type of vehicle and the system chosen by the designer. Many cars tend to use rack and pinion steering mechanisms; these are shown later in this chapter. The steering wheel is situated inside the car. This is mounted on the steering column that passes through the bulkhead and connects to the other components that are underneath the front of the car. The job of the steering linkages is to convert the movement of the steering wheel into movement at the road wheels.

When the steering wheel is turned, this turns the steering column that operates the mechanism of the steering box. The cross-shaft is the output from the steering box, which moves the drop-arm. This pushes or pulls the drag link, which operates the steering lever, which is attached to the offside stub axle. The offside wheel is thereby moved in the required direction. The wheel hub is mounted on the stub axle, which pivots on the king pin in the beam axle. The track rod is attached to the offside steering lever, so that it moves transversely when the offside wheel is turned. This moves the nearside track rod end (TRE) which is attached to the nearside steering arm that steers the nearside stub axle and the nearside wheel.

Steering box

The steering box is bolted to the chassis. The steering column is attached to the steering box; the steering wheel is attached to the inner part of the steering column. The cross-shaft is attached to the steering linkage underneath the car. The steering box does a number of different jobs, the main ones are:

- Turning the drive through a right-angle (90 degrees) between the steering column and the cross-shaft
- Giving a reduction gear ratio of about 14:1 so that the turning force applied by the driver to the steering wheel is increased at the cross-shaft. The movement of the cross-shaft will be reduced by an equivalent (14:1) movement ratio
- Stops bumps caused by road surface irregularities being passed on to the driver

With worm and peg steering, the peg is at the end of the rocker shaft. The tip of the peg sits in the worm. The worm is fitted on the end of the steering column. When the steering column is turned by the steering wheel the worm turns too. The peg follows the helical thread of the worm; this moves the rocker shaft that is attached to the cross-shaft. The cross-shaft turns move the steering linkage.

It is important that the steering box is kept properly lubricated; usually a gearbox type of oil such as SAE 80 EP is used. The oil is added through the filler plug, which is also a level plug. That is, oil is added until it reaches the lower part of the filler hole. Some steering boxes have a separate grease nipple for the end bearing. The steering box can be adjusted to compensate for wear. There is an adjusting screw for wear in the peg and the worm. The bearing adjustment is by shims under the end plate.

The normal service interval for checking the steering box oil level and the adjustment is 15,000 km (10,000 miles).

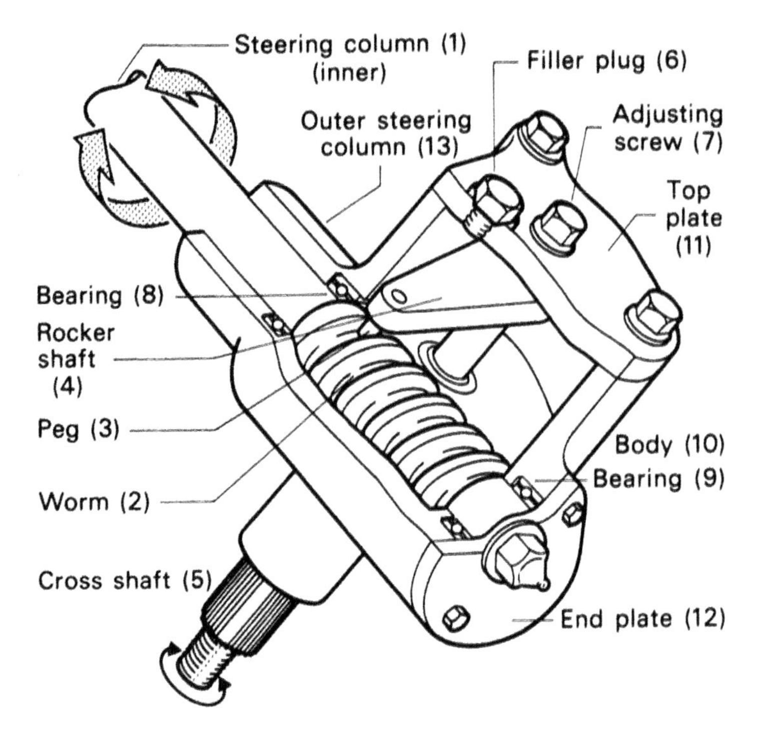

Figure 5.19 Steering box

Racer note

To check for the correct adjustment to the peg and the worm follow this simple procedure:

- Jack the front of the car up and support it on axle stands.
- Set the steering wheel to the straight-ahead position. Move the steering wheel about 90 degrees in each direction; you should feel a slight tightness in the middle (straight-ahead) position.
- If this cannot be felt screw the adjusting screw down until it can, taking note that some adjusting screws have lock nuts.

To check the end bearing this is what you do:

- Sit inside the car and see if the steering wheel will move vertically on the steering column.
- If there is excessive vertical lift then remove shims from between the end plate and the steering box body.

Rack and pinion

The rack and pinion assembly does the same job as the steering box and track rod combined together. The rack and pinion is a long and thin tubular looking arrangement. It is used on most cars because it has the advantages of being both light and cheap.

The rack and pinion is made in one unit, which is bolted to the bulkhead with U-bolts. The steering column connects to the pinion. The track rod ends (TRE) are screwed onto the track rods. The track rods are attached to the rack by ball joints. The TREs connect to the steering arms on the hub carriers using taper fit pins.

The pinion is attached to the lower end of the steering column so that it is rotated when the driver turns the steering wheel. The pinion meshes with the rack, so that turning the pinion moves the rack to one side or the other. When the rack moves the track rods, steering arms and then the road wheels move.

Wheel bearings

Hubs usually run on pairs of ball bearings. Larger heavier vehicles may use taper roller bearings. The bearings may be either pre-loaded, or they may be adjustable.

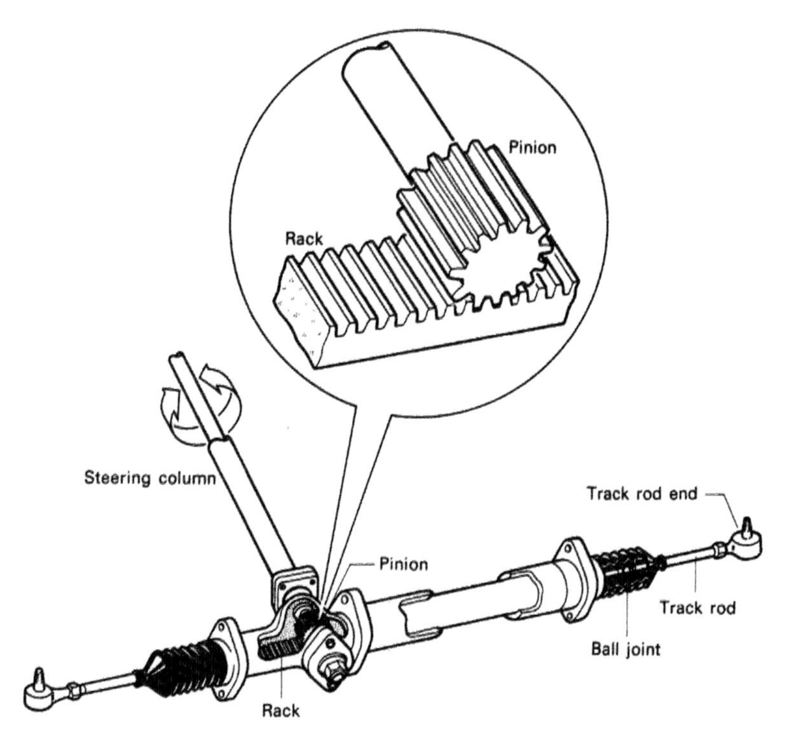

Figure 5.20 Rack and pinion steering

Checking wheel bearings

With the car supported on axle stands, there are two checks to be made. One is for bearing free play; the other for noise. To check for free play, hold the top of the wheel with one hand and the bottom with your other hand. Try to move the wheel from side to side. If it moves more than the smallest amount, then the bearing needs either adjusting or replacing. To check for noise, spin the wheel by hand and listen; a good bearing should spin freely and quietly.

Checking steering and suspension joints

Two of the main points of wear in the steering system are the ball joints and the track rod ends. To check these joints for wear, the two parts that the ball joints connect should be pulled in a direction that you would expect them to part. Using hand pressure, maybe with a small lever, a joint in good condition should show no signs of free play at all.

Adjusting wheel alignment

The wheel alignment, toe-in or toe-out, should be checked every 20,000 miles (30,000 km). This is carried out using optical gauges. The optical gauges may use a laser beam and may have a digital readout, depending on the level of technology and sophistication.

Figure 5.21 Suspension top joint

Figure 5.22 Suspension bottom joint

The wheel alignment optical gauges are fitted to the front wheel, the laser beam is focused and the reading can then be taken. The reading scale may be in mm (inches on older gauges), or in degrees. Most wheel alignment gauges are supplied with a chart, or table, so that degrees can be converted into mm for most standard size wheels, that is 10–20 inch.

Rally car specifics

Most rally cars are based on road cars; the changes to the suspension depends on the specific competition regulations. Typically, the changes will involve the springs and the dampers. The springs are chosen to give the suspension setting needed for the type of rallying – typically they are stiffer. The dampers are likely to be adjustable to allow variations in suspension behaviour – firmer, or harder, dampers are used to reduce suspension bounce. Lots of rallies are run off-road, using special stages, or white roads – so the suspension has to be robust and able to deal with hitting bumps at high speeds. To prevent damage, additional bump or rebound stops may be fitted.

Race car specifics

Unlike rally cars, race cars usually run on smooth circuits; however, the circuits tend to be tighter and the bends are taken at very high speeds. So, the amount of suspension travel is likely to be very small – as low as 3 mm (1/8th inch).

However, the suspension movement will be very rapid and dampers with additional cooling will be required as well as fine adjustment for stiffness. The suspension on open-wheel cars is very light and may be made of carbon fibre composites. The springs are very stiff and little compression is likely.

Most single seat race cars use inboard suspension so that the weight is taken on the chassis.

Race cars tend to use negative camber to keep the tyre treads on the road; and very little castor so that the car is not easy to keep in a straight line, but will corner without effort.

Wheels and tyres

Key points

- The wheels and the tyres must be perfectly round, rigid and correctly balanced.
- There are rules relating to wheel and tyre fitment and maintenance that must be followed; for example the minimum tread depth on road cars is 1.6 mm.
- Wheels may be made from steel or aluminium alloy; with either a well base or detachable flanges for fitting the tyre.
- The tyre and the wheel diameter are measured where the tyre fits the rim; the width is measured between the flanges.
- Different types of tyre construction and tread designs are used for different purposes.
- Specialist machines are used for changing tyres and balancing the wheel and tyre.

Functions of wheels and tyres

The wheels and tyres have a number of jobs to perform, namely to:

1 allow the vehicle to freely roll along the road
2 support the weight of the vehicle
3 act as a first step part of the suspension
4 transmit to the road surface the driving, the braking and the steering forces

Requirements of wheels and tyres

For the wheels and the tyres to be able to carry out their functions efficiently they must be made and maintained to the following basic requirements:

1 They must be perfectly round so that they roll smoothly.
2 They must be balanced so that the steering does not shake.
3 They must be stiff to give responsive steering and smooth running.

Wheels

There are many different types of wheels in use. We'll look at the most common ones – that is steel well based, **aluminium alloy**, **wire-spoked** and **two-piece**.

Safety notes

- Always use axle stands with the vehicle on a flat and level surface when removing wheels.
- Never exceed the maximum tyre pressure given by the manufacturer.
- Always use a torque wrench to correctly tighten the wheel nuts.

Basic construction and sizes

Basically, all wheels comprise of a **rim** and a **wheel centre**, which are attached together in some way. The rim is the part to which the tyre is fitted.

The rim has a **flange** to hold the tyre in place, a seating part to seal the tyre bead against and retain the air and a well section so that the tyre can be fitted and removed from the rim.

The wheel centre is the part that is attached to the hub. The wheel centre usually has four holes for the **wheel studs**. The back of the wheel centre has a flat section to make contact with the hub.

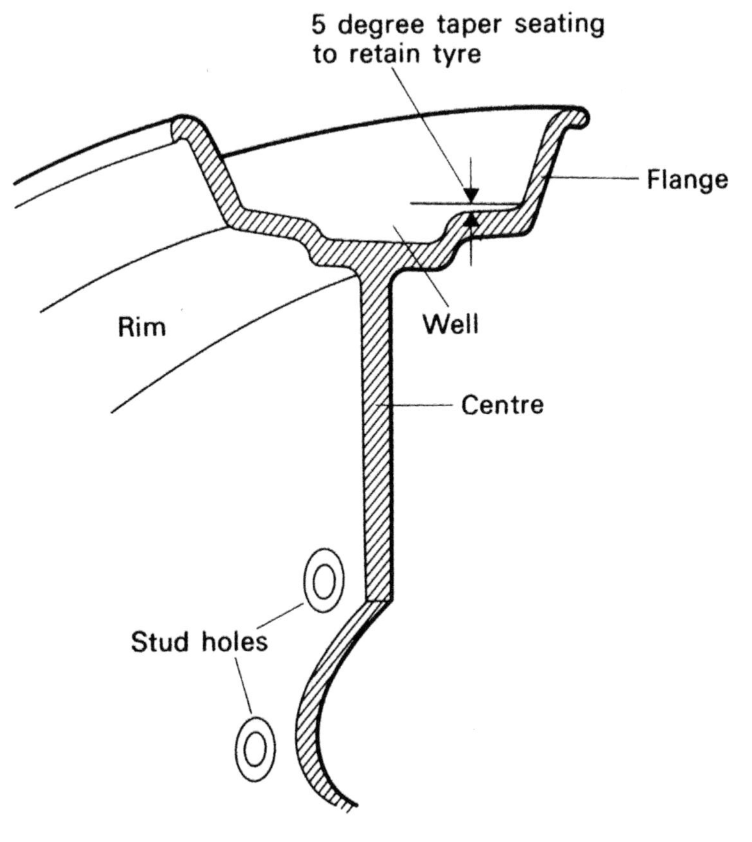

Figure 5.23 Part of wheel

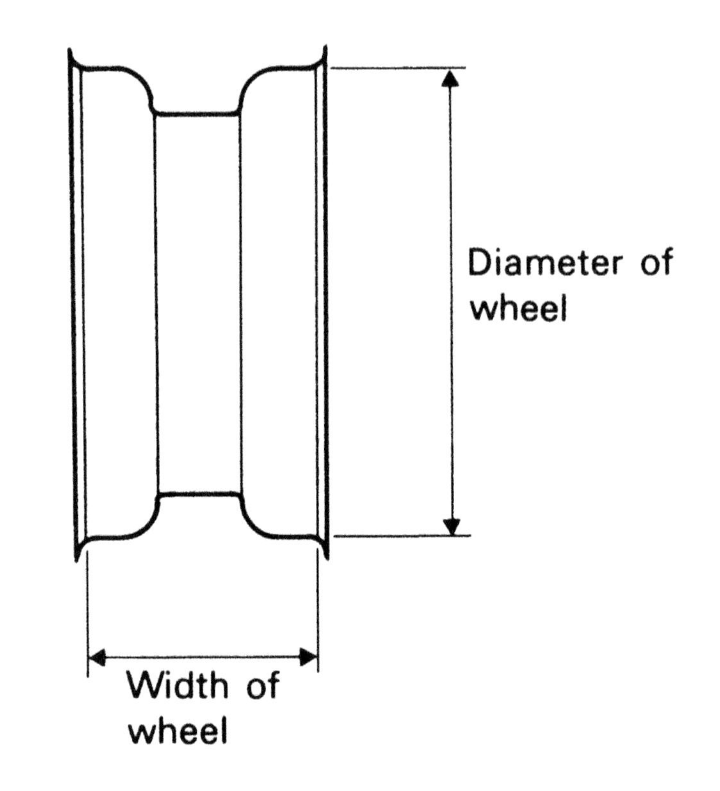

Figure 5.24 Wheel size measurements

The **wheel diameter**, which is also the tyre size, is measured at the tyre seating part of the rim. You should note that the flange extends beyond this part of the wheel. The **wheel width**, which is also the equivalent of the nominal tyre width, is measured between the inside faces of the flanges.

> **Racer note**
>
> When fitting wheels it is very important that they are tightened correctly. If they are not tight enough they may come loose; if they are too tight it can cause damage. To set the tightness correctly, you must use a torque wrench. Check the workshop manual, data book, or wall chart for the torque setting for each specific vehicle and type of wheel.

Steel wheels

The rim and the wheel centre of the steel wheel are both made from pressed low carbon steel. The rim is spot welded to the wheel centre. Steel wheels have the advantages of being cheap, tough and reasonably light in weight. Steel wheels are ideal for off-road events such as 4 × 4 cross country and rallycross as they are tough, so they will withstand a limited amount of impact damage.

Aluminium alloy wheels

Aluminium alloy wheels were originally developed for aircraft; they give a combination of extreme lightness and high stiffness. The aluminium is alloyed with silicone for wear resistance, copper for hardness and magnesium for easy casting. Aluminium alloy wheels are usually cast in one piece; but some specialist wheels are made in two parts, which are held together with a ring of bolts. This construction allows the rim to split into two parts for easy fitting of the tyre. Two-piece alloy rims are flat across the whole section – that is they do not have a well section; it is not needed.

The disadvantages of alloy wheels, as they are referred to, is that they are expensive, brittle and have a limited life span. The brittleness is a problem if the wheel hits a kerb or other hard object (called kerbing); in this case the rim is likely to chip or crack, obviously the tyre will soon deflate.

Of course, alloy wheels can be made in a variety of styles, so they are available with looks to suit the car and its owner. Good looks are in the eye of the beholder!

Wire spoke

The wire-spoked wheel is only used on a small number of sports cars. Unlike the other wheels, which are attached to the hub with studs, the wire-spoked wheel is attached to the

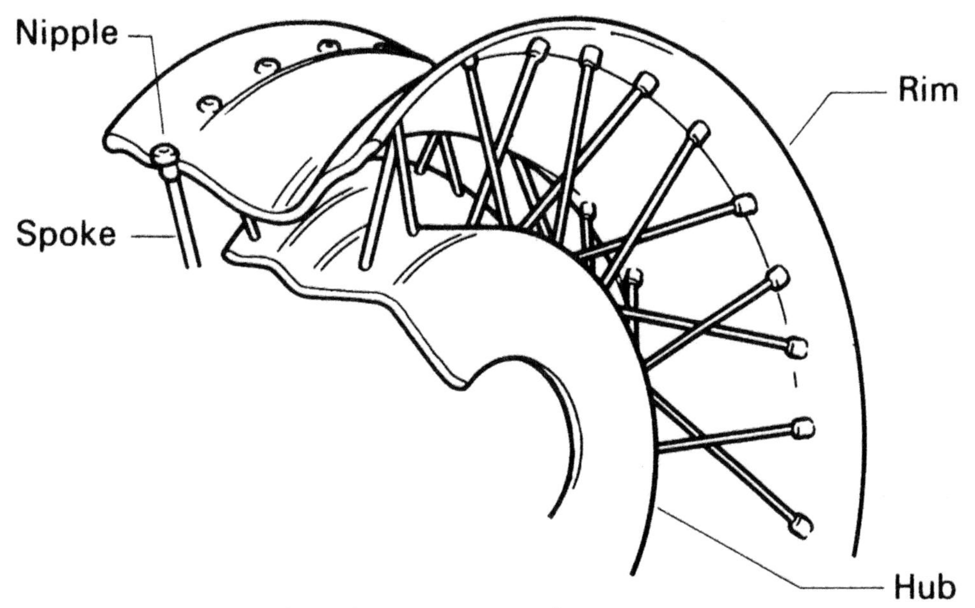

Figure *5.25* Wire-spoke wheel

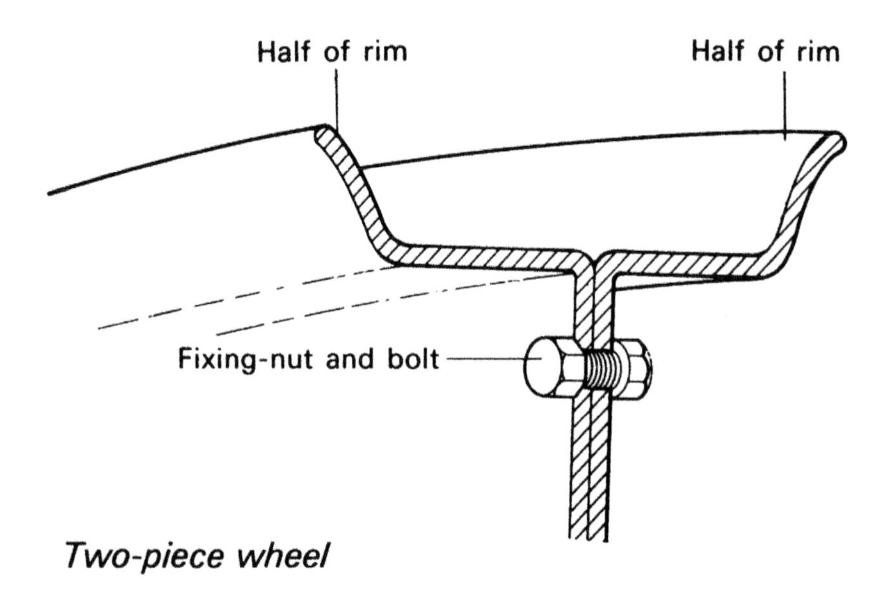

Figure 5.26 Split or two-piece wheel

hub with a splined section and a single **rudge nut**. The wire-spoked wheel is slightly flexible and springy. Spoked wheels, as they are usually abbreviated, have the advantages of being light, good looking and allowing the cool air to pass through to cool the brakes.

The disadvantages are that they are easily buckled, especially if the spokes become loose, and they need an inner tube to seal in the air.

Tyres

Construction

The basic construction of all vehicle tyres is very similar. The main components are the **tread**, the **casing**, the **wall** and the **bead**. The wire bead forms the shape and the size of the tyre. The textile plies and the rubber covering runs from one bead to the other. The two main different types of tyre construction are radial ply and diagonal ply.

Radial and diagonal ply

Radial ply tyres are so called because the plies run in a radial manner. The plies of **diagonal ply** tyres usually run at an angle of about 45 degrees to the tyre radius.

Radial ply tyres roll more freely than diagonal ply tyres; this gives both better fuel economy and longer tread life. However, diagonal ply tyres are quieter and less inclined to make the suspension or the steering knock or vibrate if a road bump is hit. For this reason diagonal

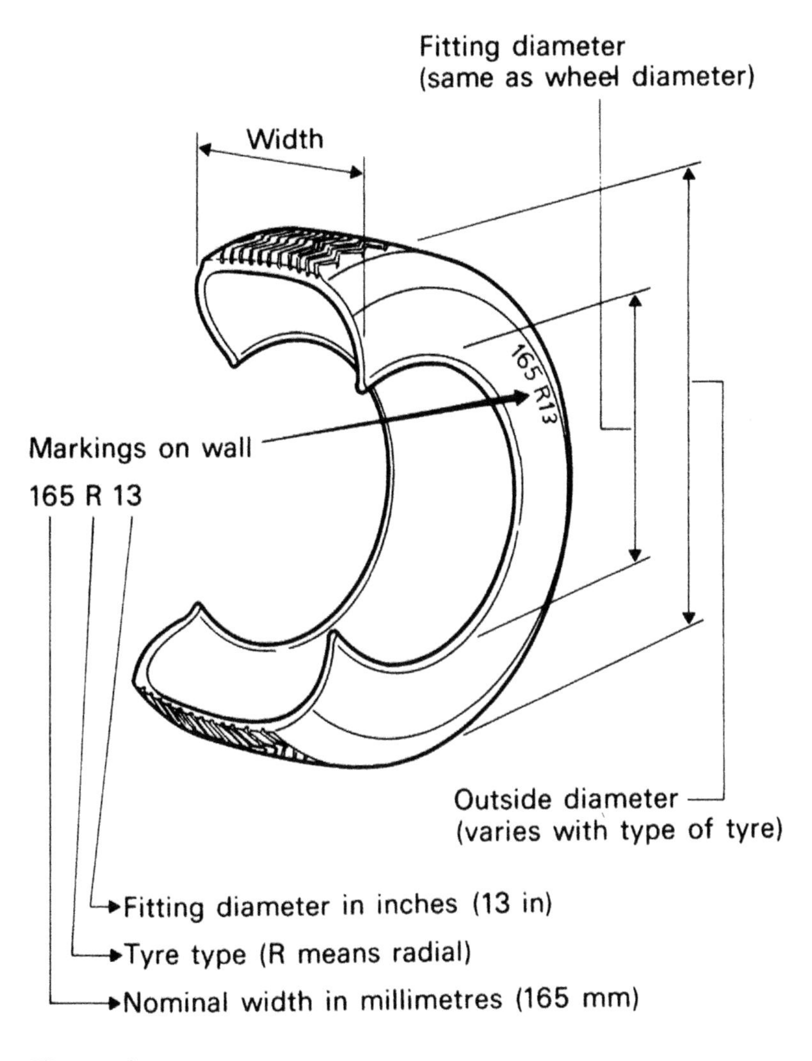

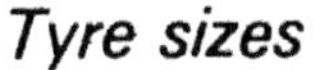

Figure 5.27 Tyre identification marks

ply tyres are used on some American vehicles – those likely to be used on poorly surfaced roads, or off-road.

Because of their constructional differences radial ply and diagonal ply tyres behave differently; for this reason, they must not be mixed on the same axle. If fitted in pair, the radial ply tyres must be fitted to the rear wheels.

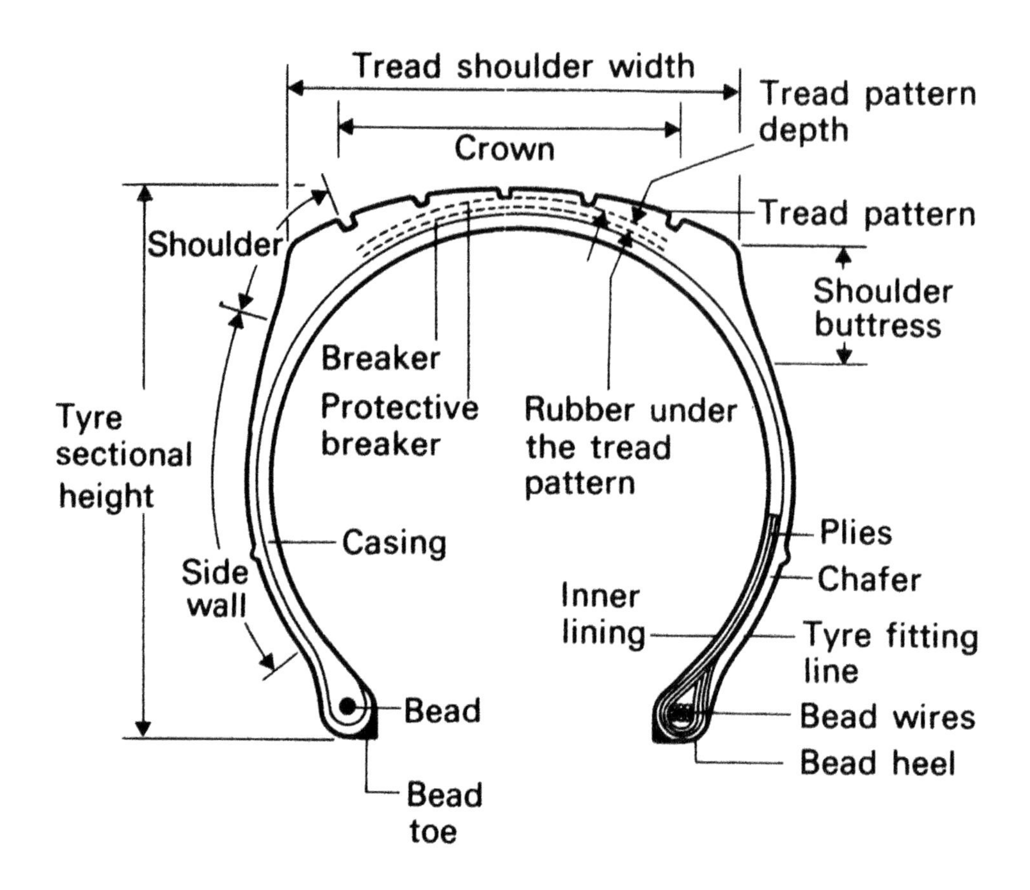

Figure 5.28 Tyre part names

> **Nomenclature**
>
> The term diagonal ply refers the fact that the plies run from bead to bead in a diagonal or angular manner. Previously this type of construction was more commonly referred to as cross-ply; this name is still used by older mechanics and tyre fitters.

Tyre treads

Different types of tyre treads are used for different purposes. The tyre tread patterns are designed for the different applications of the vehicle. Typical examples are ordinary **multi-purpose** car tyres (budget), **truck tyres**, **high-speed** car tyres, **off-road** vehicle

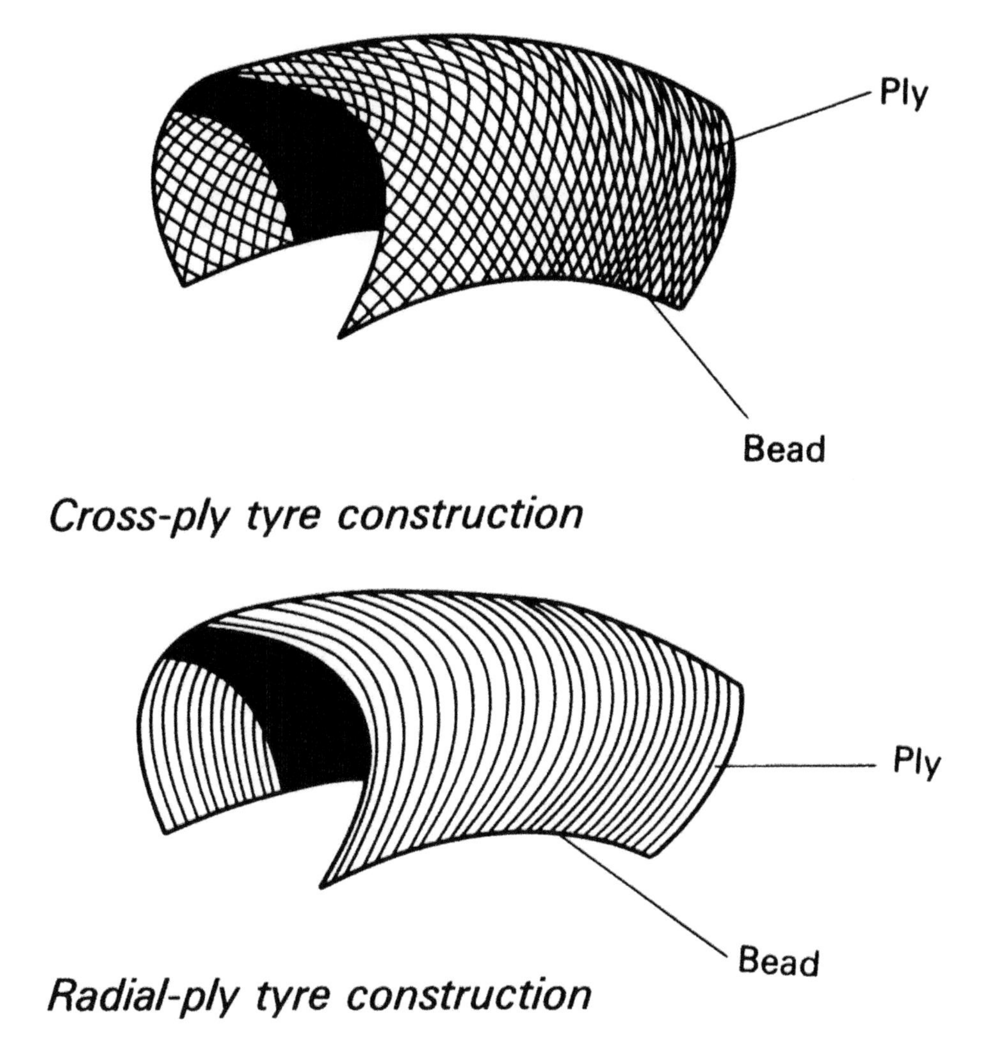

Figure 5.29 Diagonal and radial ply construction

tyres, **asymmetric** treads, **mud and snow** (M&S), **wet race**, **intermediate** and **dry race** – **slicks**.

The purpose of the tyre tread is to **discharge water** and enhance the grip, on average tyres the tread is about 7 mm (1/4 inch) deep when new. M&S tyres need to grip into the soft surface, so the tread is almost double the depth of a typical tyre. High-speed tyres have treads designed to give the maximum grip and run very quiet at high speed with minimum heat generation. Race tyres usually use a much **softer rubber compound** for the tread to give better grip – this means that the tread will not last as long as for road tyres.

Radial tyre tread for bad weather conditions

Radial tyre tread for high speeds

Figure 5.30 tyre tread patterns

Tyre sizes

The size of the tyre depends on the size of the wheel. The **diameter** of the tyre is measured across the bead; it is the same as the diameter of the wheel rim. Most tyre diameters are given in inches (in); increases in size are in one-inch increments. A few specialist tyres are measured in millimetres (mm) – called **metric tyres**; these sizes always correspond to half-inch sizes between the inch increments of regular tyres. The reason for this is so that metric tyres cannot be fitted to regular rims and vice versa.

The tyre width on radial ply tyres is given in millimetres; on diagonal ply tyres it is given in inches.

For example:

Width		*Diameter*	
165	X	13	*Radial ply*
5.60	X	15	*Diagonal ply*
160	X	313	*Metric radial ply tyre*

Other markings

Between the width and the diameter numbers you will find a double figure number and a pair of letters. The double figure number is the aspect ratio of the tyre – that is the height expressed as a percentage of the width. You will find aspect ratios between about 50 and 80%. Two typical letters that you will find are SR; the S shows that the tyre is suitable for speeds up to 180 kph (112 mph), the R shows that it is a radial ply tyre.

For example:

175/65SR13

175	Width
65	65% aspect ratio
S	Use up to 112 mph
R	Radial ply tyre
13	Wheel diameter

Other information that you will find on the side of a tyre includes country of origin, maximum tyre pressure, maximum load-carrying capacity, and compliance with European regulations.

Changing a tyre

Jacking up to change a wheel – when jacking up the car to change a wheel the following procedure must be observed:

- place vehicle on level ground
- ensure that the hand-brake (parking brake) is fully on
- place chocks behind the wheels, which are to remain on the ground
- slacken wheel nuts about a quarter of a turn
- place jack either under the chassis or a suitable suspension member – you should look at the workshop to locate a suitable place

When refitting the wheel you should use a torque wrench to ensure that it is correctly tightened.

Racer note

Over-tightening the wheel is just as bad as leaving it loose. Over tightening will also cause loss of time at a tyre change. Always use a torque wrench.

Use of tyre machine

To remove and replace tyres on wheels it is usual to use a tyre machine. The operation of tyre machines varies, and you should read the instruction booklet or have a short training session before operating the tyre machine.

Tyre pressure

The tyre pressures should be checked regularly. Tyre pressures must be kept within the limits specified by the manufacturer. In either hot or cold weather the tyre pressure may vary. The tyre pressures also vary with the use of the car. The best time to check the tyre pressures is before the car is driven. You should not alter them in the middle of a journey or event unless it is essential for a specific reason.

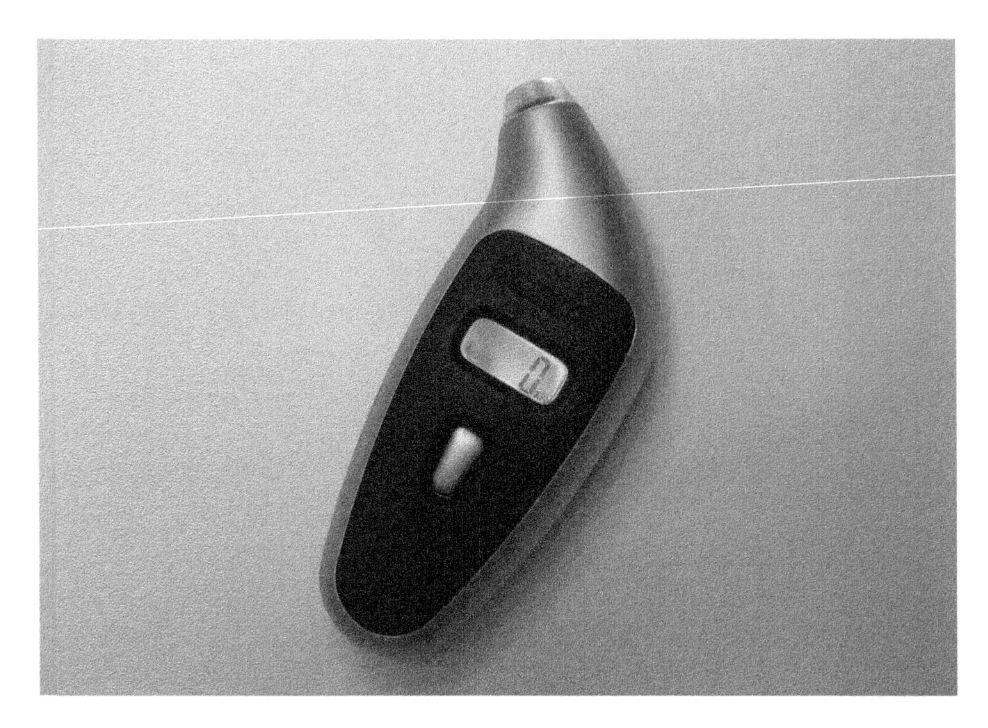

Figure 5.31 Digital tyre pressure gauge

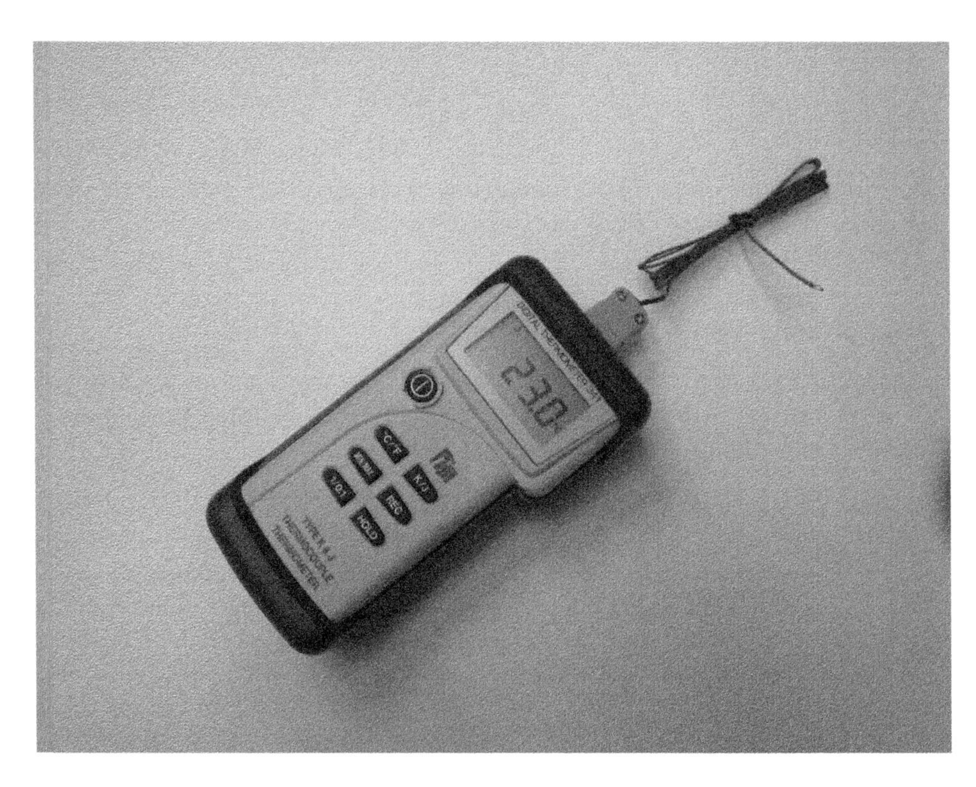

Figure 5.32 Digital tyre temperature gauge

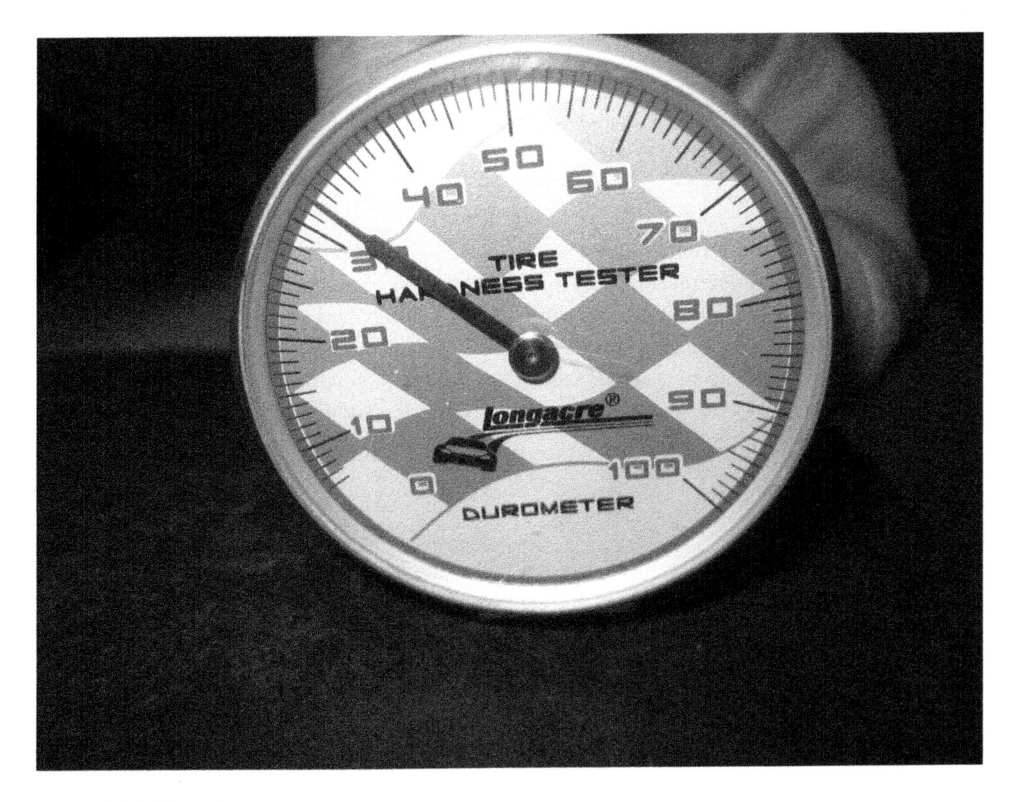

Figure 5.33 Tyre durometer

The use of an accurate tyre gauge is essential; the pencil type gauges are usually very accurate, for race use digital gauges give the best consistency of readings. Calibrated and certificated gauges are available for use with race and high-performance vehicles.

> **Racer note**
>
> Ideal tyre pressures are usually established during practice from lap times and alignment of the cross-tread temperatures.

Tyre tread depth

The tread depth is set by European Regulations; currently the minimum depth allowed is 1.6 mm. This may be changed at any time.

Because of the inconvenience of changing a wheel after a puncture, and the risk of an accident or indeed closing a traffic lane, you are advised to regularly check all the wheels and tyres, and if a tyre fails fit a new one.

Tyre wear

If the tyre pressures are set incorrectly the tyres will wear unevenly. If the pressure is **too high** the tread will wear in the **middle**; if the pressure is **too low** it will wear on the **outer edges**. Faults in the steering and suspension can also cause uneven tyre wear.

Inner tubes

These are circular hollow rubber rings with a valve. The inner tube is used on vehicles fitted with wire-spoked, three-piece and two-piece wheels to prevent air leakage. The inner tube is inflated inside the wheel and tyre assembly. With wire-spoked wheels a **rim tape** is fitted between the inner tube and the rim to prevent puncturing by the sharp edges of the spokes. In an emergency, punctures in inner tubes can be repaired with patches, these are larger versions of bicycle patches. However, this practice is not advised and should not be carried out on high-speed vehicles.

Tubeless tyres

Most cars, and a large number of trucks and motorcycles, use tubeless tyres. That is they do not have inner tubes – the tyre bead forms an **airtight seal against the rim**. Punctures in tubeless tyres can be repaired by inserting a special **rubber plug** into the hole. This must not be done on the tyre sidewall. The plug can be inserted without removing the tyre from the rim. High-speed tyres should only be plugged as a temporary repair and the vehicle driven at a reduced speed.

Tyre valves

Tyre valves hold the air in the inner tube, or they are fitted to the rim in tubeless set-ups. The most popular type is the **Schrader valve**. When air is pumped into the tyre the valve core is forced downwards to allow the air to pass. When the intake of air is stopped the pressure of the air in the tyre helps to keep the valve closed.

Tyre damage and repair

Punctures by nails and similar items in the tread of tubeless tyres can be plugged using special plugs and insertion tools without removing the tyre from the wheel. However, this is not recommended for high-performance vehicles, or where high-speed motorway driving is likely.

The maximum length of cut allowed by law is 25 mm (1 inch).

Inner tube punctures can be repaired by patching; but this is not recommended.

Wheel balancing

It is essential to balance the wheels and the tyres as a unit to prevent wheel shimmy – that is unbalanced wheels may cause the road wheels to shake, which will cause the steering wheel to shake from side to side. Wheel shimmy on out-of-balance wheels is usually noticeable at between 30 and 40 mph. Wheel balancing is carried out using a special machine; you will need a short training session for any specific machine.

Tyre fitting regulations

The laws on tyre fitting and usage in the UK and most of Europe can be summarised as follows:

1. Radial or diagonal ply tyres can be fitted to all vehicles.
2. If only two radial ply tyres are fitted, these must be fitted on the rear wheels.
3. Radial ply and diagonal ply tyres must not be mixed on the same axle.
4. The tyre pressures must be kept within the manufacturer's recommendations.
5. The tread must be not less than 1.6 mm deep for the entire circumference over all the tread width.
6. The tread and the sidewalls must be free from large cuts, abrasions or bubbles.

Wheel rotation

To ensure even wear on all the tyres of a vehicle it is good practice to move the tyre positions; this should include the spare wheel if possible. It is important to note that all the wheels and tyres must be of the same size and type.

Braking system

Key points

- Brakes on road vehicles must comply with MOT regulations
- Brakes work on the principle of converting kinetic energy into heat energy
- Disc brakes are less prone to brake fade
- A system of compensation is needed for even braking
- Hydraulic systems use the principles of Pascal's Law
- Special care must be taken with brake dust and brake fluid

The main function of the braking system is to stop the vehicle. The braking system also has two less obvious functions; these are to be able to control the speed of the vehicle gradually and gently, and to hold the vehicle when parked on a hill or other incline.

The Highway Code gives a guide to typical **stopping distances** from different speeds. Road and weather conditions affect stopping distances greatly – in wet weather it may take twice as long to stop as in wet weather.

MOT requirements

To comply with the requirements of the **Vehicle and Operator Services Agency (VOSA)** the braking system must fulfil these basic minimum requirements:

- Comprise two **independent and separate systems** – usually this is interpreted as the **foot-brake** and the **hand-brake**
- **The main braking system** – foot-brake – operating on all four wheels, must give a retardation of **50%**

- **The secondary braking system** – hand-brake – operating on only two wheels must give a retardation of **25%** and hold the vehicle in a parked position. This is also referred to as the emergency brake
- Braking on each axle must be even

Nomenclature

VOSA is the current name of the part of the UK government agency that administers the policy on operating vehicles; other countries have similar bodies. Their names change from time to time; people often just say Department of Transport. The term MOT – meaning Ministry of Transport (now defunct) is still used on official documents. In the USA the equivalent department is referred to as DOT.

Vehicles over three years old must pass an **MOT test** each year. One of the tests is that the braking system complies with the above minimum requirements. Typically, the foot-brake must record above 80% retardation and the hand-brake 50%. Hand-brake travel is usually a maximum of three 'clicks' of the ratchet; foot-brake travel is usually about 25 mm (1 inch).

The percentage braking figures are read out from the dials of the **brake testing rolling road**. They relate to **deceleration** (opposite of acceleration) expressed as a **percentage of the acceleration due to gravity** (G). The value for this is 9.81 m/s/s (32 ft/s/s).

Racer note

Brakes on motorsport vehicles need to be perfect in operation – there is no room for error.

Friction

Friction is the resistance of one body to slide over another body. It is only dependent on the surface finishes of the materials; size of contact area does not affect the friction. When two areas are in contact and force is applied to hold them together the friction generates heat.

The braking system uses friction to convert the **kinetic energy** of the vehicle into **heat energy** that is dissipated into the atmosphere.

Nomenclature

Dissipate is a science word for dispelling, getting rid of, or spreading about. The heat from the brake pads and discs is dissipated into the atmosphere so that it does not build up and allow the brakes to get hot.

The amount of **heat generated** by the brakes to stop a racing car from say 200 mph (320 kph) will be the same as that generated in the engine to accelerate the car to the same speed.

The amount of friction depends on the materials of the **friction surfaces**, in other words the **pads** and the **discs**, or the **drums** and **shoes**.

The heavier and faster the vehicle, the bigger the brake components will need to be to dissipate the greater amounts of heat generated.

Racer note

Kinetic energy is the energy of motion – the faster a vehicle is travelling and the heavier it is the more kinetic energy it possesses – so the hotter the brakes will get on a twisty circuit, or a downhill section where the brakes are in constant use.

Mechanical brakes

Safety note

Brake dust may contain asbestos; in any case it can cause breathing problems. So use a breathing mask when appropriate and clean down brake assemblies with a proprietary product.

Mechanically operated brakes are used on vehicles such as **motorcycles** and **karts**. The **hand-brakes** on most road cars are also operated mechanically.

A rod or a cable is used to transmit the effort from the lever, or pedal, to the brake shoes. The mechanism used to move the shoes is usually a simple cam arrangement. When the cable, or rod, is pulled the cam is turned so that the brake shoe is pressed against the drum.

Mechanical brakes are very simple; but they are subject to a number of problems. These are:

- Cable stretch
- Wear of the connecting clevis pins and yokes
- Need to adjust each cable run separately

Compensator

Because it is essential to apply equal stopping force to each wheel on an axle, whether with mechanical brakes on an old car, or kart, or the hand-brake on a current vehicle, some form of compensation device is used in the system. That is a mechanical device that distributes the force evenly between the brake assemblies (drum or disc) so that the vehicle will stop evenly without skidding, or hold the hand-brake evenly on both wheels.

There are two main types of compensator. These are: **swinging link,** also called **swivel tree**, and **balance bar**.

The swinging link has three arms and is mounted on the axle casing. The longitudinal cable pulls on the longer arm, the shorter arms pull at right angles to the longer one that is turning the force through 90 degrees. This changes the direction to transverse and gives a mechanical advantage (leverage) by the difference in arm lengths. It is important to check

$$\text{Pressure} = \frac{\text{Force}}{\text{Cross-sectional area}}$$

the difference in cross-sectional areas. Piston B is twice the area of piston A, so the force on piston B is twice as great.

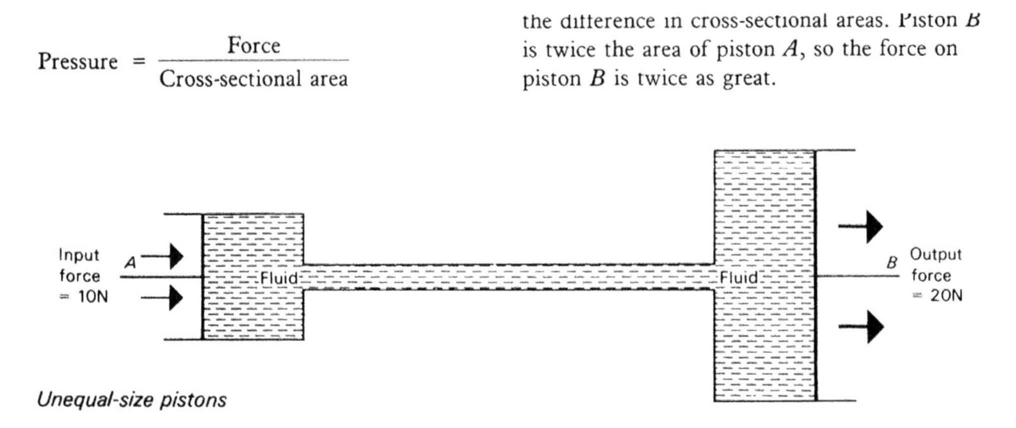

Figure 5.34 Hydraulic gain

that the mechanism moves freely – old racing cars may have a grease nipple on the swinging link to encourage lubrication.

The balance beam is used on most current car hand-brakes, forming the lower part of the lever. The centre of the beam, which is a very short metal component, is free to twist in the hand-brake; the two cables, one to each wheel are attached to each side of the beam. This allows for slight variations in cable length caused by cable stretch, or uneven adjustment between the two brake assemblies.

Hydraulic brakes

There are many types of hydraulic brake systems, and variations within those systems. The advantage of hydraulic brakes is the hydraulic system is self-compensating – there is no need for mechanical compensation systems and the attendant adjustments to be made. This is explained by Pascal's Law.

Pascal's Law

Pascal was a French scientist who lived between 1623 and 1662; he spent a lot of time studying why wine bottles broke at the bottom when the cork was pushed in the neck. His discovery, that **a liquid cannot be compressed, and that any pressure applied to a fluid in one direction is transmitted equally in all directions**, lead to the invention of hydraulic brakes about 300 years later.

Let's have a look at this in more detail. Pascal's theory is applied so that when the driver presses the brake pedal, a much larger pressure is transmitted to the brake pads and shoes:

Pressure = Force / Cross-sectional area

Force = Pressure * Cross-sectional area

For example, if the driver presses on the brake pedal so that a force of 400 N is applied to the master cylinder, which has a cross-sectional area of 50 mm^2, what force will the wheel cylinder piston exert on the brake shoes if its cross-sectional area is 100 mm^2?

$$\text{Pressure at master cylinder} = \text{Force} / \text{Cross-sectional area} \quad (12.1)$$
$$= 400 / 50$$
$$= 8 \text{ N/mm}^2$$
$$\text{Pressure at wheel cylinder} = \text{Force} / \text{Cross-sectional area}$$

Applying Pascal's Law, the pressure at the wheel cylinder is the same as that at the master cylinder; but the area is different, so we can insert the numbers that we know. Therefore:

$$8 \text{ N/mm}^2 = \text{Force} / 100 \text{ mm}^2$$

Transposing the formula:

$$\text{Force} = 8 * 100$$
$$\text{Force} = 800 \text{ N}$$

As you can see, doubling the cross-sectional area doubles the force. So, if the brakes were made using four-wheel cylinders, or disc brake pistons, each of which is double the cross-sectional area of the master cylinder, then the force applied by the driver will be multiplied by eight at the brakes.

Layout of simple system

The simple hydraulic brake system comprises a **master cylinder** and **wheel cylinders** and **drums** – as found on very old racing cars and racing motorcycles, or **callipers** and **discs** (called **rotors** in the USA) on later ones. The master cylinder is connected to the wheel cylinders by narrow bore **brake pipes**, also called **brake lines**, or **Bundy tubes**. Remember from Pascal's Law – the pressure applied at the master cylinder by pressing the brake pedal will also be applied all along the brake pipes and at the wheel cylinders. When the brakes are fully applied the pressure is typically 50 bar (750 psi), although it may be as much as 150 bar (2,250 psi). When the brakes are released the residual pressure is about 0.25 bar (4 psi). **Flexible brake hoses** are used to connect between the hard lines attached to the body/chassis and the moving components such as the brake callipers.

Nomenclature

Pipes, hoses, Bundy, lines are all terms that are used and misused in the automotive industry. The same applies to couplings, fitting, connectors, brake nuts and pipe ends. You should try to use the technically correct terms, as always. However, when building a braking system on a car, or bike, you will find reference in the workshop more to the manufacturer, or type – such as Goodridge, AP, and Bembo.

Safety note

The pressure of the brake fluid with the pedal depressed is higher than that of the air from the compressor – and you know that must be handled with care – so make sure that the pressure is released before working on the hydraulic system.

Hand-brake

The hand-brake, also called **parking-brake**, may be hand operated, or foot operated. On large automatic cars (US standard saloons) the **parking-brake is typically applied by the driver's left foot** (the right foot operates both the accelerator and foot-brake pedals). The **parking brake is released by a separate hand-operated lever** underneath the dashboard.

On most vehicles the hand-brake operates on the two rear wheels; however, some specialist FWD vehicles have the hand-brake on the front wheels to give more emergency stopping power. **Trials cars** have separate hand-brakes on each of the two rear wheels; these are called **fiddle brakes** as the drivers fiddle with the two separate levers to maintain traction up steep off-road sections.

Racer note

Sporting trials are run by the **Motor Cycling Club** (MCC) – the oldest motorsport club, which caters for both motorcycles and cars. They use tracks, or off-road sections, where it is necessary to use the fiddle brake to lock one rear driving wheel to maintain grip on the other to climb the hill, or take a very sharp corner by driving one wheel around the other.

The hand-brake is held on by a ratchet and pawl mechanism. To check the hand-brake for correct operation on a car without using a rolling road, jack up the rear of the car and apply the hand-brake one notch (click) at a time. Progressively each wheel should become harder to turn. Both wheels should be fully locked after about three clicks.

The hand-brake cables should be visually checked for **fraying** or corrosion – old vehicles may have grease nipples for ease of lubrication; most current hand-brake cables have nylon linings to save the need for lubrication – but be aware of corrosion, or damage, to the small uncovered section – especially on rally cars where they can become caked in mud. When checking the cables inspect the **clevis pins**, connecting **yokes**, and retaining **split pins** – the application of water proof grease to the bare cable and linkage is often helpful on rally cars.

Drum brakes

Current cars tend to use drum brakes only on the rear with disc brakes at the front.

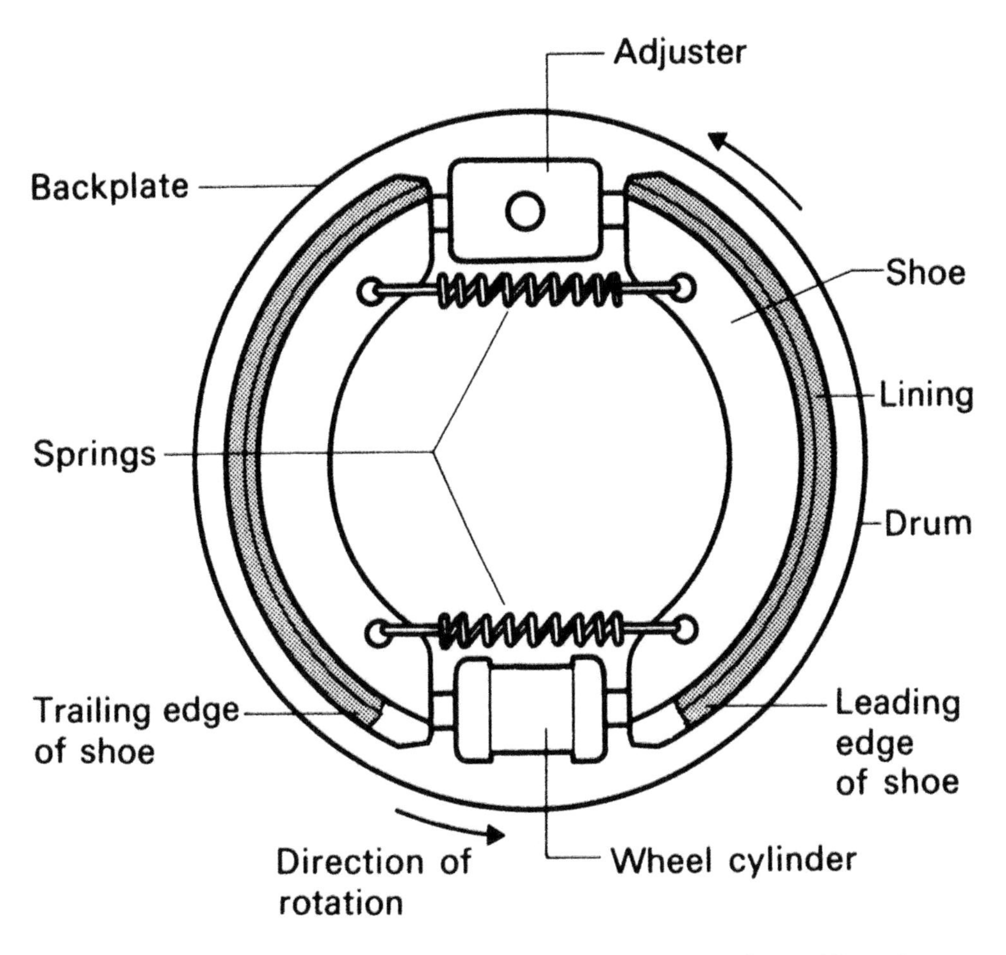

Figure 5.35 Drum brake

Master cylinder

This is operated by the driver's right foot. There are two main types; these are:

- **Single piston** type – used on some single-seater cars and older single circuit brakes
- **Tandem master cylinder**, using two pistons – used on all current dual circuit brake systems

The operation of both types is very similar. There is a **reservoir** that keeps the main **cylinder** (or chamber) full of fluid. The pedal pushes the **piston** up the cylinder to displace the fluid into the braking system through the **delivery/return valve**.

On a tandem master cylinder there are **two pistons** and **two concentric reservoirs**. Each piston displaces the fluid to its own hydraulic circuit. So that should one brake line fracture and leak out the fluid, the vehicle can still be stopped using the remaining parts of the system.

> **Racer note**
>
> In this text you will find reference to older motorsport vehicles – there is more money spent every year, and more races for, historic and classic vehicles than modern ones – and that include Formula 1. This means that there is more employment in this area of work.

Drum brake shoe layouts

Current vehicle rear drum brakes use **leading and trailing brake shoes**; older race vehicles fitted with drum brakes at the front usually have **twin leading shoes** at the front.

The twin leading shoes give the maximum braking power; the leading edge of the shoe tends to dig into the drum and give what is referred to as a self-servo action. When stopping hard, the front brakes do the bulk of the work – typically the front brakes do 70% of the work compared to 30% at the rear. However twin leading shoe brakes are only efficient when the vehicle is going forwards; so they are not useable for a hand-brake where the vehicle may be parked on a hill pointing either up or down. So, for hand-brake operation a leading and trailing shoe arrangement is needed.

Wheel cylinder

There are two main types of wheel cylinders. These are:

- Single piston – used on twin leading shoe brakes where one cylinder operates each shoe
- Twin piston – used on leading and trailing brakes so that the wheel may be held in either direction of rotation; this is needed to prevent roll back on inclines

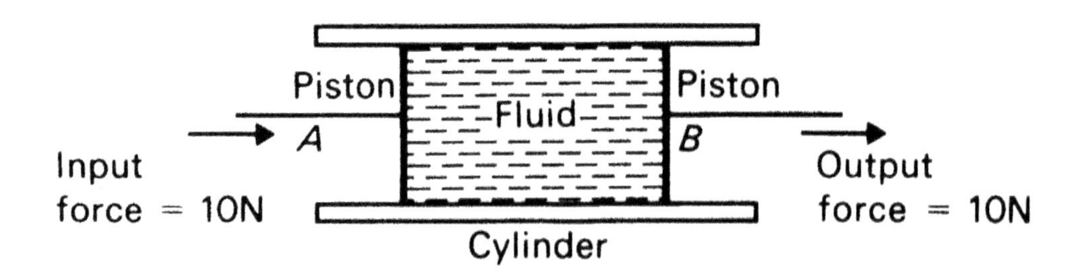

Figure 5.36 Hydraulic wheel cylinder

Brake shoes

The **brake linings** are attached to the brake shoes by either rivets or a bonding process (glue). The shoes on ordinary road vehicles are usually fabricated (welded) from plain steel; on race vehicle they are made from aluminium alloy for lighter weight and better heat dissipation. Usually the brake shoes are held in place against the wheel cylinders by strong springs.

> **Racer note**
>
> Before stripping drum brakes it is prudent to either sketch or photograph the layout of springs on shoes.

> **Safety note**
>
> Brake shoe springs are very strong. Do not put your fingers where they might get trapped or hurt by the springs slipping.

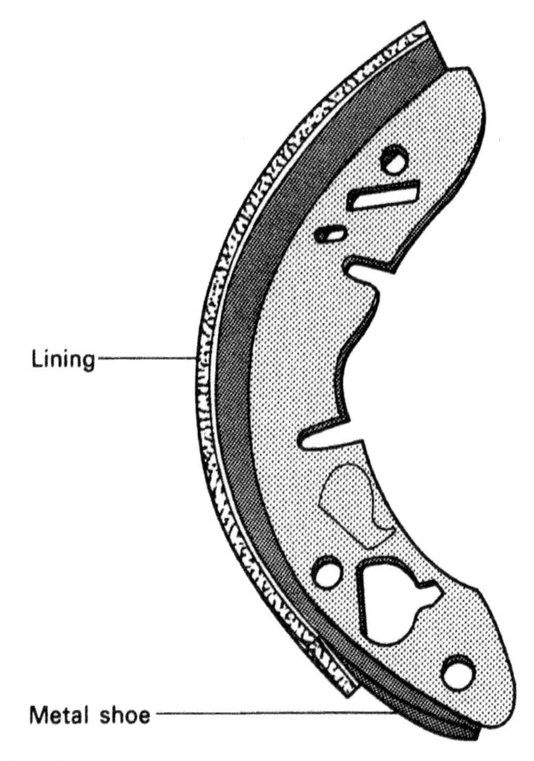

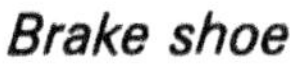

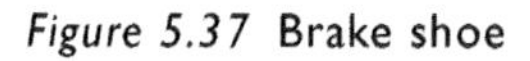
Figure 5.37 Brake shoe

At every major service the brake linings should be checked for wear; the workshop manual will give a minimum thickness figure – typically 3mm (1/8th inch). If the linings are riveted in place the rivet heads must be well below the surface of the brake shoe.

Drums

Brake drums may be made from cast iron or aluminium alloy with a steel insert for the friction surface. The aluminium alloy ones are much lighter than the cast iron ones – possibly only one-third of the weight. They may also have cooling fins cast on them to help prevent brake fade.

The brake drums are usually held in place on the hub by two small countersunk set-screws. When inspecting the brake pads the drums should be inspected – look for score marks, cracks and ovality.

> **Nomenclature**
>
> Ovality means oval or, in the setting of brake servicing, not properly round. This problem causes brake grab.

Disc brakes

Disc brakes have several advantages over drum brakes, namely:

- **Less susceptible to brake fade** – that is a reduction in braking efficiency through an increase in the temperature of the friction surfaces, usually after several successive brake applications
- Open to the air and therefore kept running **cooler**
- **Easy to change pads**
- **Greater braking effort** for size and weight with the aid of a brake servo
- **Self-adjusting**

On current popular cars it is normal to have front disc brakes with rear drum brakes. High performance and race vehicles typically have discs both front and rear; the reason for this is that these vehicles usually have even weight distribution giving even braking both front and rear.

> **Racer note**
>
> Jaguar's racing reputation was built on winning the Le Mans 24-Hour Race many years ago. They did this by using Dunlop disc brake to enable them to brake later and harder into each corner – thus increasing their lead on each lap.

Callipers

Callipers are the disc brake equivalent of the drum brake wheel cylinders. The fluid moves the pistons to press the **pads** against the **disc**. There are many variations of callipers: from

Figure 5.38 Disc brake set up

single piston callipers used on popular small cars to ones with six pistons used on race and high-performance vehicles.

> **Nomenclature**
>
> You will hear terms such as **four-pot** callipers: this means that each calliper as four pistons – two on each side.

As the pads wear the pistons will expand out to take up the wear. To return the pistons in some cases it is necessary to use a special tool to turn the piston whilst pushing it back into the calliper – look in the workshop manual for the specific application.

Discs

The **pads** act on the discs (rotors in the USA) to give the necessary friction. Plain discs are simply cast iron; but most discs are coated with some form of surface finish to improve braking and resist corrosion. The discs are bolted to the hubs.

When overheated, discs can warp, that is like a buckled wheel on a bicycle. This **warping** will give uneven braking and can often be felt at the brake pedal. When replacing pads it is a good idea to check the discs for warping or, as it is called, **run out**.

Special discs

Discs may be manufactured in a number of different ways – mainly to improve cooling, some examples are:

- **Vented discs** – a gap between the two sides of the disc with air vents fitted
- **Cross-drilled discs** – as its name says, drilling across the disc
- **Wavy discs** – the wavy edge increases the number of leading edges and therefore improves braking

On very advanced race and high performance car **carbon discs** are used; these are usually used as multi-floating disc – a system used on large aircraft.

> **Safety note**
>
> When replacing brake pads, do one side at a time; and do not press the brake pedal to ensure that the seals are not broken and no fluid is lost.

Brake pad thickness

As the brakes are used the **friction material** wears and there is a point at which it become necessary to replace the pads. This is when the thickness of the friction material reaches a minimum amount – the vehicle manual will state this figure and how to **measure** it.

Brake pad grade

A variety of different materials are used for brake pads. Basically these fit it to two major classifications. These are:

- **Soft pads** give more friction; but generally they create more heat and are therefore prone to fade
- **Hard pads** give less friction, less heat and last longer; hard pads are used for race cars

Brake lines

Brake lines connect the brake components, transmitting the fluid pressure according to Pascal's Law with a small fluid displacement. There are a number of different materials used for brake lines. On road cars they tend to be an **alloy steel**. On historic racers **copper** is frequently used to look authentic and for its ease of bending. Single-seater cars tend to use **stainless steel** for its high value of **hoop stress** – that is its resistance to expand under pressure.

Brake fluid

Brake fluid is a special type of oil developed to give the specific properties needed by the braking system. These are:

- **High boiling point** to reduce the risk of brake fade
- **Non-corrosive** to rubber and the other materials used in the braking system
- **Lubricating** properties for the mating parts
- Will not fail under very **high pressure**
- **Low viscosity** for rapid response to pedal operation

When topping up, or replacing, brake fluid **always use the recommended brake fluid; brake fluids must not be mixed**.

Safety note

Remember brake fluid removes paint – be careful when you are using it.

Racer note

Hose spanners are available to reduce the risk of damaging brake hose ends.

Brake adjustment

When the friction surfaces wear, adjustment is needed to prevent excessive brake pedal travel. There are two main types of adjuster; these are **wedge** type and **snail cam** type.

On current cars the rear drum brakes are **self-adjusting**.

FAQs What are self-adjusting brakes?

There are two main types of self-adjusting brakes: disc brakes are self-adjusting by piston movement in the callipers; rear drum brakes are self-adjusting by using a ratchet mechanism on the hand-brake mechanism inside the drum.

Fluid venting

This is also referred to as **bleeding the brakes**. That is the process of **removing air** from the braking system after repairing, or replacing, one of the hydraulic components. There are two main ways of doing this:

- **Manually** – a rubber tube is attached to the **vent nipple** (bleed nipple) by pushing it over the nipple to give a firm fit. The other end of the tube is submerged in a container of clean brake fluid. The nipple is slackened about one turn. When the brake pedal is

Figure 5.39 Brake venting

depressed, fluid will flow and bring with it any trapped air. The master cylinder reservoir must be kept topped up throughout this job. When it is seen that air bubbles have stopped flowing, the nipple is retightened and the pedal tested for a firm feel.

- **Using air pressure** – a device is attached to the master cylinder reservoir, which supplies fluid under pressure using air pressure from either a hand pump, or a compressed air supply. The nipple is attached with a tube in the same way as in the manual method. There is no need for anybody to press the pedal. When the nipple is opened fluid will flow – simply close the nipple when the air bubbles stop.

Single-seater brakes

On open wheel, or single-seater, cars it is normal to use **two master cylinders** with a **balance bar** between them to compensate between the front and rear systems. The vehicle in effect has two separate braking systems – one for the front wheel, which does most of the work, and one for the rear wheels. They do not have hand-brakes. The pedal pushes on a balance bar, a short rod, like a playground see-saw; if the brake pushrod is dead central between the cylinders the force will be equal to each one. If the balance beam is moved so that the force is more towards one of the master cylinders – this is done by screwing the threaded part of the balance bar in a threaded part of the pedal (usually there is a flexible cable with a knob below the dashboard to allow this to be done easily) the force will be different between front and rear. Adjustments are made to suit the driver and circuit, for instance to prevent rear wheel lock-up under heavy braking, or to bring the rear round on very tight corners.

Racer note

Brake fluid is hygroscopic; in other words is attracts water from the atmosphere. It should be replaced at least every season to keep the system fresh.

Electronic controls

The braking systems of all new road cars have electronic controls such as an **anti-lock braking system (ABS)** and traction control. On all wheel drive (AWD) vehicles these systems can automatically apply the brakes when sensors indicate that an accident is likely – such as when entering a corner too fast, or driving on ice.

Chassis fault diagnosis

Vehicle identification

It is wise to always check that the vehicle is what you think it is. The first stop is the **vehicle identification number** (VIN).

There are two standards in use, both very similar. These are the ISO standard and the USA standard developed by the Society of Automotive Engineers (SAE).

The standards are based on 17 digits. The first three digits identify the country and the manufacturer. The forth to ninth digits are the Vehicle Identifier Section (VIS). The tenth to seventeenth digits are the Vehicle Identifier Section (VIS) – or serial number.

With the US system the ninth digit is a check digit – that it is a digit calculated from a formula based on the other digits in the sequence. So if somebody changed one of the digits, for instance the engine capacity, they would also need to change the check digit. This is to prevent car crime.

Before the introduction of the VIN, and in some cases overlapping this, you will find **chassis numbers** and separate **body numbers** too. On formula cars you will find a chassis number or a **serial number**. Motorcycles will be found with a **frame number**.

You will also need to exercise some care in verifying the various numbers when dealing with certain cars. **Heritage organisations** supply certificates to prove genuine certain documented vehicles. Companies such as British Motor Heritage Ltd produce complete body/chassis units for many specialist vehicles such as the ever popular MG and Mini; these incorporate improvements of design and manufacture.

Try this

Look at the VIN plate on any vehicle and write down its identity.

Steering and suspension terminology

This section sets out to describe and define a range of suspension terms and nomenclature relevant to the design of motorsport vehicle suspension.

Anti-dive and anti-squat – anti-dive is achieved by inclining the upper and lower wishbones so that their axes intersect at a point behind the suspension. This gives the equivalent of a leading arm suspension so that when the brakes are applied the braking force is applied in the anti-clockwise direction, which lifts the front suspension against the weight transfer that is trying to depress the suspension in the opposite direction. The rear suspension is treated in the same way but the angle is reversed so that the point of intersection is in front of the suspension.

Axle tramp – is the sideways movement of the axle relative to the chassis, which can occur on live-axle cars when cornering under power. Often this take the form of an oscillation as the free play is taken up in the suspension joints. It can also combine with wheel patter to cause the loss of traction.

Camber and swivel axis inclination – on traditional suspension systems the camber of the wheel and the swivel axis inclination (also called kingpin inclination – KPI) were designed to intersect at the road level, called centre-point steering. Most current systems have negative scrub radius. That is the intersection of the camber and swivel pin axes is above the road surface. The reason for this is that with centre-point steering, if a front tyre blew out there would be a steering torque about the steering axis. The intersection point would no longer be at a centre point, but instead would be below the road surface; the rolling resistance of the wheel would generate a torque about the steering axis. With negative scrub radius the intersection point is above the road surface so that in the case of a tyre failure the change in the intersection point would default to centre-point steering. This prevents the vehicle veering to one side of the road or another in the case of a tyre becoming deflated. The distance between the tyre contact patch and the point of intersection between the steering angle and the road determines the amount of torque steer.

Variations in the camber angle can be brought about by changes in the swivel pin position. With modern multi-link systems this variation is very complicated; a method of analysis has been developed at the Hyundai Motor Co.

Castor angle – is the amount of trail between the wheel contact patch and the point where the steering axis intersects the road; this gives the steering wheels their self-aligning torque, so that the wheels return to centre after a corner and maintain stability at high speed on straight roads.

On a level road the torque is a function of the castor angle, that is the rearward inclination of the (imaginary) swivel pin, ($\angle_c$), usually between 1 and 7 degrees. Large values of castor angle are sometimes used to give stability and high self-aligning torque; this is quite effective but can have an adverse effect on the handling. When a vehicle is heavily laden, such as with the addition of rear seat passengers, or a boot full of luggage, the rear suspension is usually the more compliant. The dipping of the body at the rear as the springs are compressed will increase the castor angle; this will, in turn make the steering more heavy. The swivel pin inclination angle ($\angle_{kp}$), that is the angle by which the swivel pin (imaginary) leans transversely inwards at the top to give either centre-point steering or a negative scrub radius, can also affect the amount of castor. That is, during cornering the swivel pin inclination may move because of two reasons. One is the movement of the theoretical centre of the pivot during steering; the other is the lateral forces that alter the shape and the position of the tyre contact patch. If the centre of pressure (CP) is in front of the centre of gravity (CG) then a side wind will exert a lateral force on the steering, which will cause the steering to be turned in the same direction as the wind pressure. This can be overcome by fitting tail fins, which move the CP rearwards, or the forward positioning of the engine and gearbox to move the CG. An alternative is to design in negative castor, but the steering will feel somewhat twitchy. The 1934 Auto Union 16-cylinder racing car is a good example of extreme positioning of the CG, which gave such a variation in handling characteristics that the only driver to control the vehicle was Hans-Joachim Stuck, who broke seven speed records with her in the first season of competition.

Co-ordinate systems – vehicle movements and motions can be described in terms of co-ordinate systems. There are two systems, one that relates directly to the vehicle, the other that relates the vehicle to the earth – or its travel on the road.

Vehicle fixed co-ordinate system

The vehicle motions are defined with reference to a right-hand orthogonal co-ordinate system by SAE conventions. The co-ordinates originate at the centre of gravity (CG) and travel with the vehicle. The co-ordinates are:

x – forward and on the longitudinal plane of symmetry
y – lateral out the right side of the vehicle
z – downward with respect to the vehicle
p – roll about the x axis
q – pitch about the y axis
r – yaw about the z axis

Earth fixed co-ordinate system

Vehicle attitude and trajectory through the course of a manoeuvre are defined with respect to a right-handed orthogonal axis system fixed on the earth. It is normally selected to coincide with the vehicle fixed co-ordinate system at the point where the manoeuvre is started. The co-ordinates are:

X – forward travel
Y – travel to the right
Z – vertical travel, positive is downwards
Ψ – heading angle
ν – course angle
β – sideslip angle

A similar co-ordinate system has been devised by the SAE for vehicle tyres.

Driving characteristics

Think safe

Only test drive on the track on test days.

Driving and braking forces – the suspension must transmit between the chassis and the road the driving and braking forces. The rolling radius of the road wheel affects the amount of tractive effort (TE) which can be transmitted. At this point the co-efficient of friction between the tyres and the road should also be considered. Race car drivers enjoy being able to spin the wheels, so this must be taken into account. Indeed, wheel spinning is often used to increase the co-efficient of friction (μ); on dragsters the tyres are 'lighted' by spinning the wheels so that they get hot and momentarily set on fire to melt the outer rubber. The suspension mounting points must be capable of transmitting large amounts of driving torque. The driving and braking forces are responsible for large amounts of weight transfer and squat and dive actions.

Lateral friction – when travelling in a straight line with no lateral, or side, forces on the vehicle, the tyres have a true rolling motion over the road. When cornering, centrifugal force exerts a lateral force on the tyres; this in turn gives rise to slip angles. That is the difference between the line of the wheel and the actual line of travel. Side winds and variation in the steering angles can also generate slip angles. If both front and rear slip angles are equal then the vehicle has neutral steer characteristics. If the slip angles are greater at the front than at the rear then the vehicle will under steer. If the slip angle is larger at the rear than the front, then it will over steer. Under steer is a stable condition and therefore desirable in a road car, but not necessarily a race car. Variables which bring about changes to the lateral friction are:

- load distribution – a heavy load at the rear will increase the rear wheel slip angles and cause over steer
- tyre aspect ratio (height/width) – tyres with an aspect ratio < 0.70 allow high cornering speeds
- road conditions – typical co-efficient of friction for a good road is 0.80–0.85
- changes to the track width brought about by suspension movement can change the slip angles
- changes in the load distribution, such as brought about by heavy breaking, alter the vertical forces on the tyre. This alters the amount of lateral friction and hence the vehicles stability
- camber changes, both wheel and road camber, alters the lateral friction

Load-carrying capacity – before calculating the load-carrying capacity it is important to define kerb weight (KW) and gross vehicle weight (GVW). Some literature uses the more technically correct term vehicle mass; for all intents and purposes the two terms are synonymous. The use of the term weight is unlikely to go out of colloquial use in this country or America. By definition of EU and DIN standards the kerb weight includes: the chassis; body and trim; engine; gearbox; all ancillary items such as starter, alternator, battery, fuel system, exhaust system and cooling system; optional items such as sunroof and air conditioning; all the oils and fluids including 90% of the maximum fuel; all the mandatory safety items, namely, jack and brace, spare wheel, warning triangle, first aid kit and spare bulbs.

EU-type approval regulations, which do not currently apply to one-off cars, require that the maximum mass of drivable vehicle (MDRV) must be stated; this includes an allowance of 68 kg for the driver and 7 kg for luggage.

The load-carrying capacity, or payload, is the difference between the permissible gross vehicle weight, as determined by the vehicle design, and the kerb weight.

Roll – is the difference between the overturning moment and the righting moment. The speed at which a vehicle will overturn depends upon the height of the centre of gravity, the track width and the radius of the corner. The roll centre of the vehicle may be above or below the road surface; this depends on the layout of the suspension mountings and the angle of the suspension arms.

Sprung and unsprung mass – the sprung mass (SM) is that mass above the springing medium; the unsprung mass (USM) is that below. Part of the mass of the suspension components may form unsprung mass, the remainder sprung mass. It is desirable to keep the unsprung mass as low as possible in relationship to the sprung mass so that the amplitude of the force generated when the wheel hits a bump is kept to a minimum and cannot therefore cause driver or passenger discomfort. Three concepts related to the sprung to unsprung mass ratio that affect the ride quality are:

Wheel hop – this is usually a resonant condition of the unsprung mass which is at a different frequency to the frequency of the body/chassis. On classic cars this can be felt through the steering. Makers of the Vincent, a replica of the classic 1930s MG, fit FX taxi wheels to an otherwise lightweight suspension system to generate an amount of wheel hop, sufficient in their words to give a vintage feel to the suspension. The wheel hop must not be such that it adversely affects the handling of the vehicle. With a live rear axle, tramp can develop when cornering at speeds exceeding the design figure. This feels like wheel hop; indeed it is a sideways form of wheel hop generated by the driving forces.

Wheel patter – the low amplitude wheel hop, particularly noticeable on washboard type surfaces, such as concrete sections of motorway.

Boulevard jerk – is caused by the **stiction** in the suspension system. This gives a jerky ride on a fairly smooth surface. On a bumpy road this cannot be felt because of the suspension action in dealing with the bumps.

Track width – (TW) is the distance between the centre of the tyre tread contact patches on each axle. It is common for there to be a small difference between the front and rear track widths. The TW has a major influence on the vehicles cornering behaviour and the amount of body roll. The TW should be as large as possible within the constraints of the body/chassis width. The **overall width** (OW) is the width of the outside bodywork of the car at its widest point. The track width ratio (k_2) is an indication of how well the suspension fits the body chassis and is calculated by: $k_2 = TW/OW$.

The Motor Vehicle Construction and Use Regulations (C&UR) require that the bodywork completely covers the wheels and tyres; therefore it is impossible for the k_2 figure to be equal to, or greater than (=>) unity. The wheels and tyres need running clearance to cope with the build-up of mud or snow and allow for minor variations in tyre sizes when replacements are needed. In bump and rebound conditions the TW may vary, so there must be body/chassis clearance to allow for this.

On many independent suspension systems the bump and rebound movement of the suspension causes changes in the TW. This is because of the geometry of the suspension pivots. The problem is that variation in track width causes lateral forces which give rise to variations in slip angle (α). This leads to an increased rolling resistance and adverse affects on the vehicle's directional stability, which may include steering effects. That is **bump steer**. The loss in performance and the detrimental effects on the steering at high speed are to be avoided. Changes in track width also cause tyre noise and increases tyre wear. The tyres used on motorsport vehicles are expensive, so excessive wear generating geometry should be avoided, and for environmental reasons vehicle noise should be kept to the minimum.

Wheel alignment – this is the relative position of the road wheels. The wheels are usually set with a predetermined value of static toe-in or toe-out so that under dynamic cruising conditions (30–70 mph or 50–110 k/h) the wheels will become parallel to each other and so have no excessive wear on any one edge. Conventional layout vehicles, that is ones with a front-mounted engine driving through a gearbox and propeller shaft to a rear axle, usually have a small amount of toe-in, typically 2 or 3 mm measured at the wheel rim. The positive wheel camber tends to make the front wheel want to peel outwards at cruising speed, so that the static toe-in is changed to a parallel position. With radial ply tyres the toe-in can be set to a minimum value; the final recommended setting would depend on tyre-wear observations.

Figure 5.40 Corner weight – like a bathroom scale

Figure 5.41 Corner weight set-up

Figure 5.42 Corner weight readout

To ensure that the wheels have true rolling motion on corners, the Ackermann principle is usually designed into the steering arm layout. That is so that the inner wheel turns through a greater angle than the outer; typical figures are that the inner wheel turns through 22 degrees when the outer is turned through 20 degrees. This difference is referred to as toe-out on turns.

On vehicles with independent rear suspension (IRS) a small amount of toe-in is given to the rear wheels to allow for suspension movement caused by positive rear wheel camber.

Wheelbase (WB) – is measured from the centre of the front axle (imaginary) to the centre of the rear axle. The overall length (OL) is measured from front bumper to rear bumper. Both the wheelbase and the ratio of the wheelbase to the overall length (k_1) k_1 = WB/OL are important variables in suspension design. A long wheelbase relative to the overall length of the vehicle allows for the accommodation of passengers between the axles, so the floor can be flat in the foot well and the seat cushion height has less special constraints. It reduces the affect of load positioning in the vehicle; this includes the position of the engine and gearbox. It reduces the tendency to pitching, especially on undulating country roads, which in turn allows the use of softer springs that tend to give passengers a more comfortable ride. With the reduced overhang there is less **polar inertia**, which improves the **swerveability** of the vehicle. The length of the wheelbase affects the turning circle for any given input of steering angle. American car companies have set ratios for wheelbase and overall length; this is to give a particular aesthetic to their full size cars – for example the Lincoln. On the compacts and sub-compacts (ordinary European-style cars) this does not apply.

Try this

Drive two different vehicles and compare the feel of them on the road using the terms you have learnt about.

Suspension types

Beam axle

The beam axle type of suspension (also called rigid axle) is the simplest and the oldest type. It is frequently supported using semi-eliptic leaf springs. Beam axles are often used at the rear of front-engined race cars in the form of a live-axle, that is one which transmit the power through half-shafts inside. Beam axle front suspension was used on cars of the 1930s and 1940s, but it is not a viable option for front suspension on a race car because of the effect that hitting a bump has on the steering at high speeds. Racers used to cut the front beam axles in half and convert them to swing axles by the addition of a central pivot point. Interestingly, this was used by Ferrari as well as on a great number of Ford E93 based kit cars. The problem with beam axles is that if one wheel hits a bump both wheels are tilted, so the caster, camber and kingpin inclination are all altered. A live rear axle is a possible choice for the rear suspension of a race car as it allows the use of a variety of engines and gearboxes with a conventional layout chassis. Also sufficiently strong rear axles are available for transmitting the 300–400 bhp that is needed to get the quarter mile in well below 10 seconds.

Independent suspension

The beam axle is non-independent; most other suspension types, where each wheel can move separately, are referred to as independent suspension.

MacPherson strut

MacPherson strut suspension is a concentric arrangement of coil spring and damper unit, which is also the suspension upright member. It is used on many cars for both front and rear suspension. It allows for a large amount of suspension travel and can therefore provide a soft ride with a long stroke damper action. The mounting parts are well apart and so the load can be spread across the body/chassis components. This allows plenty of space between the struts for either an engine or luggage space; on the negative side a considerable height is needed to accommodate the struts – this constrains the vehicle height to a minimum figure. MacPherson strut suspension, which by its nature needs to be mounted with a small angle of inclination to the vertical, tends to give large amount of track width variation on bump or rebound. This creates a lateral force which is taken by the plunger arm such that it is susceptible to wear and may develop a coarseness in the steering. An anti-dive action is not usually incorporated.

Vertical pillar strut

Vertical pillar strut suspension is initially similar in both action and appearance to the MacPherson strut design, but the sliding member is such that the action is truly vertical so that the tyre maintains a constant contact patch and track dimension. It is also designed so that the damping, suspension springing and track control actions are separate. This suspension is usually only to be found on Morgan sports cars.

Wishbone

The usual arrangement is a double wishbone, the upper wishbone being shorter in length than the lower one. Anti-dive is achieved by inclination of the arm axis. This design requires separate dampers and springs. The position of the mounting point is critical in terms of vehicle roll and other suspension and steering actions. The space between the inner pivots of the wishbones can constrain the engine bay space; the height is such that it is usually no higher than the road wheels.

Multilink

This is an adapted version of the MacPherson strut, which controls the wheel movement to give a constant track and a more robust mounting arrangement. It is a feature of many Japanese or other far Eastern saloons.

Trailing arm

Trailing arm suspension can be used on both the front and the rear, but it is normally only used on the rear. The problem is that the arm tends to lift under hard acceleration and traction is lost. Vehicles that drive through rear trailing arm set-ups have very poor grip on snow and ice. Trailing arms are now used only on dead (no drive) rear axles. They ensure that there is no change to either the track width nor the camber angle.

Try this

Make a list of motorsport vehicles and identify the types of front and rear suspension used – which is the most popular?

Suspension mediums

A number of different mediums, or materials, can be used to form the suspension. The displacement of the medium is the action of the suspension. Of importance is the amount of displacement and the rate of displacement; this depends on the elasticity of the medium and a number of other factors related to the design of the suspension. On a rally car it is necessary to have a moderately large amount of displacement in the form of available suspension travel because of the size of the components and to give reasonable ride height and comfort over a variety of road surfaces. On a circuit car the suspension movement may be as little as 2 mm. The rate of displacement, measured in terms of suspension rate, is the load necessary to compress the medium by a unit distance; in race car terms this is usually referred to in pounds per inch (lbf/inch).

Hydrolastic

The Hydrolastic (registered trade name) fluid is used to transmit the forces from the suspension arm to the thick rubber diaphragm that actually does the job of springing. The shock absorber function is built into the Hydrolastic displacer unit. The usual arrangement is to connect the displacers front to rear on each side. This reduces pitching and the bump effect when one wheel hit a road irregularity. Hydrolastic suspension could be used on a race car, but it is generally restricted to lightweight (class A) saloons.

Hydropneumatic

The suspension medium is gas, which is compressed inside a spherical chamber by means of a diaphragm against which the oil based fluid is pressed. The fluid transmits the force between the suspension arm and the diaphragm. The hydropneumatic suspension may be fitted independently to each wheel; or on light vehicles the suspension is often connected between the two sides on each axle. That is they are connected transversely. On more sophisticated cars, like the Citroen, all the suspension units are connected through a valve system. The fluid is pressurised by a pump so that the suspension can be raised and lowered and constant body height may be kept when the vehicle is fully laden. On pumped systems, the length of suspension travel can be quite long.

Torsion bar

The torsion bar is favoured for its light weight; but it needs a long installation length. The length factor is to some extent compensated for by the minimum amount of height that is needed. Torsion bars generally have only small amounts of suspension travel available; they are not, therefore, suited to custom cars. They can be found on small Fiats and the now classic Morris Minor. The mass of the torsion bar is entirely sprung weight as it is mounted at each end to the body/chassis.

Rubber cone

The rubber cone is used solely on the old Mini as a main spring, and occasionally on other vehicles as a secondary spring. In its secondary spring role it is favoured by caravan towers to prevent the rear suspension from bottoming and provide a variable rate spring for a fully laden vehicle. The load-carrying capacity is limited to a small figure and the suspension travel distance is very small.

Coil spring

Helical coil springs can be made to allow for a long length of suspension travel. They can be made in a range of spring rates including variable rates. Coil springs can be fitted to suspension arms in a large variety of ways including concentricity with the shock absorbers.

Leaf spring

Leaf springs can also take on the role of suspension member as well as that of spring medium. They may have one leaf or a combination of different length leaves laminated together to form the spring. The leaf springs are usually made from medium-carbon steel, but plastics and composites are optional materials. Leaf springs may be used in either semi-elliptic or quarter-elliptic forms. Leaf springs are easy to attach to both the axle and the chassis; they also may provide a medium range length of travel.

Ladder bars

The ladder bar is a ladder-shaped linkage used between the chassis and the live rear axle. It operates in a similar way to a trailing arm. This set up is often used on dragsters as the length can easily be changed and mounting to the axle is simple.

Suspension components

Four bar

These have four separate linkage bars between the chassis and the live rear axle. They operate in a similar way to a trailing arm, but the mounting points are spaced apart to transmit the driving a braking forces more evenly. They also offer a high resistance to the axle, twisting under heavy acceleration.

Watt linkage

Watt linkage is a centrally pivoted linkage system that locates the wheels transversely; this is used with IRS on some vehicles.

Panhard rod

This is a single transversely mounted rod used on both live and dead rear axles to absorb lateral forces and reduce the risk of axle tramp.

Steering geometry

Non-steered wheels – on a live, or a dead, axle the wheels will both be perpendicular and with no toe-in, nor toe-out.

On non-steered independent rear suspension (IRS) a certain amount of camber (positive or negative) may be incorporated to maintain maximum tyre contact on cornering – especially short circuit cars. Toe-in or toe-out will therefore be needed to prevent unnecessary tyre scrub.

Camber – the inclination of the road wheel when viewed from the front of the vehicles. Leaning out at the top is positive camber; leaning in at the top is negative camber. The camber angle can be 0–8 degrees. Road cars usually run small amounts of positive camber; single seat race cars are likely to have large values of negative camber – the reasons for camber are to accommodate camber changes on bump steer and maintain maximum tyre contact patch size.

Castor – this is the rearwards inclination of the suspension upright; this gives the self-centring action of the vehicle when you come out of a corner.

SAI (KPI) – the inclination of the swivel axle (better known by the old term kingpin, usually the suspension upright on a race car). This usually inclines inwards at the top end by 0–4 degrees.

Toe-in and **toe-out** – the static setting of the road wheels looking from the top. Toe-in means that they are closer at the front; toe-out is that they are wider apart at the rear. Cars that drive through the rear wheels usually have toe-in; FWD cars have toe-out. The reason being the under steady state cruising the driving forces will move the wheels into a neutral position to ensure no tyre scrub. A typical figure for toe-in or toe-out is 0–3 mm (0–1/8th inch).

Ackermann principle – this is used to reduce tyre scrub on turns, giving as near as possible true rolling motion on turns. This is achieved by lining up the centre of the track rod end (TRE) on an imaginary line between the suspension upright pivot centre and the centre of the rear axle, so that as the outer wheel turns the inner wheel turns through a bigger angle. Typically when the outer wheel turns through 20 degrees the inner wheel will turn through 22 degrees.

Toe-out on turns – this is the test for correct Ackermann:

- Place the vehicle with the front wheels on turn tables and the rears wheels on thin chocks to ensure that it is level
- Set both wheels to straight-ahead position – check the steering wheel
- Set the turn-table scales to zero
- Turn the steering to the left until the outer wheel, the one on the right, has turned through 20 degree
- Check the angle of the inner wheel, the one on the left; it should be 22 degrees or as recommended by the manufacturer

If the Ackermann angle is incorrect the vehicle will scrub off tyre tread in a similar fashion to incorrect wheel alignment. Ackermann faults can be caused by: bent steering arm, out-of-centre steering rack fitting, steering rack adjusted to one side. However, be careful and check the manufacturer's data; not all vehicles use Ackermann – they just scrub off tyres.

Try this

Carry out a steering and suspension check on a vehicle.

Slip angles – this is the difference between the direction of travel of the vehicle and the angle of the wheel.

Self-aligning torque – the torque exerted by the castor angle that tends to pull the vehicle into a straight line. On circuit cars the castor angle is kept small, typically 0–3 degrees to keep the steering light. Heavy luxury cars may have up to 8 degrees of castor to keep them on a straight line at speed on the motorway.

Over steer, under steer and neutral steer – are the three characteristic steering tendencies of vehicles. Neutral steer is the situation when you go into a corner having applied an amount of lock and the vehicle follows the planned path with no more steering changes. Under steer is when you enter a corner then have to apply more lock to negotiate it completely; in other

Table 5.1 Hydraulic power steering fault finding

No	*Test*	*Mfg pressure*	*Test pressure*	*Comment*
1	Check for fluid leaks	n/a	n/a	only use specified fluid
2	Check for noises	n/a	n/a	check for bearing wear
3	1,000 rpm full left	45 psi (3 bar)		typical figures
4	3,000 rpm full left	75 psi (5 bar)		
5	1,000 rpm full right	45 psi (3 bar)		
6	3,000 rpm full right	75 psi (5 bar)		

words the vehicle does not follow the front wheels. Over steer is when you go into a corner and then have to remove lock and the vehicle feels like it will roll over. Tyre types, sizes and pressures play a big part in this area of handling. Also, changing the load in the vehicle will alter the handling characteristics – a vehicle with four passengers will be much more difficult to drive than one with just two occupants.

Rear and four wheel steer – some vehicles have been made which have rear wheels that move a small amount when the front wheels are steered.

Hydraulic power steering – relies on hydraulic fluid pressure to move the steering rack, or in some cases a steering box. Some systems are speed sensitive – the amount of power steering assistance decreases with the road speed. In its simplest form this actually just measures engine speed, the argument being that engine speed correlates to road speed. The more complex systems pick up road speed information from the ECU (see Table 5.1).

Racer note

Driving fault-free cars helps you identify faults on others.

Electronic power steering – is controlled through an ECU which monitors:

- Vehicle speed
- Force applied by the driver
- Steering angle position
- Rate of acceleration of steering angle – how fast the steering wheel is being changed

Fault finding is usually limited to using a fault code reader.

Modifications

Suspension adjustment

Adjustable ride height – there are a number of different systems, namely:

- **Self energising** – an independent unit, like a bicycle pump, is mounted between the axle and the body/chassis. As the suspension moves it pumps up internal pressure. A set of internal valves control the pressure to match the external load and the relative ride height.

- **Pneumatic suspension with height control** – this has a height control sensor that measures the effective ride height and allows more fluid to be pumped into the suspension units to effectively raise the suspension. Under normal conditions this will maintain a constant ride height irrespective of the load. There may be some form of override control to allow different ride height for different conditions.
- **Electronic controlled suspension or active suspension**, the valve and pump mechanisms being controlled by an ECU.

Fault finding starts with using a fault code reader if appropriate, then checking for leaks, which are the usual failure. Some suspension systems need occasional repressurising, or the replacement of the diaphragm units because of internal leaks or the weakening of the rubber diaphragms.

Adjustable shock absorbers (dampers) – are available for most vehicles. Adjustment may be by a small screw at the base or twisting the complete body assembly with a 'C' spanner. Some dampers are remotely electronically adjusted by a stepper motor, or solenoid, controlled from a dashboard switch.

Rocker and bell crank systems – are used on most single-seaters to reduce unsprung weight and put the damper into a cooling airflow. It is easy to check them visually.

Push-rod and pull-rod systems – in a similar way to the bell crank system they allow the reduction of unsprung weight and allow remote positioning of components.

Adjustable axle location – the use of rose joints and threaded tie-rods allows suspension components to be adjusted to give exact set-up steering geometry.

Try this

Drive a vehicle with adjustable suspension and make some adjustments – how does its handling change? Do this on a track only.

Think safe

Legal requirements

Road safety and track safety – in the wrong hands any vehicle is a lethal weapon. By its very nature a motorsport vehicle is very dangerous, that is because of its high power to weight ratio and its twitchy handling and unforgiving controls. To put safety into perspective, consider the following facts based on typical figures for the UK:

- 70% of road accidents are by drivers aged under 24 years old
- 12,000 people are sent to jail each year for driving offences; 2,400 of these are for driving under the influence of drink or drugs
- Six people die each day on the roads, hundreds are injured

In the USA and other countries, the figures are proportionally similar. As a general guide in the UK there are about 30 million vehicles; in the USA the number is about 250 million.

The risk is not just from you as a driver, but includes other drivers colliding into your car and the risk of mechanical failure causing an accident. Think safe – the following positive actions will help to keep you alive:

- Check the vehicle over before you race, or when you take over a vehicle that you are not familiar with; this may include: a spanner check, oil/coolant/fluid level check, wheels and tyres, door/boot/bonnet catches, and general check for condition
- Carry out a dashboard check and start-up procedure for the vehicle – remember some engines need to be started in special ways
- Wear appropriate race/rally clothing – Nomex underwear, driver's overalls, driving boots, helmet, visor and gloves
- Fasten the seat belt correctly
- Follow all flag and light instructions
- Watch out for and avoid other drivers not driving safely

Lighting – on the road it must comply with the appropriate regulations. If a rally event goes into another country then it must comply with those regulations too.

Historic racer note

In 1965 the Monte Carlo winning Mini Cooper S was disqualified for having the wrong head lamps.

Tyres – if used on the road these must be road legal; you cannot use slicks on the road. For race events the tyres must comply with the race regulations. It is normal procedure for Formula-type events to have control tyres – that is one particular make, size and tread pattern. There may be two choices, namely wet weather tyres and dry weather tyres. The tyres must be fitted and inflated to meet strict guide lines. It is normal to do this with the wheels and tyres separately before scrutineering, when they will be fitted with a seal. As tyres run best when warm the use of tyre warmers is a good idea. Be aware that tyre temperatures will be very different, and therefore so will handling and lap times, in different weather conditions; February testing at Cartagena will be much faster than at Snetterton.

Racer note

Recently one Formula 1 team had a choice of 1,500 different tyres.

Steering – must be free from play, and not damaged.

Brakes – must comply with regulations and be similar to others in the same class. It is important that braking power is even for each axle set, and that front rear balance gives straight-line braking in a straight line under dry conditions. Pads and shoes must have sufficient friction material for the event.

Trials cars have fiddle brakes – that is there is a separate lever for each rear wheel; some motorcycles have inter-connected front and rear brakes.

Seat belts – for competition use it is essential to have multi-point seat belts, the current trend is for six-point mounting, although you will also find four- and five-point melting seat belts.

Emissions – there is a lot of talk and publicity about emissions. To put it into perspective, the amount of pollution generated by motorsport in the UK equates to that generated by one jumbo jet flying from the UK to Australia – and the aeroplane puts the pollution in the most dangerous place. However, we must ensure that we do not cause unnecessary pollution from our exhaust, or elsewhere. Where possible use catalytic converters, and for diesel engines use after-burners. Also, keep the noise down as much as possible. Most events have a scrutineering area for testing exhaust noise and many can test exhaust emissions. Thruxton circuit has *church hour* on a Sunday morning; no racing or testing is allowed for this hour.

Rally cars may be fitted with two engine ECUs; one giving an over rich mixture to keep the turbo-charger working for maximum acceleration on special stages, the other to give road legal emissions when driving between stages.

Historic racer note

Depending on the vehicle and its age, the regulations may vary. For example:

- Steering may have a limited amount of play in the straight-ahead position
- Brakes may have low (by comparison to current vehicle) efficiency
- Seat belts may not be needed
- Emissions limited to no visible smoke from exhaust

EuroNCAP – the European New Car Assessment Programme is an independent crash testing organisation standard. It is used by all the major car manufacturers in Europe. Australian car makers are now adopting it as are car makers in the USA. Currently the USA uses the National Highway Traffic Safety Administration (NHTSA) standards. It is not a legal requirement to comply with any of these standards; but they are an excellent measure of how the vehicle will perform when involved in a collision. Based on their performance the vehicles are given a *star rating*. Five stars is the highest rating. Formula car tubs are sometimes tested in the same way. There are four major tests: front impact test, side impact test, pole test and pedestrian test.

MOT – in the UK this refers to the MOT test certificate. Its name came from the Ministry of Transport, now Vehicle Operating Standards Agency (VOSA). However the name MOT still appears on the certificates in deference to popular demand.

The MOT is a 30-minute examination. Certification is now done online – all MOT certificates are kept on a database allowing retaxing of the vehicle over the telephone or online. Vehicle insurance is also kept on a database so that VOSA know which vehicles are taxed, have insurance and MOTs.

A similar situation exists in the USA and other western hemisphere countries. Less developed countries still require tax and insurance, but the testing and databases are not as well developed, as administration tends to be at a more local level.

Competition requirements – general competition requirements are covered in the **MSA *Bluebook***. Specific regulations (**regs**) for each class, or type of racing, are given by the organisers; usually these are published on the internet to reduce the use of paper and postage costs.

FIA – the Fédération Internationale de l'Automobile, is the world governing body for motorsport. However it is more important in Europe and the emerging motorsport countries like the UAE and Japan than in the USA. Many form of motorsport do not recognise the FIA, preferring to have independence. Sometimes this leads to arguments and conflicts.

MSA – the Motor Sport Association, is the leading motorsport authority in the UK. They are a member of the FIA. The local motor clubs form the membership of the MSA along with licence holders. The MSA is responsible for most of the racing in the UK. The MSA issues racing licences to both individuals and race circuits. It organises some premier events like the British Grand Prix. Motor cyclists have the **Auto Cycle Union** (ACU). Kart racing is administered by the MSA; but drag racing is not – it is done through the **National Hot Rod Racing Association** (NHRA). Probably the oldest club is the **Motor Cycle Club** (MCC) which may be confusing as it organises off-road car trials as well as motorcycle trials. The USA has a number of clubs, the best known probably being **National Association for Stock Car Auto Racing** (NASCAR), which operates in the UK too.

Bluebook – is published annually by the MSA and sets out the rules and regulations of racing in the UK.

ARDS – the full name, which is rarely used, is the Association of Racing Driver Schools. ARDS administers and certificates the qualification for a racing licence. The normal procedure is that before starting racing you have a medical examination, then you spend a day, and the cost of a couple of tyres, on a one-day training course with an ARDS registered racing driver school. At the end of the day you take a short written examination – multiple choice questions and a number of observed around the circuit. The purpose of the ARDS is to ensure that you will be safe on the track and observe the flags and appropriate directions. If you pass your medical and ARDS you can apply to the MSA for a competition licence.

Competition licence – issued by the MSA following your medical and passing your ARDS. There are a number of different competition licences for different classes of racing. For non-MSA events, like some drag racing, you can join the NHRA, or in some cases pay on the day for a temporary licence. The difference is that in MSA events there are usually many competitors on the track at the same time; drag races are run against the clock – the risks are different.

Scrutineering – the checking of the vehicle before an event. The scrutineer checks to see if the vehicle complies with the regulations relating to the event and this includes general safety too.

Driving on the road – to drive a vehicle on the road, both the vehicle and the driver must comply with all the regulations. For example:

- Safe legal vehicle
- MOT if needed
- Insurance
- Driver's licence
- Road tax

Driver's licence – to drive a vehicle on the road in the UK you must be over 17 years old. You can apply for your licence and your test up to three months before you are 17. In the USA the age varies in each state, typically 14 in the mid-West and 18 in New Jersey. There are also rules in the USA that require you to have a learner's permit before a full licence will be issued.

Road tax – is essential if the vehicle is being used on the road. Road tax can be bought at the Post Office, online, or over the telephone. To apply online, or over the telephone, you must have **insurance** registered on the insurance database, and if the vehicle is over three years old a new style **MOT certificate**. These are also registered on the VOSA database.

Insurance – legally required for a vehicle to be used on the road in any country; it usually does not cover any racing or rallying. Also motor trade insurance does not cover competition vehicles as standard. Special insurance is available for racing and rallying. There are different levels of cover, and these may be very specific. Typical examples are:

- Transporting to an event
- Testing
- At event (not on track, not racing)
- On-track (not racing)
- On-event (racing)

Check out the fine detail of any insurance before buying, or going to an event.

Company regulations (smoking/alcohol) – it is illegal to smoke in a company-owned vehicle in the UK, even if it is your own company. It is also illegal to smoke on any business premises. Generally speaking smoking, the use of alcohol, or drugs is banned in all motorsport premises and vehicles in the UK, the USA and most western hemisphere countries.

Highway Code – you have a legal requirement to fully understand and abide by the Highway Code in the UK. In the USA the highway regulations do vary in detail between states for things like (traffic) stop signals and lane usage; but similar principles apply – that is taking care, signalling and obeying the signage. European countries tend to follow a similar pattern.

Vehicle care – as a technician there is no need to mention that you have a duty of care of any vehicle in your charge.

Health, safety and the environment requirements

Motor racing can be dangerous, so take no risks off the track – remember to document all your actions.

Think safe

Both **employee** and **employer** are responsible for third parties. **Vicarious liability** means that the employer is responsible for the actions of the employee as well as the employee being responsible for any action – this means that all parties may be liable for prosecution.

COSHH – Control of Substances Hazardous to Health – you should have a file for each of the substances that you use in your workshop showing the hazards and how to deal with them – your suppliers will be able to provide this data; and most companies have it available on their website.

Risk assessment – assessing each action before doing it.

Disposal of waste materials – is simple if you follow a few basic rules – to be environmentally conscious is very important; doing this correctly will save costs in the long run and give a good image to motorsport in general.

Think safe

Disposing of motorsport equipment wrongly can cost you in fines and you must also be careful not to give away clues as to your racing secrets.

Never put in a land fill bin anything which can be recycled. In many cases it is against the law – it will depend on your local authority – and it costs nothing to ask for advice; but ensure that you follow it. Table 5.2 gives examples typical of most areas in the UK and the USA.

Think safe

Professionally and personally, over the past three years I have reduced my land fill waste to a quarter of what it was. As well as recycling everything I can, I also challenge the use of unnecessary packaging of items, that is asking why things are packed in a certain way and requesting changes where I can. I also look for the best choice of alternative materials being specified and reuse of materials where appropriate.

Think safe

Ensure that the workshop, the transporter and all the vehicles are equipped with the correct fire extinguishers and that they are live and tested before every event.

Table 5.2 Disposal of waste materials

No	*Item*	*Where to dispose*	*What happens to it*	*Comment*
1	old units	return to manufacturers	may be overhauled, remanufactured	
2	electrical components	specialist disposal at ERS	precious metals removed for reuse	
3	tyres	specialist disposal through supplier	shredded for use in road surfaces and safety surfaces	specialist companies will collect these
4	batteries	specialist disposal at ERS	precious metals removed for reuse	specialist companies will collect these
5	oil and fluids	specialist disposal at ERS	converted into paint stripper or burnt to generate heat	specialist companies will collect these
6	paper and card	collected by local authority	recycled into paper again	
7	wood	specialist disposal at ERS	burnt in power stations to make electricity	
8	metal items	weigh in for scrap	melted down for reuse	
9	glass	specialist disposal at ERS	melted down to make more bottles	

(ERS) Environmental recycling site

Technical information

Information sources – technical information for road vehicles is not always easy to find; for race vehicles it can be impossible. As a race engineer you should keep your own records and notes as well as those needed by the customer, team, or other service provider. The use of a digital camera is recommended so the information can be uploaded on to your computer. If you are purchasing a mobile phone with a camera, then look for one with zoom and flash built in. The macro facility too is useful for photographing small parts and pages of data. Sources of information may include:

- Manufacturer's workshop manual
- Independent type workshop manual (such as Haynes)
- Driver's handbook
- CD-based manual
- Parts manual
- Online data source (such as Autodata)
- Information from test bench manufactures (such as Snap On)
- Company records
- Data acquisition system
- Vehicle log

Maintenance of records – work on any vehicle must be documented. You must maintain an audit trail to ensure that you have both a technical and a legal record of what you have done. Depending of the type and level of racing, you are likely to keep records in one, or more, of the following ways:

- Vehicle service record
- Company job card
- Company file card
- Electronic customer record management system (CRM) also called customer relationship management
- Data log – hard or soft
- Vehicle log – hard or soft

Legal requirements relating to data – you must keep detailed records of all your work for the following purposes:

- Ensure that the vehicle complies with the relevant competition regulations
- Meet the needs of the driver and/or vehicle owner
- Satisfy the requirements of HM Revenue and Customs (UK) or other tax and VAT requirements in other countries
- Meet HASAWA criteria – in case of an accident
- Ensure compliance with the Data Protection Act (UK and most other countries)

The law requires that you keep this data safe and do not disclose it to third parties. Written (hard copy) records must be kept in a locked file – usually these should be fire resistant too. Electronic (soft) data must be password protected and only accessible to authorised parties. The Data Protection Act, which is universal in most countries, makes these stipulations, which are covered by criminal law as well as civil law.

Fault diagnosis

This topic is largely covered in Chapter 8. However, specific to the vehicle chassis the following procedures ought to be borne in mind when trying to diagnose faults:

1. Where possible, and appropriate, take readings using the fault code reader to interrogate the chassis ECU. This is always the first stop if possible.
2. Following an off, no matter how light, check the steering geometry, and the chassis alignment, and corner weights. If the car was not damaged by the off, it could have been a faulty steering, suspension or chassis component that caused the off in the first place.

For most purposes checking the toe-in/toe-out with optical gauges is good, and a manual check of castor, camber and SAI with a bolt-on gauge will suffice. The use of a complete four wheel system will show up more faults – that is relative alignment of all four wheels. The corner weights can also indicate incorrect suspension adjustment or faults.

If it is suspected that the chassis is bent then carry out either a jig test or drop test with plumb-bob and chalk.

Set-up procedures

Pit garage – using a pit garage will usually have an on cost, whether it is a track day, test day or race day. Costs vary with time of year and circuit. The advantage of a pit garage is that you will have a flat floor, shelter, be right next to the track and can usually set it out and maintain security. Depending on the size of the vehicle you may be able to accommodate more than one car. Most pit garages have transporter, or trailer, access from the non-track side.

Paper-based systems – have their place in *clubman* and *historic* vehicle events where the changes to the vehicle are likely to be small and the set-up follows pre-set patterns or methods. Then it is a matter of recording only limited variables like tyre pressure and fuel consumption.

Laptop systems – most competitive teams and individuals use a laptop computer-based system. This allows lots of data to be recorded quickly and retrieved for analysis and comparison.

The *Intercomp Race Car Management* software has six heading tabs. These are:

Geometry – details of the suspension including: castor, camber, SAI, which components and vehicle height
Set up – details of spring and damper settings
Weights – corner weight information showing CG
Track – details of track and weather conditions
Race – race times/speeds, section times and comparative information
Tyres – compound, pressure and temperature

Of course you can record this information on paper manually; but the low-cost of software and easy access to laptops, coupled with the speed and accuracy of such a system, makes it ideal.

Data logging – consists of three main areas:

- The set-up data of the vehicle and how this changes during practice and racing
- The performance of the vehicle under practice and race conditions; this will be logged against individual drivers
- The behaviour of the vehicle on the circuit

These areas may be separate, or they may be integrated.

Data analysis – once you have got the data you have to analyse it. That is, you need to break it down into small manageable chunks and try to make sense of what is going on. The big problem at the beginning of any season, or on delivery of a new vehicle, or even a new driver, is the establishment of a baseline setting. Once you have baseline data you can make changes and get faster times. Teams frequently go testing on the same day on the same track, then they watch each others' times for comparison – if they test on different days then they try to factor in changes for the weather.

To enable easy data logging most circuits have one of two systems that can interface with your vehicle.

Stack and *PI* are probably the most popular. The speedometer has a sensor that picks up a signal from the start/finish post. This allows automatic lap timing and recording in the car. Coupled with a standard built-in trip computer it can give detailed information about performance.

For kart and other club level events each competitor is given a transponder which sends a signal to the control office (usually in a high tower). This allows the organiser to plot the lap times of each car; usually this gets a special prize or points.

To log the performance of the vehicle in terms of acceleration, braking, fuel consumption, engine operation and cornering forces, the vehicle can be fitted with sensors to record this information onto a memory chip in a small ECU-like device – a data logger unit. This information can be overlaid the circuit information for full analysis.

The analysis will lead to a series of questions, and these questions will develop with your experience, for instance?

Could third gear be used at that point?

Why are the tyre temperatures so different?

Is it possible to use more revs out of that corner?

Whole vehicle set-up

Setting up a race car, or bike, should be tackled as a whole. Changing one small part can affect the rest of the vehicle. In this section we will highlight some of the main points for your further research.

Historic racer note

The first documented set-up of the whole racing car was by William Milliken (USA) and Maurice Olley (UK), funded by Cadillac and under the supervision of Cornell University.

Wheels and tyres – is probably the place to start with any vehicle; the tyres transmit the power, the braking forces and the turning moments between the vehicle and the road. The factors to be considered are:

- Rim widths, diameters and off-sets
- Different designs and sizes
- Tread compounds
- Aspect ratio
- Front/rear combinations

- Different tyre circumferences
- Track width
- Caster, camber, SAI and toe-in/toe-out

Centre of gravity (CG) – the aim is to get the centre of gravity into the position that you want. Usually it should be as low as you can get it and near the centre of the vehicle. This is measured simply using corners weights; or if you run Formula 1 or WRC you will have a more complex test rig. Standard cars are unlikely to have an accurately placed CG, so getting it accurate may mean moving heavy items like the battery to a different location or adding ballast – lead weights are the normal ballast. This needs to be done both laterally and longitudinally. Setting up the static wheel alignment also affects CG as does ride height.

Ride height and body attitude – the orientation of the body/chassis, also called in aerodynamic terms angle of attack, alters the roll and pitch and yaw. Generally the more stable position is when the car is higher at the rear than the front and the CG is in front of the centre of pressure (CP). The following points should be considered:

- Steering geometry
- Bump/roll steer
- Anti-dive/anti-squat
- Suspension travel
- Aerodynamic aids

Aerodynamic forces and moments – as we have already mentioned these are related to body attitude. In addition you must consider drag, lift, down force and vehicle movements. Often forgotten is the airflow inside and underneath the vehicle. To get this one right you will need to study aerodynamics in detail and use computational fluid dynamics (CFD) software as well as a wind tunnel.

Brakes – vehicle brakes are now well developed and available in various set-ups for most vehicles; but the same problems still need setting up for competition vehicles. These are:

- Cooling – dissipating the heat energy from braking is something that needs careful consideration – the heavier the car the more the heat developed. Consider that if you have a car developing say 250 bhp, the heat from the radiator is probably enough to heat an average house. Braking that same car is going to develop the same heat as accelerating it. Where is the hot air going?
- Brake balance, or distribution – under heavy braking conditions you will have a lot of weight transfer from the rear to the front; this needs to be managed. On road vehicles the anti-lock braking system (ABS) and possibly the electronic stability programme (ESP) will deal with this. On an open-wheel car you are only likely to have two master cylinders and a balance bar – a lot of testing will be needed to get the right choice of brake pads and balance bar position.

Historic racer note

Jaguar won their first Le Mans 24-Hour Race by having disc brakes, which enabled them to lose less time when braking for corners, so they got quicker lap times than the other cars that had drum brakes.

Drive line – adjusting the gear ratios to suit the vehicle, the power curve and the circuit needs accurate attention – the choice of the correct gear ratios will become noticeably important to handling under transient accelerating conditions. That is when cornering you should enter on the best line for a power exit; at this point it is important to have the correct gear ratio, to have the correct part of the power band available to give the required acceleration to power the vehicle into the next straight. Remember that you do not want to be changing gear on your exit from the curve; you want the car to be in a straight line for both maximum handling and maximum weight transfer to put the power onto the track. Gear ratios are further discussed in Chapter 4 on transmission.

Springs and dampers – this area also relates to ride height and body attitude as well as load transfer. The factors to be considered here are:

- Suspension type – IFS, IRS, and type of layout, and suspension medium
- Spring characteristics – these should be calculated in conjunction with the tyre characteristics
- Damper characteristics

The suspension layout will control dive and squat and give other characteristics such as the amount of movement relative to wheel movement and bump steer. On single-seater cars it is normal to have inboard mounted dampers and springs to reduce the unsprung weight, which in turn improves wheel control. On road cars softer suspension is needed – for rally cars it is normal to have variable rate springs to give some initial movement to cope with uneven surfaces.

Different front to rear spring rates will alter pitch and roll – it is normal to have slightly stiffer springs at the rear than at the front; this keeps the roll centre higher at the rear to enhance stability.

Compliance – this covers chassis stiffness, suspension stiffness and other factors of body/chassis design. For maximum stiffness many open-wheel formula cars use a composite tub construction. Rally cars often incorporate roll over bars to add stiffness as well as safety. Short circuit racers – USA Midgets and Escorts (not to be confused with UK cars of the same name) have stiff tubular steel chassis.

Driver–vehicle interface – often referred to as ergonomics, is probably the most important factor. Does the car suit the driver, or does the driver suit the car? The race car engineer can change a number of factors, such as:

- Steering gear ratio
- Brake pedal force
- Clutch pedal movement
- Steering feel – kickback and movement
- Seat position
- Instrument position
- Gear lever position – movement

Wheels and tyres

The choice of wheels and tyres is almost endless. The current trend is to go for large diameter wheels with very low profile tyres. The resultant small sidewalls allow high-speed ratings.

When fitting replacement wheels and tyres you must check the following points to avoid failure and possible damage:

- Confirm the wheel and tyre manufacturer's fitting is approved to the vehicle concerned
- Check the off-set for chassis and bodywork clearance
- Check hub/flange/stud fitting and bearing load
- Check the speed rating
- Set the pressures accurately
- Torque up correctly using the designated nuts and studs

Steering

Rack – the normal modification is to fit a quick-rack, which is one that requires less turns from lock to lock than the standard one.

Geometry – this may be modified in a number of ways – changes to the castor, camber, SAI (kpi) and toe-in. On some vehicles there is a facility for adjustment. On track cars the use of an adjustable rose joint is the usual method. To improve handling the use of negative camber is common.

Brakes

The braking system converts kinetic energy into heat. The limiting factor with braking systems is the maximum rate of heat dissipation – how fast they can conduct and radiate the heat away. On top flight race and high-performance vehicles, the brake components are made largely from carbon. This is not, however, suitable nor affordable for most motorsport vehicles. Typical modifications of the main components are:

Discs – replace with ones that are ventilated, cross-drilled, or larger diameter. In all cases the aim is better cooling.

Callipers – replace with ones that have more pistons to increase the applied pressure.

Hoses – typically add braided flexible hoses and stainless-steel lines to give less hoop (that is outward like a hoop) expansion when maximum pressure is applied.

Fluid – use of a high boiling point fluid to prevent brake fade. Remember to test brake fluid for water content – the hygroscopic test.

Questions and skills

1. Using the diagram of the Ackermann angle as a guide, make a model of the steering linkages. Turning the inner wheel in steps of 5 degrees measure how many degrees the outer wheel turns. If possible try this on a car using turn tables.
2. Draw up a table of the steering angles for a range of different vehicles and see if there is a particular pattern to them.
3. Obtain either an old steering rack or steering box and strip it down using the workshop manual – can you name all the parts.
4. Make a list of your favourite vehicles and draw up a table to show the diameter, width, aspect ratio, speed rating, tyre type and tyre pressures that are used.
5. Investigate the mechanical brakes used on motorcycles, karts, or bicycles. Some are internal expanding; others are external contracting, see if you can identify the different ones.

Chapter 6

Electrical and electronic systems

Figure 6.1 Dashboard of JCB land speed diesel car

Electrical and electronic systems

Key points

- The battery is the central store of electricity for the vehicle.
- Battery electrolyte is very corrosive – avoid splashing it on your skin or the vehicle.

- The alternator is driven by the crankshaft to keep the battery charged – not all race cars/bikes have them.
- Many vehicles use the metal body as earth return.
- Race cars often use a separate trolley battery for starting.
- Vehicles may have several ECUs – engine on road/engine on special stage/body controls/transmission controls and other separate systems.

Safety note

Electrical short circuits are the main cause of vehicle fires. When working on an electrical system it is easy to cause a temporary short circuit by two terminals, or cables, touching. For this reason you *must* always *disconnect the battery* before carrying out any work on the electrical circuit.

Battery

The battery is the **central store of electrical energy** for any vehicle. Its purpose is to store electrical energy in chemical form. Most vehicle batteries operate at a nominal 12 volt. Batteries are also available as 6 volt and 24 volt. To control weight distribution, or give added power, or reliability, vehicles may use more than one battery. Two 6 volt batteries may be connected in series to give a 12 volt supply; two 12 volt batteries may be connected to give a 24 volt supply. A 12 volt battery may have three terminals to allow both 6 volt and 12 volt supplies.

Nomenclature

The word battery means a group of things and came into use in the seventeenth century as a place where guns and soldiers of the English Civil war were located – in Hampshire there is a village called Oliver's Battery. The technically correct name for a car battery is an accumulator; you may come across this term when you are working at an advanced level with your studies.

Casing – this is made from a **non-conductive** material, which can withstand low level shock loads. Inside it is divided into compartments, or cells.

Cells – each cell is made up of **positive plates** and **negative plates** divided by **porous separators** and submerged in **electrolyte**. There is always an odd number of plates – there is one more negative plate than positive plates to make the maximum use of the positive plates.

The **nominal voltage** of each cell is 2 volts, so six cells are needed to make up a 12 volt battery. We say nominal voltage, as it is not the exact voltage. The voltage given by each cell varies with the state of charge of the cell, and therefore the battery. A fully charged cell will produce 2.2 volt, meaning that the fully charged battery will produce 13.2 volt.

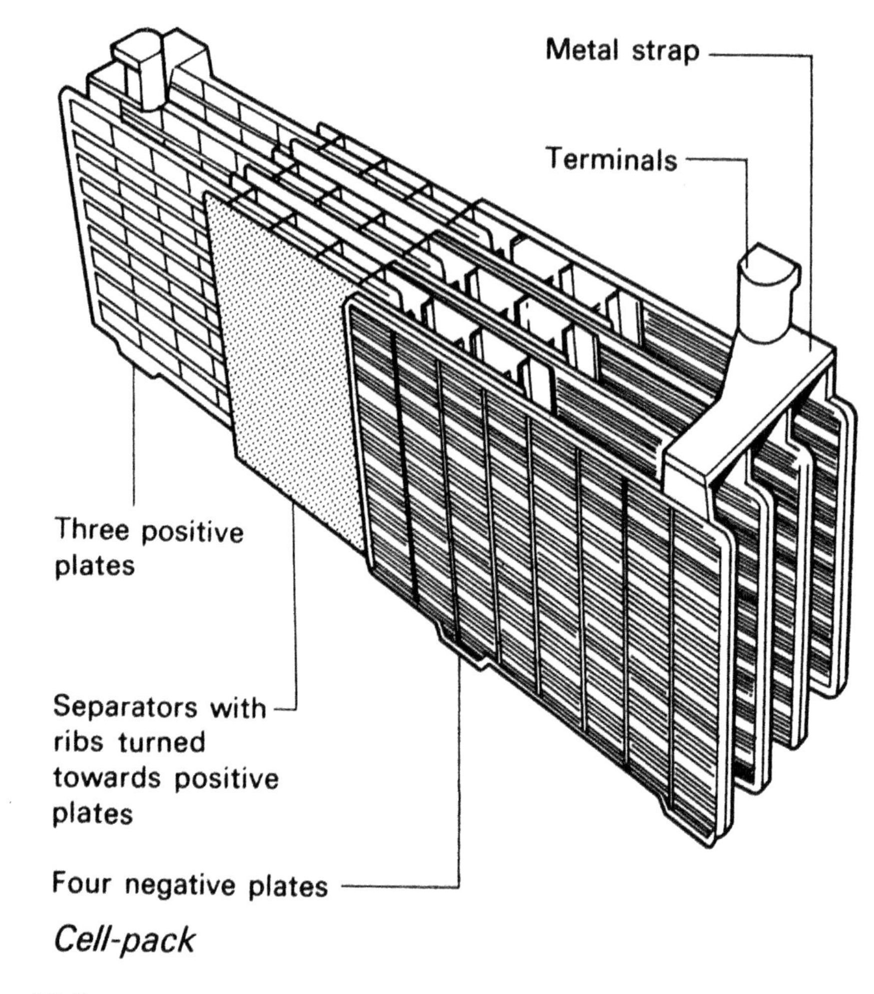

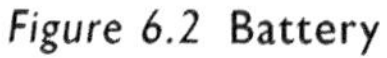
Figure 6.2 Battery

Electrolyte – is a solution of **sulphuric acid**, it is **highly corrosive, treat it with care**. Batteries may come **dry charged**, in which case the electrolyte is added only when the battery is purchased – this aids storage and transportation; or it may be **wet charged** – the battery is charged after the electrolyte is added.

The electrolyte in motorsport vehicle batteries is in a **gel** (jelly) form to prevent spillage in the case of an off.

Safety note

Electrolyte, usually called **battery acid**, is very **highly corrosive**; it will burn the skin off your hands and make holes in your overalls. Read the COSHH sheet and take extreme care – **treat batteries with respect**.

Table 6.1 Battery chemistry

	Fully charged	*Discharged (flat)*
Positive plate	Lead peroxide (PbO_2)	Lead sulphate ($PbSO_4$)
Negative plate	Spongy lead (Pb)	Lead sulphate ($PbSO_4$)
Electrolyte	Strong sulphuric acid ($2H_2SO_4$)	Dilute (weak) sulphuric acid ($2H_2O$ with a percentage of $2H_2SO_4$)

Battery charging – batteries are kept fully charged when on the vehicle by the alternator (or dynamo on older vehicles). When not on the vehicle a battery charger is needed. It is wise to keep a battery fully charged. The battery off a race car for example should be fully charged before it is stored for the off-season period. **Maintenance free batteries** must only be charged at low amperage to avoid damage to the plates; check that the battery charger is the correct type for use with maintenance-free batteries. There are three types of battery charger in use. These are constant current charging, constant voltage charging and taper charging.

Charging the battery alters the chemical structure in the plates and the electrolyte acidity (see Table 6.1).

Relative density (specific gravity)

> **Nomenclature**
>
> Relative density (RD) is the technically more correct term for specific gravity (SG); in automotive engineering they are used to mean the same thing. When water is at its maximum density, its temperature is 4°C, and a litre will weigh 1 kg. Acid has a greater density, so a litre of acid will weigh more.

The RD or SG of a battery electrolyte is an indication of the state of charge of a battery (see Table 6.2). This is checked using a hydrometer; this is a glass tube with a float inside it. The denser the liquid the higher the float will be in the electrolyte, the float may have a numbered scale, or simply a° coloured scale to indicate the state of charge.

Maintenance – the following points should be checked:

- **Electrolyte level** – **top up with de-ionised water** as needed, sealed maintenance-free batteries should never need this activity

Table 6.2 Relative density readings

Colour	*RD or SG*	*State of Charge*	*Comment*
Green	1.280	Fully charged	Leave battery to cool after charging before testing
Yellow or orange	1.160	Half charged	
Red	1.040	Discharged (flat)	May vary with temperature

- **Keep batteries full charged** – if the vehicle is laid up for the off-season keep it charged with a bench charger
- **Use proprietary protector** (petroleum jelly or similar) on battery terminals to prevent corrosion
- **Keep terminals and connectors clean and tightly fastened**

About half of all breakdowns are caused by battery faults; the AA and the RAC frequently publish statistics on causes of breakdowns.

Starter pack (or **power pack**) – for breakdown and recovery work a special battery in the shape of a small brief case (or hand luggage) is used. This is simply charged from the mains with its own cord and plug (230 volt UK; 110 volt Europe, USA and most other countries).

Race battery – batteries for race vehicles are very light and usually of the gel type. They do not need to start the vehicle as a portable trolley battery – a large heavy duty battery mounted on a trolley – is used for starting. Or a starter pack system may be used, depending on the size of the engine.

Racer note

Keep your trolley battery fully charged and remember which side of the vehicle it plugs into.

Alternator

The alternator is driven by a 'V' belt from the engine front pulley; it produces **alternating current** (AC), which is converted into **direct current** (DC) to charge the battery. The alternator is made up of a **rotor**, which spins inside a **stator** and is encased in adjoining **front and rear casings**. The electricity is generated by the movement of the rotor inside the magnetic stator. The electricity leaves the rotor through two slip rings and brushes. The rear casing houses the

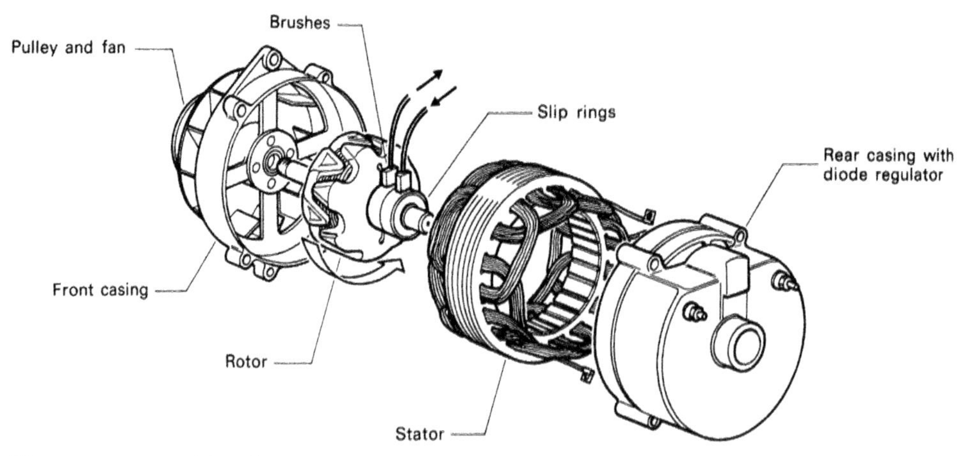

The components of the alternator

Figure 6.3 Alternator

rectifier and the **regulator**. The rectifier converts the AC current into DC current; the regulator uses a set of **diodes** to control (regulate) the amount of current supplied to charge the battery. Too much electricity would cause the battery to boil and become useless; too little and the battery would become flat. Typically it takes 20 minutes of running the engine to recharge it to replace the electricity used in starting the car from cold. Much less electrical energy is used starting the engine from warm than starting from cold. Maintenance of the alternator is minimal – check the 'V' **belt tension**; there should be about 12 mm (½ inch) free play on the longest section; **inspect it for cracks or fraying**, and **read the output** using the engine **diagnostic tester**.

Dynamo – before the alternator was the dynamo. This ran at much slower speeds because the commutator's soldered segments would break away above a specific speed – around 6,000 rpm.

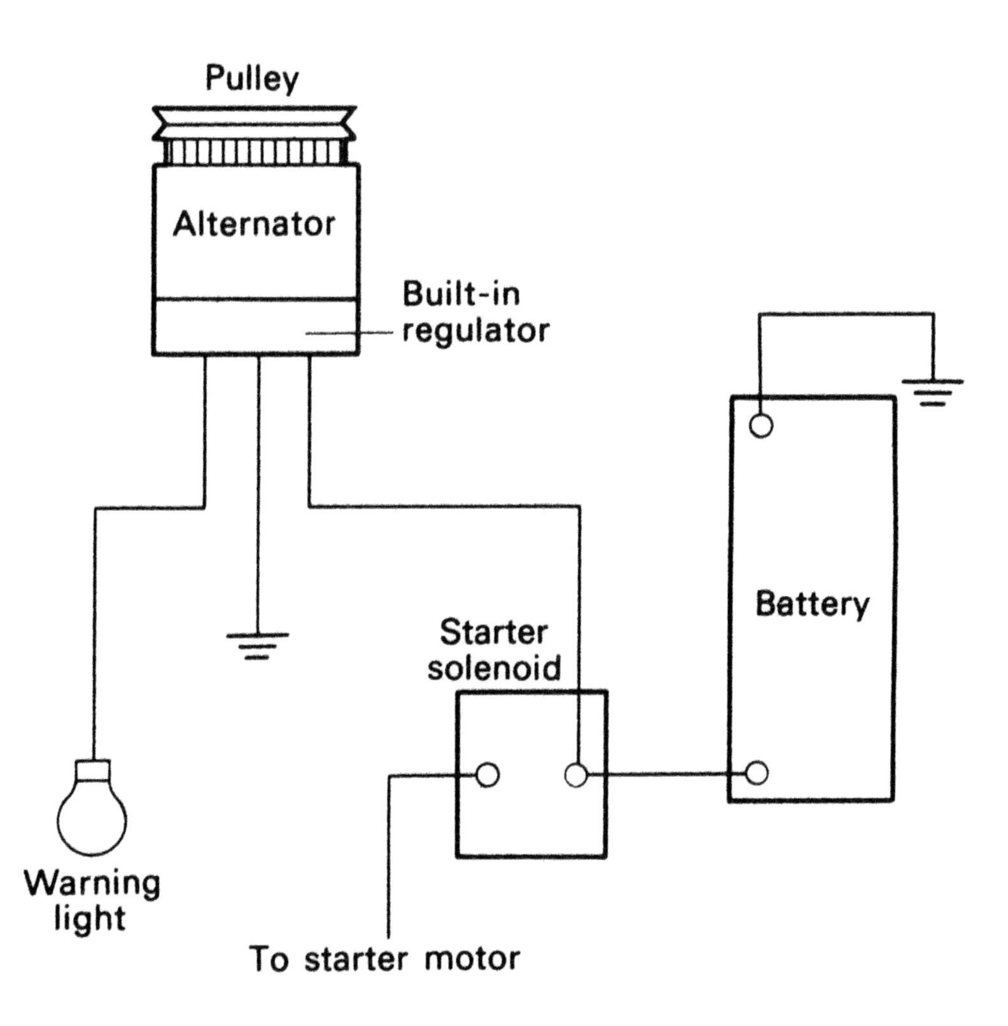

Figure 6.4 Charging circuit

Magneto – this is used to generate HT for the ignition circuit on small capacity motorcycles (and some historic vehicles). On vehicles, usually motorcycles, where a power supply is also needed, the magneto is combined with a dynamo to make a **mag-dyno**. The mag-dyno is often used on sporting trials motorcycles; the power for the lighting (this is the only use of electricity apart from the ignition) is only provided when the engine is running, and is also dependent on engine speed.

> **Nomenclature**
>
> Alternator, dynamo and mag-dyno are all referred to as *generators* as they all generate electricity.

Starter motor

There are two major types of starter motor. These are:

- **Inertia type** – this uses an inertia drive from the motor to the flywheel so that a mechanical linkage is made only when the starter is turning fast enough
- **Pre-engaged type** – this has a solenoid to engage the gear on the starter motor with the flywheel ring gear before the starter motor turns

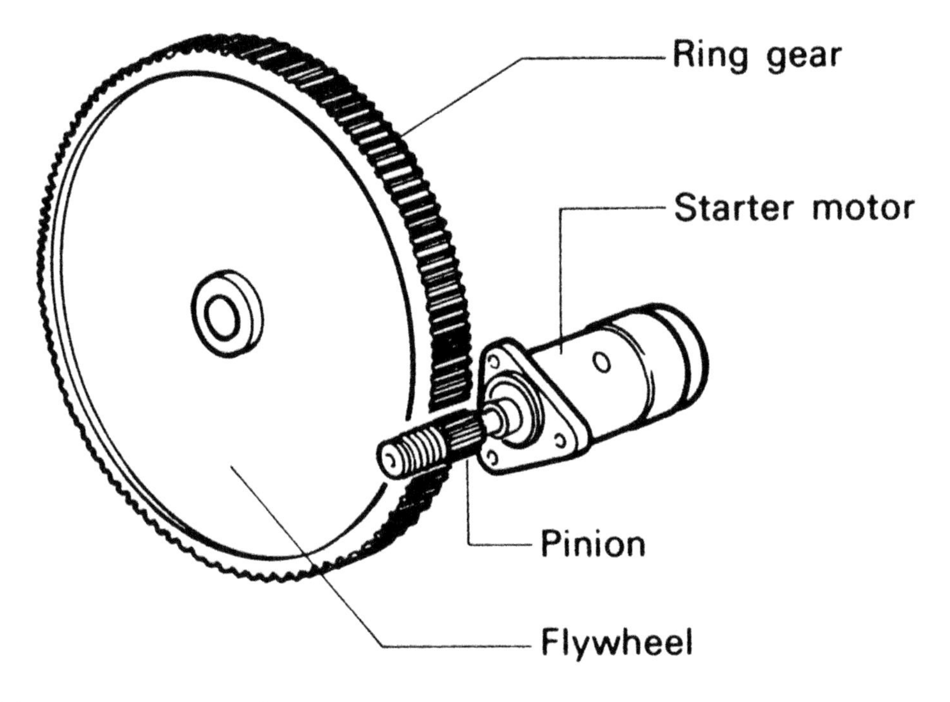

Figure 6.5 Starter motor and flywheel arrangement

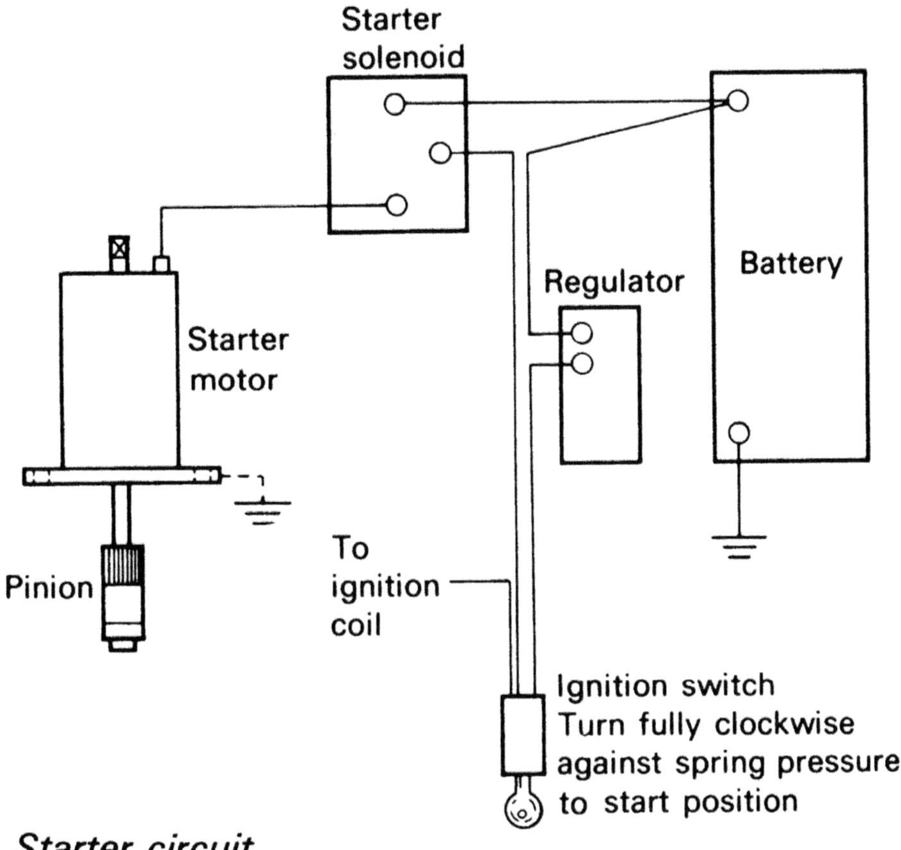

Figure 6.6 Starter circuit

The actual motor part is similar on both types of starter motor. The **armature** is turned inside the **field coils** of the motor by the effect of supplying a large electrical current to the starter motor brushes. Because of the heavy current involved a special switch called a **solenoid** is used. This is operated by power supplied to it when the key is turned to the start position. On race cars a separate starter button may be used, the ignition being controlled by a flick switch with on or off only.

As a guide, a starter motor takes about 180 amps to turn an engine from scratch; engines typically need to rotate at 120 rpm to initiate combustion. Some engines can achieve this in less than half a turn of the crankshaft.

Safety note

If you are testing a starter motor off the vehicle do so in a suitable rig – not loose on the bench top as the current required could cause a burn, or start an explosion.

Vehicle circuits

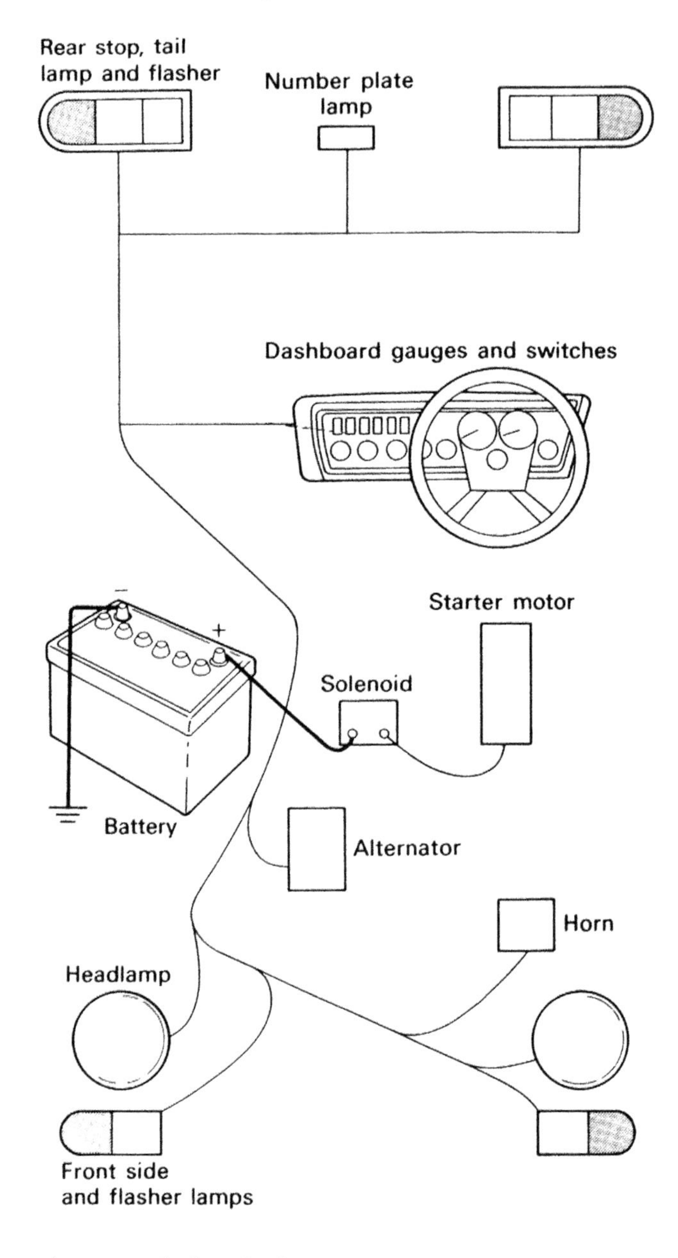

Figure 6.7 Wiring schematic

Electrical symbols to make electrical circuit diagrams easy to understand and small enough to be printed on A4 paper, a system of electrical symbols is used. There are variation between countries and manufacturers; but it is usual for them to be easily identified even if you see them for the first time.

Circuit diagrams – most vehicle manufacturers provide circuit diagrams. The student should seek out those of vehicles that are of interest.

Racer note

Beware when using circuit diagrams – they are like the London underground map; the position of a component on the diagram is not an indication of its position on the vehicle.

Chassis earth this takes large amounts of current, especially when starting the engine (typically 180 amp). It is essential to check all chassis earth points. There is a major lead between the battery and the chassis, and a similar lead between the chassis and the engine.

Racer note

It is prudent to check all earth connection for tightness as part of your spanner check procedure – also use a suitable proprietary anti-corrosion coating to keep them in good condition.

Cables and connectors – all cables on current road cars and motorcycles use crimped ends into semi-locking plastic connectors. The type used varies with the manufacturer. The student is advised to investigate those of the manufacturer that is of interest.

On older motorsport vehicles mainly copper wire is used – as compared with the aluminium or gold-plated wires of current high-performance vehicles. Copper wire is usually soldered to brass, or tin-plated, connectors on older motorsport vehicles. The soldering is necessary to give good electro-mechanical connection and prevent the wires coming off the terminal.

Racer note

Soldering is a good skill to learn if you intend making up some of your own cable and connector with high electrical and mechanical integrity.

Lights

Sidelights are fitted to indicate the size and position of the vehicle to other motorists; in the UK they must be white to the front and red to the rear. Other countries use amber lights at the

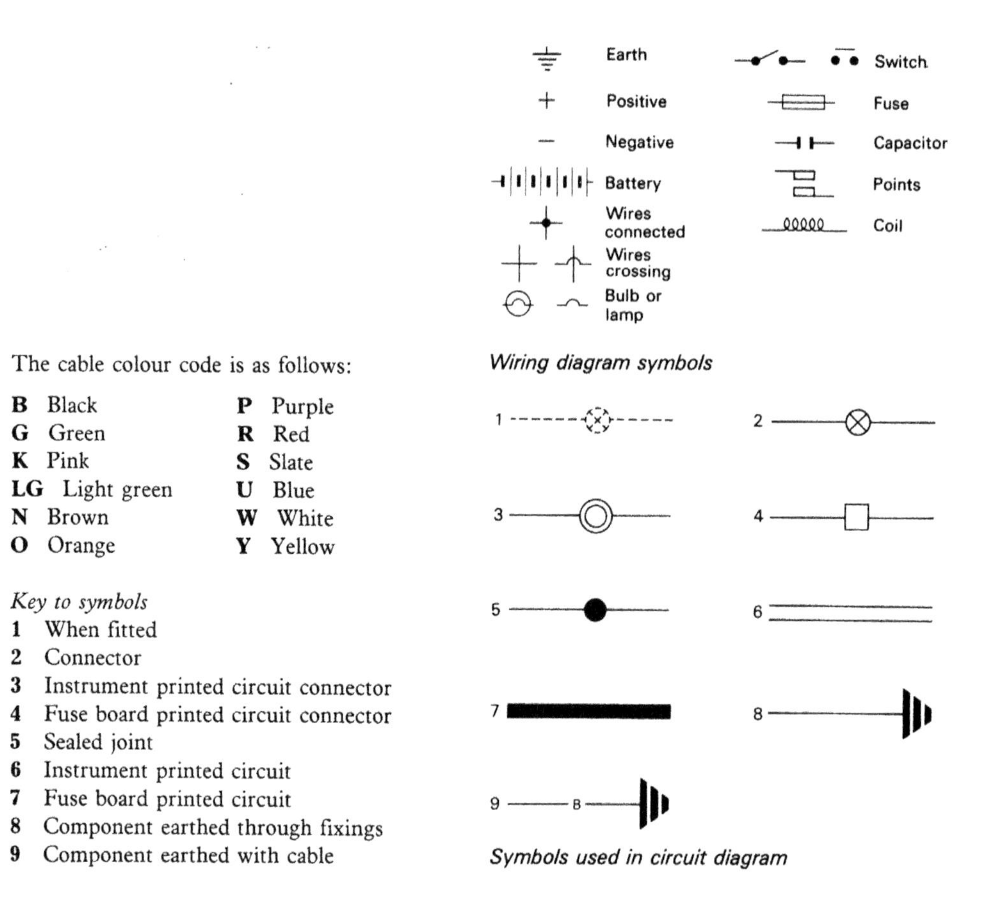

Figure 6.8 Wiring symbols

front. There are regulations as to position on the vehicle and minimum size. Until recently a 5 watt lamp was compulsory, now LED lights are permitted.

Headlights are fitted to give the driver clear vision in darkness. The law sets regulations on lamp position and beam placement – the rules are devised to prevent them from dazzling other motorists. Headlights may be a number of different colours; but white is normal with blue (to look like police cars) the second most popular.

Spot and fog lights – used as their name imply. They should only be used in appropriate conditions.

Direction indicators must show amber in colour and flash at a rate of 60 flashes per minute. Their position is also subject to regulations and side repeaters may be used too.

Brake lights are operated by the brake pedal. The minimum is two rear brake lights – 12 watt bulbs, or equivalent LEDs. High level brake lights are normal on current vehicles using an array of LEDs.

Running lights – on US vehicles and in some other countries too. These may be the equivalent of the side lights – usually amber in colour.

Race car rear lamps – in bad weather race cars are required to show a red light to the rear. These are almost universally LEDs as they use very little current.

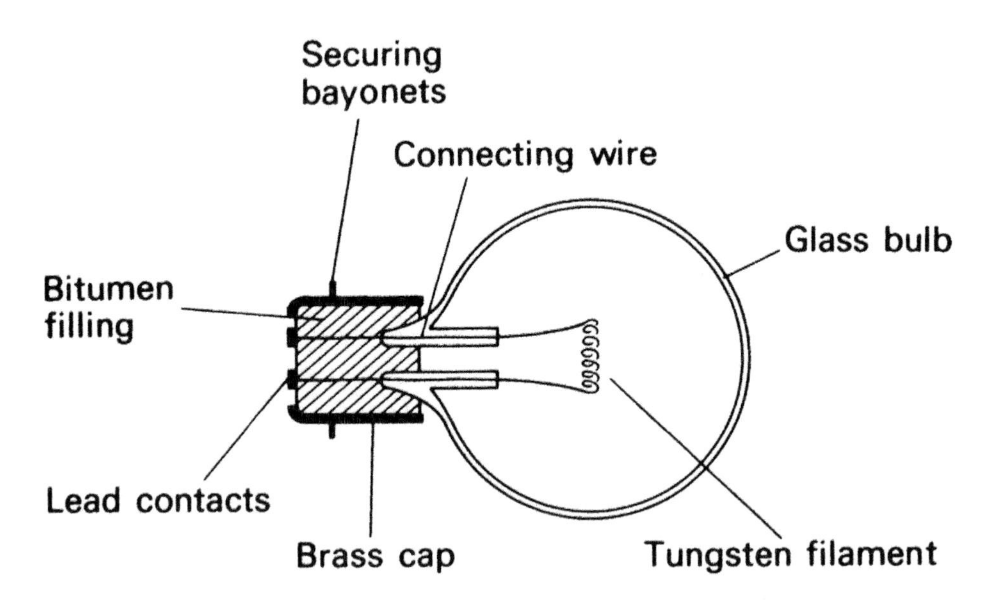

Figure 6.9 Pin bulb

Courtesy – an interior light that illuminates when the doors are opened, or in some cases when the remote locking doors are unlocked.

> **Racer note**
>
> Check operation of *all* lights before taking the vehicle to scrutineering.

Body electrics

Door locks – remote locking (central locking) uses electronically controlled solenoids to operate the mechanical locking mechanism.

Electric windows – these are usually operated by a small electric motor, which uses a pulley mechanism to move the window up and down; or turn a cable drive to do the same thing.

Electric mirrors – a small stepper motor moves the mirror within its housing to give accurate adjustment; also it is usual to have a heating element at the back of the mirror glass to clear condensation or frost.

ECU

Integrated circuit (IC), also called a chip. This device performs functions like the SIM card in a mobile phone. The functions are to make decisions based on information from sensors to control the ignition and fuelling.

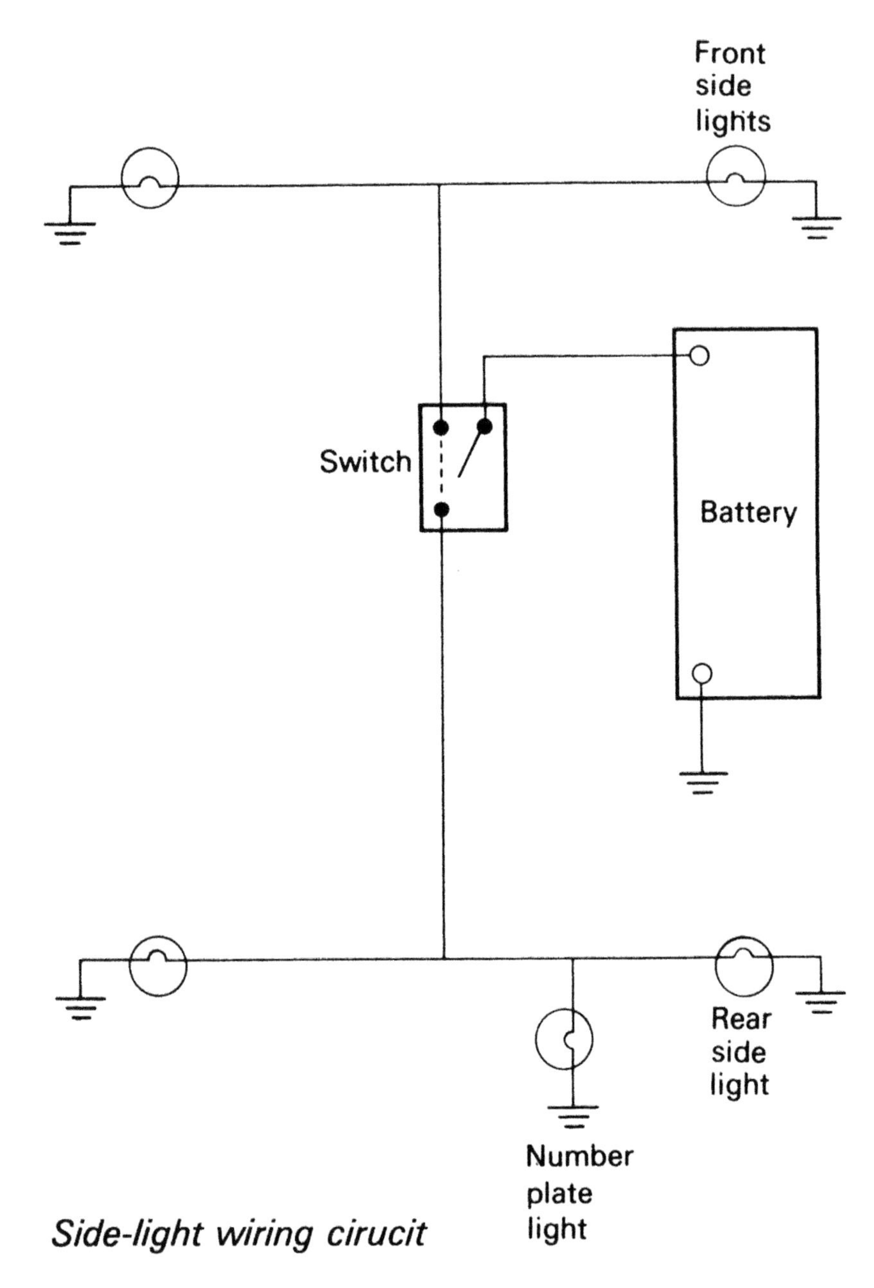

Figure 6.10 Sidelight circuit

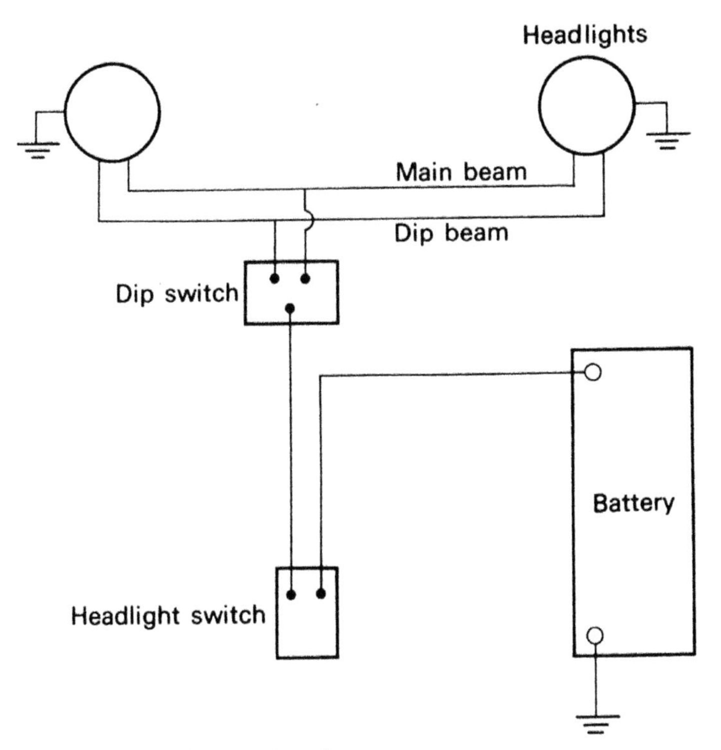

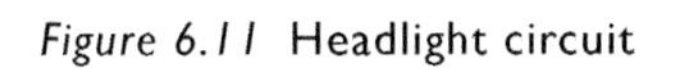

Figure 6.11 Headlight circuit

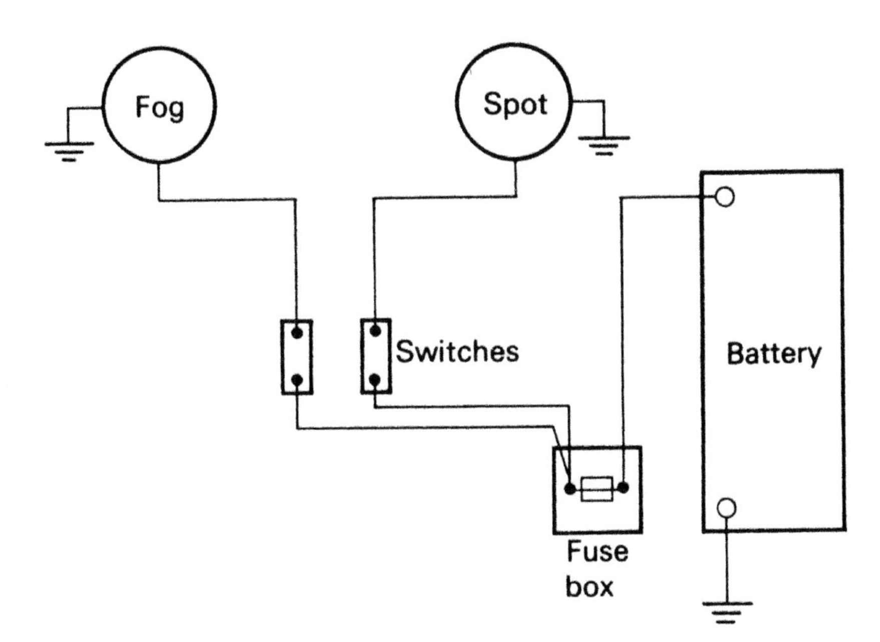

Figure 6.12 Spot/fog lamp circuit

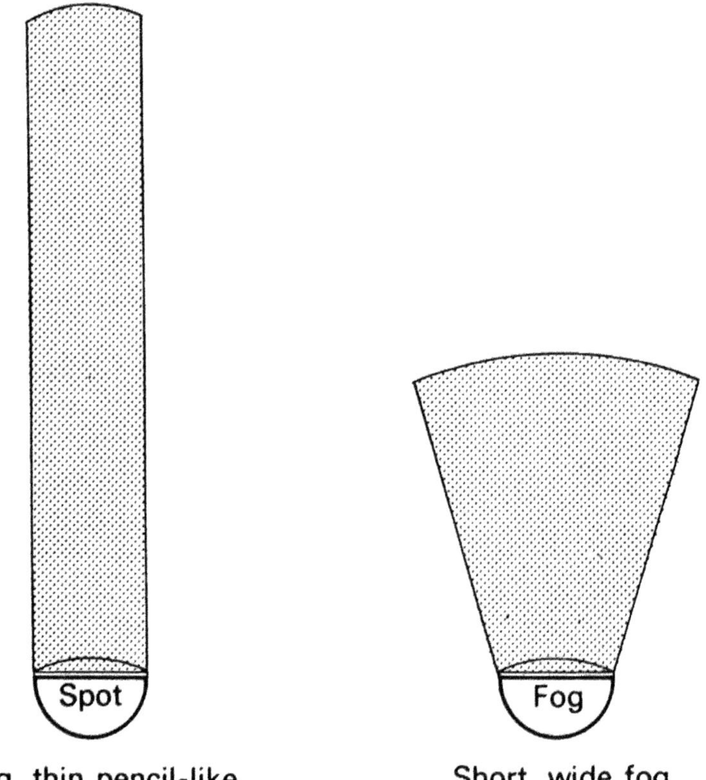

Figure 6.13 Spot/fog beam patterns

Electrical and electronic principles

The electrical and electronic systems on motorsport vehicles fall into a number of different and overlapping classifications. These are shown in Table 6.3.

Safety – working with electrical and electronic systems is not easy for many motorsport engineers, the reason being that you can't see electricity except when it sparks – and often that is too late, the damage is done. The following points should guide you safely when working on any vehicle:

- If any fault is suspected, the first job is to use the on-board diagnostics. If this shows a problem than you have the answer.
- Before you start to disconnect any wiring, or component, disconnect the battery (often this can be done by turning the isolator switch), then wait for two minutes for any stored capacitance to discharge. If you are working on a SRS (airbags) then you must be very

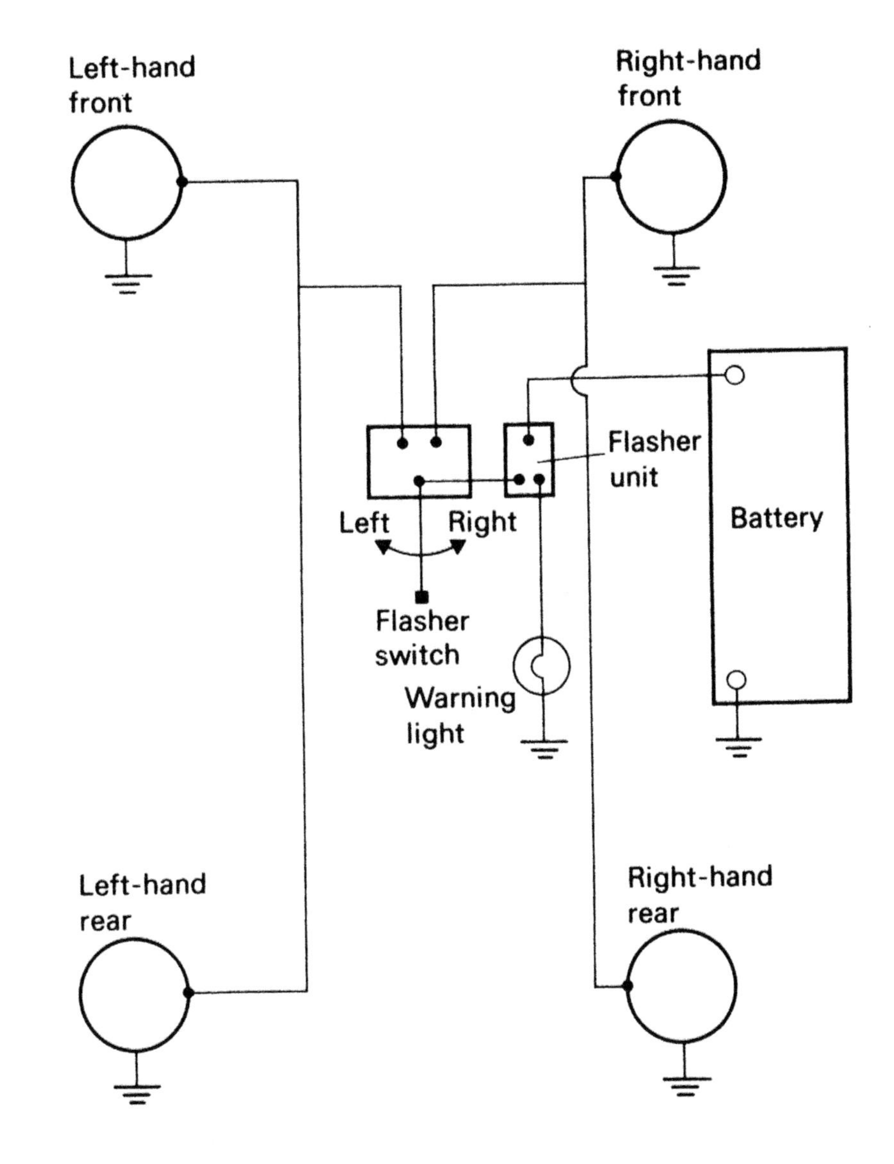

Figure 6.14 Flasher circuit

Table 6.3 Vehicle system identification

No	Type of vehicle	Basic lighting	Advanced lighting	Engine electronics	Chassis electronics	Transmission electronics	OBD	Data logging
1	F1		X	X	X	X	X	X
2	Historic	X						X
3	Le Mans		X	X	X	X	X	X
4	Motorcycle – trial	X		X				
5	Motorcycle – circuit			X	X		X	X
6	Rally car – Club		X	X	X		X	X
7	Rally car – WRC		X	X	X	X	X	X
8	Single seat	X		X			X	X
9	Sports		X	X	X	X	X	X
10	Touring car		X	X	X	X	X	X
11	Trial car	X		X			X	X

careful as you need to wait for two minutes for the current to leave the circuit or the airbag may discharge. Treat the anti-submarine seat belt pre-tensioner in the same way.

- Data logging systems should be downloaded onto your laptop before you carry out any work on the electrical system for fear of losing the information in them – rapid discharges of electricity or other electro-magnetic interference can cause a loss of stored memory.
- Do not use a mobile telephone near a memory device – this too can destroy the memory.
- Keep HT (ignition) devices separate from LT and memory devices to prevent interference.

General principles

In this section we'll recap on some of the electrical terms and formulae.

Volt (V) is the measure of electrical force (E); potential difference (PD) is measured in volts as is volt drop and electromotive force (EMF). Volt drop allows us to calculate the resistance of wiring or an electrical component.

The nominal voltage of most vehicle circuits is 12V. If the battery is fully charged it will read 14.2V, less than 12.6 and it is discharged and less than 10.5V and it probably has a fault. ECUs tend to run at 5V, dash instruments at 9V or 10V and some electronic components at 2V. HT spark is 5kV to 10kV on single coil systems, 40kV or more on systems using ballast resistors, electronic boosting or multi-coils.

Ampere (A) is the measurement of the quantity of electricity, called current (I). Batteries are rated in the number of amps they can provide over a time period (see battery). Ignition systems use about 0.5 amps, lamps about 5 amps each, the starter motor needs about 170 amps to turn over the engine from cold.

Resistance (R) is measured in Ohms; it is the slowing down of the flow of electricity. Ohm's Law states that current flow (I) is equal to the electrical energy (E) divided by the resistance (R):

$$I = E / R \left(\text{also written } I = V / R\right)$$

So that: $V = RI$

Electrical power (P) is given in watts (W), or kW (1,000W):

$$P = EI(\text{also written } P = IV)$$
$$\text{Watts} = \text{Volts} * \text{Amps}$$
$$\text{As} \quad V = RI$$
$$P = I(RI)$$

So it can also be written:

$$P = I^2R$$

In **series circuits** the resistance is additive:

$$\text{Total Resistance} = R1 + R2.................$$

The voltage across each resistor is proportional to the resistance, using Ohm's Law. The sum of the voltages across each resistor is the total circuit voltage:

Kirchhoff's voltage law

$$\text{Total Voltage} =$$

If one resistor in a series circuit fails, the entire circuit becomes dead. So you can carry out an Ohm's Law calculation for each resistor and add them up to get your total circuit voltage and resistance – occasionally useful if the bulbs can be unscrewed.

In **parallel circuits** adding resistors reduces the total resistance; but the voltage remains constant.

Kirchhoff's current law tells us that the current entering a junction equals the current leaving, so:

$$\text{Total Current} = I1 + I2..................$$
$$\text{As } I = V/R,$$
$$V/\text{Total } R = V/R1 + V/R2............$$

If we divided throughout by V:

$$1/\text{Total } R = 1/R1 + 1/R2.................$$

Parallel circuits are used for lighting circuits as all lamps will give equal brightness and if one fails the other will remain lit. The lamp that is out is obvious and so no need for calculations; however it can be useful with a voltmeter to trace faults on items like printed circuit board dashboards or rear lamp assemblies.

Figure 6.15 Race car rear light for bad weather use

Figure 6.16 Modified Mini dashboard

Figure 6.17 Home-made dashboard in Locust race car

Electrical and electronic components

Battery – on a race car these are normally gel batteries, that is the electrolyte is a gel (think of jelly) so that when inverted the electrolyte will not leak out. The plates are calcium based so that no topping up of the electrolyte is needed.

Once the electrolyte is added to the battery it will start to have a chemical effect on the plates; in other words the plates will start to deteriorate. Under these conditions batteries have a limited life, so they are usually stored without electrolyte. To make up a gel battery the electrolyte is added to the cells in a liquid form, then put on a slow charge for six or seven hours so that the liquid turns into gel. The charging must be done in one session with the battery off the vehicle using a variable voltage charger.

Gel batteries only have a limited ampere/hour capacity and discharge rate. That is, they are only suitable for powering electrical systems on the running vehicles, not starting. There are some exceptions, such as starting motorcycle engines. So to start a race car it is normal to have a battery on a trolley, which plugs into a socket on the vehicle, or a plug-in starter pack. The jack plug may be positioned left, right or rear depending on circuit direction. Most racing circuits are by convention driven in a clockwise direction, so the starter battery plug-in socket is placed on the right-hand side of the car to keep the mechanic safely out of the line of other cars, should a re-start be needed.

The best way to test a gel battery is by measuring the voltage. Do not use a high-rate discharge tester or other heavy load devices.

Rally cars and more robust race cars may use heavy duty lead-acid batteries. In the case of rally cars the battery must be mounted in a separate sealed battery box in case of a roll; this will prevent electrolyte from spilling on the driver and the navigator. On rear wheel drive cars it is a good idea to mount the battery near the rear axle to improve traction and weight distribution.

Ampere hour capacity (AmpHr) is the number of amps a battery will supply for one hour. However, the normal test is over a period of 20 hours. So a 40 AmpHr battery will supply 4 amps for 20 hours. For rally cars, use at least 1.5 times the recommended capacity – the extra weight does not outweigh the risk of a flat battery following stalling on a night stage.

Reserve capacity is the time in minutes that a battery will supply 25 amps at 25 °C, Typically this figure is about 60 minutes. This figure may replace the AmpHr capacity.

Cold cranking is a measure of the ability to provide a current of 170 amps at –18 °C (0 °F) at greater than 8.4 volt. There are SAE, BSI and DIN standards for this. Generally, if you take the race vehicle out for February testing and it won't turn the engine over, it is an internal battery problem. If before you refitted the battery you had slow-charged it overnight and cleaned and put petroleum jelly on the terminals, the plates have sulphated – that is the filler paste has turned solid and won't pass electricity. Don't waste testing time (and track costs) messing with the battery; fit a new one.

Sensors – are used to gather information for transmission to the various ECUs. See Table 6.4.

Table 6.4 Sensors and locations

No	*Sensor*	*Location*	*Comment*
1	acceleration/ deceleration	scuttle or chassis	for data logging
2	ambient temperature	front valance or mirror	may 'ping' warning on dashboard
3	cabin temperature	up to four, foot wells	with automatic temperature control
4	coolant temperature	radiator and /or block	to sense cold starting, or restarting with electronic engine management
5	deceleration	front valance and/or scuttle	two sensors with SRS
6	door open	'A' post	also for boot and bonnet
7	engine knock	engine block	piezo electric
8	engine speed		speed counter and TDC locator may be separate
9	fuel pressure	fuel system	high pressure side
10	gear selector	gear lever	or gearbox
11	induction vacuum	inlet manifold	may have pipe to ECU
12	lambda	exhaust front pipe	may have before and after catalytic converter – two for each pipe
13	lap time	instrument panel	signal from start finish line
14	lateral acceleration	scuttle or chassis	for data logging
15	oil pressure	main oil gallery	may also be on cooler
16	oil temperature	sump or oil cooler	may have thermostat on cooler circuit
17	outside light level	windscreen	
18	rain	windscreen	

No	Sensor	Location	Comment
19	road speed	gearbox output shaft	
20	seat belt	connector	
21	throttle position	throttle butterfly	
22	transponder	upper right of roll cage	for race control
23	turbo pressure	turbo outlet	
24	tyre pressure	valve and wheel arch	
25	vehicle direction	as road speed	
26	vehicle lights	lamps	
27	wheel speed	suspension upright brake disc	for ABS and tyre pressure

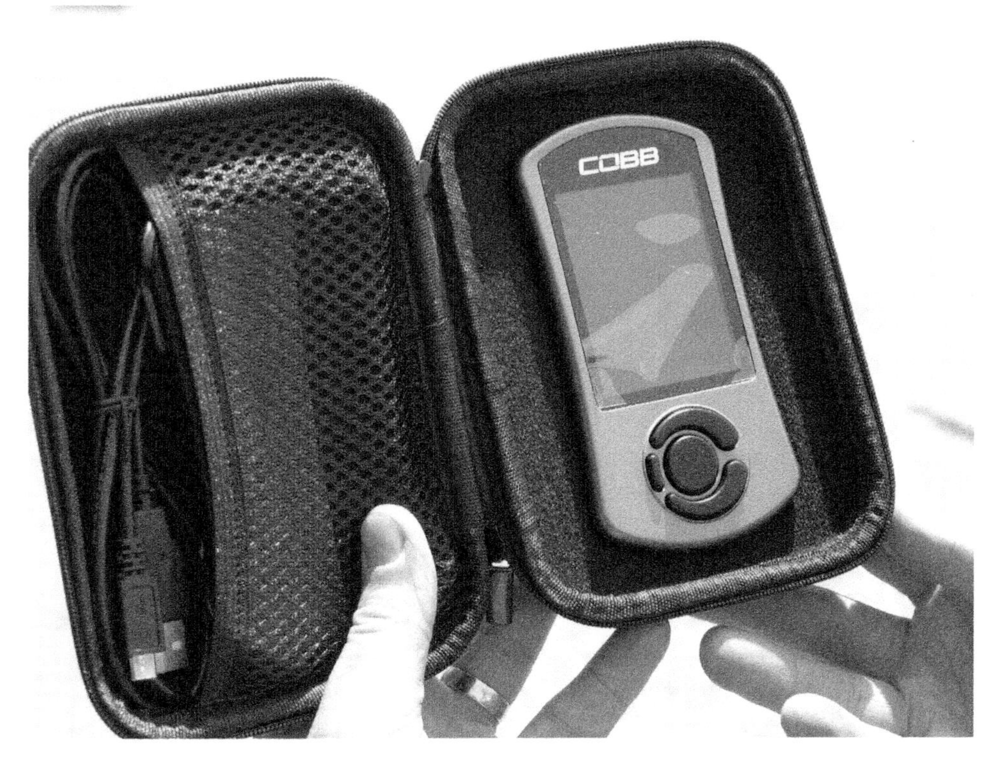

Figure 6.18 Cobb programmable dashboard mountable data logger

All sensor faults are found using the system diagnostic socket. As they are mainly solid state digital electronics they are robust and unlikely to fail unless abused or misused. The technician is advised only to test them following the manufacturer's methods and procedures. Do not apply any 12 V supply or use a multi-meter on ohm setting, as this is likely to damage the sensor. Operating voltage is in the region of 2–5V; with the current in low milliamps.

ECU – the electronic control unit is based on one or more integrated circuit (IC) devices (referred to as chips). The ICs are about 40 × 20mm (¾ inch by 1½ inch) and contain detailed circuits. There are a number of ICs in the ECU with appropriate circuits and a connector. The ECU typically operates at 5V.

There are a number of different types of ICs. Random access memory **(RAM)**, read only memory **(ROM)**, electronically programmed read only memory **(EPROM)** and electronically erasable read only memory **(EEPROM)**.

The principle of operation is that of logic. The IC gathers information from a number of sensors and makes an on or off decision. The basis of digital electronics is 0 or 1, off or on. The simplest system uses logic gates. For ignition and fuelling it is normal to use look-up tables. In this case the EPROM is programmed to give a specific output for given inputs rather than making a series of on-off decisions. There may be ECUs for:

- Fuel only
- Ignition only
- Ignition and fuel (engine)
- Transmission
- Braking only
- Chassis – braking and other system controls
- Data logging

Think safe

Only test ECUs using the diagnostic socket.

Actuators – are components which carry out a mechanical function in response to an electrical signal. As cars increasingly move to drive by wire; the use of actuators increases (see Table 6.5).

Table 6.5 Actuators

No	*Actuator*	*Location*	*Type*	*Comment*
1	ABS	scuttle or chassis	spool valves	from ECU
2	choke control	side of carburetter	stepper motor	
3	cruise control	throttle control and gear change		constant speed with maximum fuel economy
4	fuel cut-off	fuel line	inertia switch	
5	gear change	gearbox	solenoid controlled valves	
6	solenoid	scuttle or chassis	single movement of armature	inertia type starter motor
7	solenoid	starter motor body	two stage movement	pre-engaged starter motor

No	Actuator	Location	Type	Comment
8	SRS	airbag and anti-submarine device	pyrotechnic	
9	stepper motor	carburetter, light position	motor moves in 15 degree steps (or as programmed)	precise control of shaft movement
10	temperature control	air conditioning controls	solenoid or stepper motor control of valves	
11	throttle movement	throttle butterfly	stepper motor	

Actuator faults are tested through the system diagnostic socket. As they work at low voltages with low currents do not test by applying battery voltage unless the manufacturer's procedure advises you to do this.

Racers note

When on-event you may not have access to a full range of fault code readers. Experience should suggest to you which sensors and actuators are likely to fail, so it is a good idea to carry spares that can be substituted quickly.

Cables and wiring

The purpose of the electrical cables is to get electrical energy from one part of the vehicle to another. On a competition vehicle this need to be done with the minimum amount of weight and the least loss of power.

Racers note

The most common cause of electrical system failure is a dry joint – the cable does not join the connector properly and fails to pass a current.

A number of different types of cables and connectors are used. The simplest and cheapest connectors are snapping connect plastic ones – used on production vehicles with either copper or aluminium cable. The cables may be soldered to the connectors or crimped on the end. Soldering gives lower electrical resistance and can be used with either copper or aluminium cable. Crimping is cheaper and quicker; but likely to separate if subjected to vibrations and, if wet, corrosion between the cable and the connector may prevent the flow of current.

F1 cars use small screw-together connectors to give weatherproof sealing and low risk of separating. The cable is very low resistance so that a complete wiring loom is about the same diameter as one large cable on a standard salon. The cable and connectors may be gold coated. Gold is a noble metal and will not corrode under most conditions as well as being the best electrical conductor. Formula 1 and similar professional wiring harnesses with connectors may cost the same as a complete Formula Ford car.

Racer note

Noble metals include: gold, silver, platinum, iridium and rhodium. You can wear these as jewellery without your skin going green!

To make up a **loom** for a racing car the best plan is to set out the components, or dummy components, on a clean surface and measure the relevant distances; remember changes in height too. You can then make up a CAD drawing from your measurements. This drawing can be printed out full size as a template. Or, using sheets of plywood make a dummy of the chassis and place pins in it where the cables are going; again remember the changes in height. Then using cable off reels, thread it out and bind it into looms. For this you can use a wide tape or a shrink-wrap material. Make the loom as small and tight as possible. Leave free ends sufficient for termination.

The choice of cables will be dictated by the current that they need to carry – generally apply a safety factor of two. Colours should follow a typical colour code. For example: supply brown, post-ignition white with tracer colour, earth black, auxiliary green with tracer colour, lighting red and or blue. Data logging – follow the diagram supplied with the data logging kit.

Multiplexing is a system of transmitting data in a shared environment, at the same time connecting up the components in a circuit. In computer terminology it is called a local area network (LAN). There are several different types of multiplexing systems, as you would expect, from different manufacturers. Examples are:

- Controller area network – CAN-DATABUS (Bosch)
- Body electronics area network – BEAN (Toyota/Denso)
- Universal asynchronous receiver transmitter – UART
- Audio visual communications – local area network – AVC-LAN

Most systems work with a twisted pair of cables (like telephone systems) to reduce the possibility of interference. The working voltage is generally 5V with a baud rate of 10–100 Kbits/second. All the ECUs are connected to form a daisy chain. Sensors and actuators are connected to the network.

Fault finding is by fault code reader in the first instance. Damaged cabling will not show a fault code and will be found by inspection, usually after a serious incident or an off. Looms cannot be repaired and must be replaced in complete sections – they are very expensive.

Electronic brake control

Anti-lock braking systems (ABS) and anti-skid control systems (electronic stability control) work through the same ECU. The ECU constantly monitors the speed of each wheel through the wheel speed sensors. If under braking, one wheel slows more than the other, then it is likely that the greater slowing wheel is on ice or other slippery surface. If it were to lock then the vehicle would skid. Within about 2 milliseconds (0.002s), faster than you or I can blink, the ECU sends a signal to the ABS actuator telling it to reduce the brake fluid pressure to that wheel; this prevents the wheel from skidding. This activity is continuous all the time that the brake pedal is depressed. You can feel the pedal going up and down as this is happening; the dashboard warning light may illuminate. ABS does not work on soft snow or gravel, as there is no positive traction with these surfaces. In these instances you should switch the ABS off. **Electronic stability control** systems prevent wheel spin by doing exactly the same as the ABS; but in the opposite way. That is, if one wheel spins more than the other the brake is applied to the spinning wheel to slow it down. This also works when cornering to prevent the vehicle from rolling over.

Ignition system

Since about 1990 all production cars have been fitted with electronic ignition systems. On classic cars before this date you may find mechanical contact breaker points in a distributor with coil ignition. In the 1960s DIY electronic ignition became available and was used extensively on motorsport vehicles.

Racer note

The conventional cb points and coil ignition was invented by Dr Kettering working on the racing Cadillac of 1905 – becoming almost universal for nearly 90 years.

Spark plugs – race plugs differ from standard road plugs in that they will withstand much higher temperatures and pressures for much longer periods. The spark plug in a race engine must be capable of providing a spark to ignite the mixture at almost double the pressure of a road vehicle engine, twice as many times per minute and hence about 400 °C hotter.

Before replacing any spark plugs check:

- Reach
- Diameter
- Thread type
- Seating or sealing type
- Socket fit
- Electrode type
- Heat range

Racer note

Engine speed: standard 8,000 rpm; race over 18,000 rpm

Compression ratio: standard 9:1; race 16:1

Ignition fault finding is mainly down to the fault code reader. However, if you are **tuning** a competition engine you may wish to investigate a few of the areas and make adjustments. Some of the adjustments may mean altering the ignition IC programme – see **chipping**.

- Plug temperature – plugs will run hot if the mixture is too weak; you can spot this if you look at the insulator core. If it is too hot it will look a very light brown, or white, and look burnt. Dark, or black, means too rich or oily. Look out for bits stuck too it too. Retarded timing also causes overheating.
- Burn time – you can get this from your engine test bench oscilloscope or digital test kit.
- Dwell angle.
- HT value – most important is that they should all be equal. If one is too high, or too low, suspect a plug fault too.
- Ignition timing – especially at high revs.

Fuelling system

From the point of view of electronics, fuel systems have had a number of different systems up to the current electronic fuel injection (EFI). As a motorsport technician you will come into contact with many old systems as well as the current ones. Looking at some of the more common ones involving electronics:

Electronic carburetter – developed by SU, uses a stepper motor to move the jet to control the mixture for cold starting. The mechanism between the stepper motor and the jet is subject to wear. Many were converted to a mechanical choke.

Single point fuel injection – this is a step up from the carburetter. It is fully electronically controlled through the engine ECU. It has two fuel outlets, one for normal running and one that cuts in on cold start. Maintenance involves mainly changing the air filter and the petrol filter. However, the throttle potentiometers are prone to failure – this causes intermittent surging. The fault can easily be verified by the fault code reader.

Electronic fuel injection (EFI) – by this we mean multi-point, either individually supplied by fuel, or **common rail** which is almost universal. The pressure pump sends fuel to the injectors at a pressure of between 4 and 8 bar (60–110 psi) depending on the vehicle and the regulator setting. The amount of fuel going into the cylinder depends on this pressure (which is pre-set) and the amount of time that the injectors are open. The longer the injectors are open the greater the amount of fuel that will be injected. The ECU controls the open time depending on engine requirements. These include:

- Load conditions
- Engine speed
- Engine temperature
- Ambient temperature
- Lambda feed back
- Throttle position

This information is fed to the ECU by the sensors; see **ECU** and **sensors**.

The injectors are held open by a **duty cycle**, not a continuous current. The time period is between about 4 and 8 milliseconds (0.004–0.008s). Doubling the time at a constant pressure will double the amount of fuel injected.

The flow of fuel is also controlled by the size of the injector orifice. Standard injectors are often referred to as 'grey' because of their colour code. For high-performance engines 'green' injectors are used – these have a bigger orifice and can therefore deliver more fuel.

Fault finding on EFI systems should start with the fault code reader. The air filter and the fuel filter should be changed regularly. The injectors are tested by removing them and observing their spray pattern on a test rig. The main problems are usually related to sensor failure – typically throttle position sensor because it moves and wears, lambda sensor because it gets hot and deteriorates.

Racer note

Ultrasonic cleaning using a system such as ASNU will save the cost of new injectors.

Modifications

Chipping

The operation of the ignition and the fuelling systems depends on the **mapping** of the ICs (chips). For race and rally cars different ICs are required to suit the application. Fitting a specially set-up IC to the ECU is called **chipping**. It is possible to both buy ready set-up ICs for most popular applications (Ford, BMW, Peugeot, Vauxhall and similar) and to set up the chip for an individual application. When a chip is changed it is normal to check the power before and after on a rolling road dynamometer (dyno). If an individual chip is being set up then expect to spend two or more days on the dyno with a lot of test equipment.

The chip is set as a contour map. Parameters for engine speed on the x axis and load on the y axis are converted into ignition setting and fuelling settings on the z axis. This is read on a laptop computer as a look-up table. Each square on the table has a value. Reading the table rows from left to right gives the figures for increasing load, going down the columns the engine speed is increasing. The figures in the fuelling table correspond to the opening times of the injectors – figures such as 0.004s (4 milliseconds). With a suitable interface between the laptop and the chip, and access software, you can alter these figures and check the changes to the power output on the dyno. As you can see, if the table has 40 rows and 20 columns that is 800 individual boxes to check.

Lighting

Headlamps and rear lamps used in Europe must comply with EU regulations and carry the 'e' or 'E' mark. In the UK they must dip left; in continental Europe they must dip right.

Racer note

To use some UK cars at night in continental Europe it is usually necessary to place specially shaped black tape over the lamps to prevent dazzle to the oncoming cars.

All complying lamps – head, tail, indicators, stop and fog – carry a set of digits as well as the EU mark. These digits are a code that indicates the position and type of lamp.

LED lamps have now become popular. Their slow introduction was because of Construction and Use Regulations, which required a light filament. This regulation was withdrawn. LEDs have the advantages of:

- Using less power
- Being quicker in reaction
- Being more robust and longer lasting
- Allowing the use of different shapes

Gas discharge lamps (GDL) – this is a form of arc light. The lamp is about the same size as a normal halogen one made of quartz glass. The high voltage arc is provided by an electronic control unit from the 12V system. It has become popular with scooter riders as it uses only a low current. The amount of light given is about three times that of a tungsten bulb for the same current usage.

Ultra violet or **blue lights** are sometimes used as one pair on multi-lamped vehicles. UV light will light up road marking and be almost invisible to oncoming traffic. They do not give back reflection in fog and snow, therefore appearing to see through them. As UV rays can be dangerous, a two-layer filter is fitted to remove UVB and UVC – the dangerous ones, leaving UVA. The usual light source is a GDL.

Power supply

Alternator – rally cars and off-road vehicles usually need larger output alternators to power the additional electrical equipment.

Battery – rally and off-road vehicles are usually fitted with larger capacity batteries to provide a bigger reserve of electrical power for the additional electrical equipment. Often the electrical equipment will be draining the battery when the engine is not running, or running slowly so that the charging rate is below the drain of the electrical equipment.

Where motorsport vehicles are likely to roll, that means most of them, then the use of gel batteries is recommended so that there will not be a loss of electrolyte. Also ensure that the battery box complies with the appropriate competition regulations – normally this means some sort of fully enclosed box with a breather to the outside of the vehicle.

Auxiliary systems

Motorsport vehicles may be fitted with a variety of auxiliary systems, typically data logging, navigational aids and communication systems. When fitting these ensure that the instructions are followed in detail and that any legal requirements are fully met.

Lighting

Additional lighting when fitted must comply with legal and competition regulation requirements.

Figure 6.19 Door-mounted speakers

Questions and skills

1 Find any faulty electrical component and open it up to see what is inside, make notes and draw a diagram to show all the parts.
2 Under the supervision of one of your tutors, take voltage, current and resistance reading of a variety of components – note these for future reference.
3 Obtain the circuit diagram for a vehicle of your choice and then draw it out as a large-scale plan view as the components appear on the vehicle.
4 Choose one of the systems that is controlled by an ECU on a vehicle of your choice, then using a workshop manual to help you draw out a chart to show the connections.
5 Investigate what is meant by a look-up table and how this can be used with the ignition system.

Chapter 7

Body and chassis

On most popular cars the body and the chassis are made one and the same. Motorsport vehicles may use either separate chassis or a combined body/chassis arrangement. The chassis is the part to which the engine, gearbox, suspension and other components are attached. The **body** is the covering for the components, the passengers and the load. The **chassis** is load bearing, being made from strong steel. The body does not carry a load and may be made from aluminium alloy, or some form of plastic, as well as the more usual steel.

Figure 7.1 Cross-bar strengthening tops of suspension struts

Key points

- Most vehicles are front wheel drive layout; alternative layouts are: conventional, mid-engined and rear-engined.
- The chassis is load bearing and must be free of corrosion.
- Jacking points and seat belt mounting points are specially reinforced.
- Airbags must only be handled following a special procedure.
- SIPS and crumple zones are sometimes added to give extra passenger protection.

Vehicle layout

By vehicle layout we mean the position of the engine and gearbox on the chassis in relation to the driving wheels.

Conventional layout means that the engine is mounted at the front with a gearbox behind it and a propeller shaft takes the power to a rear axle so that the rear wheels are driven.

Front wheel drive (FWD) is the most common. The engine and the gearbox are mounted at the front of the car and short drive shafts take the power to the front wheels.

Rear-engined vehicles drive the rear wheels from a rear mounted engine (RWD).

Figure 7.2 KTM front end

Figure 7.3 KTM front end

Figure 7.4 Low-cost kit car in racing trim

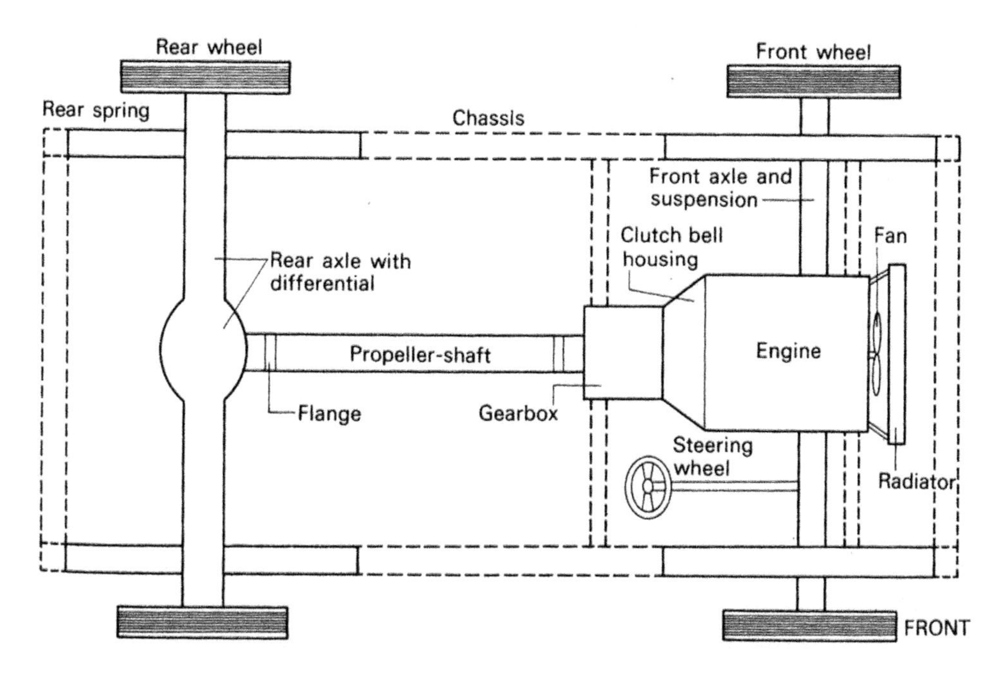

Figure 7.5 Conventional drive layout

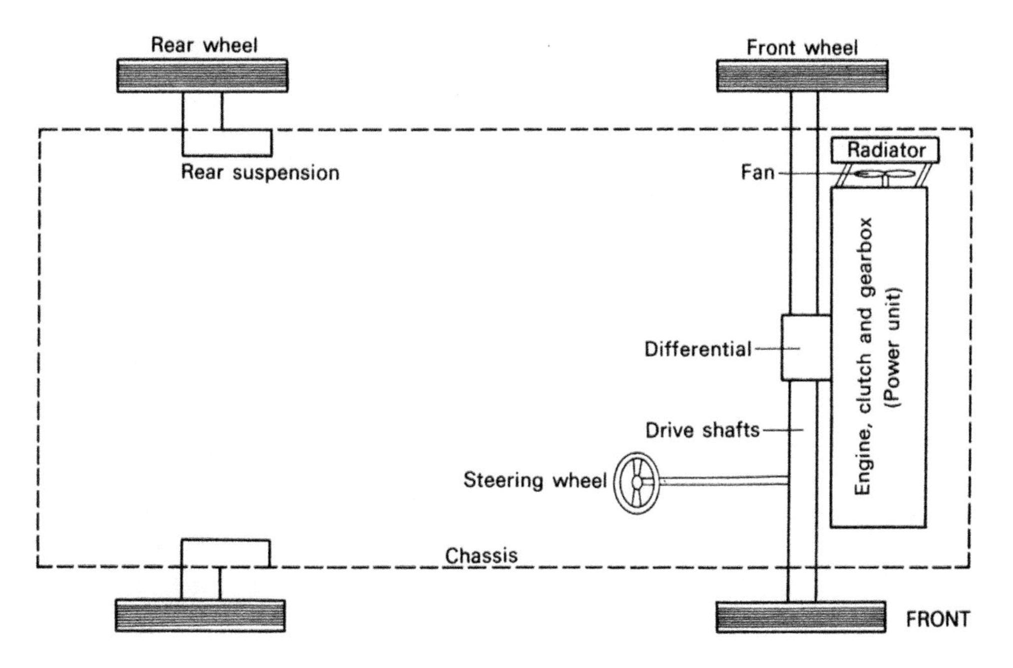

Figure 7.6 Front wheel drive layout

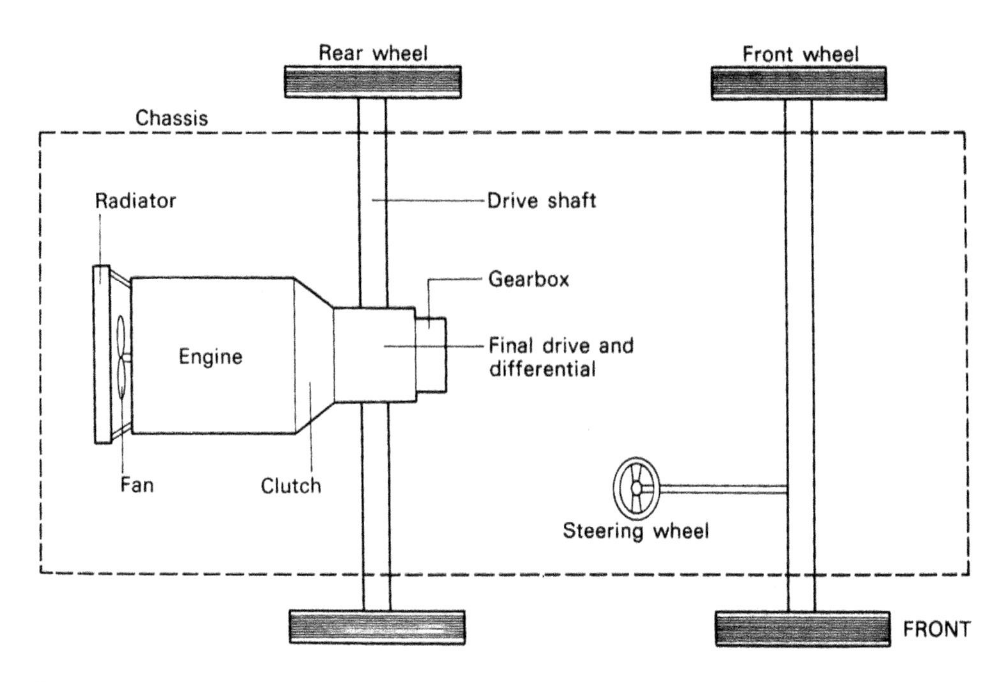

Figure 7.7 Rear wheel drive layout

Increasingly **mid-engine** set-ups are becoming more popular in high-performance cars. The engine and the gearbox are mounted in the middle of the vehicle and the rear wheels are driven.

For off-road use **four-wheel drive** gives better grip, this can be with front- or mid-engined layouts.

Nomenclature

4x4 means that the vehicle has a total of four wheels and that it is driven by all the four wheels. 4x4 vehicles may also be called all wheel drive (AWD), off-road, or all-terrain vehicle

Advantages and disadvantages of different vehicle layouts

Each type of layout has certain advantages and disadvantages when compared to the others (see Table 7.1).

Chassis

The chassis is the **load-bearing** part of the vehicle. That is to say it carries the weight of the **load** and the **passengers** and locates the engine, transmission, steering and suspension. On

Table 7.1 Comparison of typical vehicle layouts

Type of Layout	*Advantages*	*Disadvantages*
FWD	• good traction • maximum passenger space	• noise from transmission • complicated drive shafts
Conventional	• easy to repair • high level of safety	• passenger space limited by drive tunnel
Mid-engine	• best weight distribution • good for 4x4 transmission	• limited passenger space • difficult to access engine
RWD	• engine noise behind the driver • good traction	• long control cable/rods needed • Luggage space limited – between front wheels and location of fuel tank is difficult

most popular cars the chassis and the body are one and the same; but on specialised cars and goods vehicles separate chassis are used. There are three main types of chassis; these are **ladder chassis**, **cruciform chassis** and **backbone chassis**.

Racer note

If you look underneath a popular car you will see square sections, like little box girders, connecting the main suspension and transmission parts. These are called the chassis sections; although the vehicle does not have a separate chassis these give extra strength. It is important that the chassis sections are in good condition for the car to be safe.

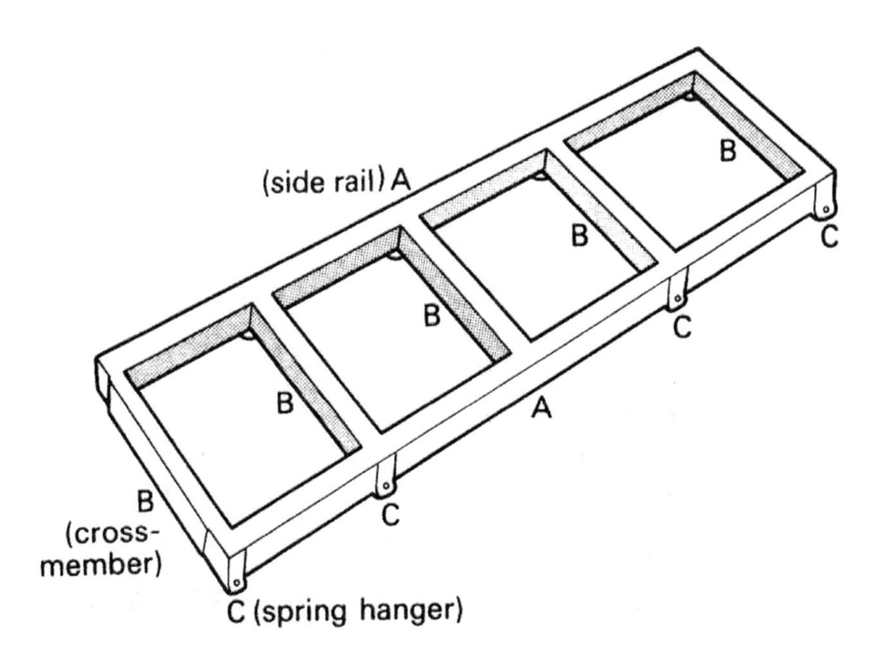

Figure 7.8 Ladder chassis

Ladder chassis – it is called this because of its shape, it looks vaguely like a builder's ladder. It has two side rails connected by cross-members. Ladder chassis are used on trucks and buses as well as vehicles like Land Rovers and some kit cars. The rails, which run lengthwise, are called **longitudinal** members; those that go across the vehicle are **transverse** members. Generally the suspension is attached to the longitudinal rails and the engine will sit between these rails. The gearbox tail housing is attached to a transverse member.

Cruciform chassis – it is cross-shaped in the middle to give resistance to twisting. This type of chassis is used on some old rare sports cars such as Lea Francis.

Backbone chassis – it looks roughly like a person's skeleton – a backbone with arms and legs. Lotus and Mazda use this design on their small sports cars (Elise and MX5); the propeller shaft can fit through the middle of the hollow backbone section.

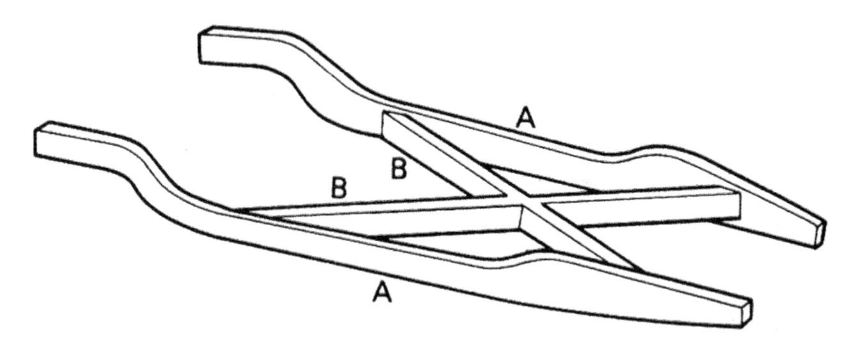

Figure 7.9 Cruciform chassis

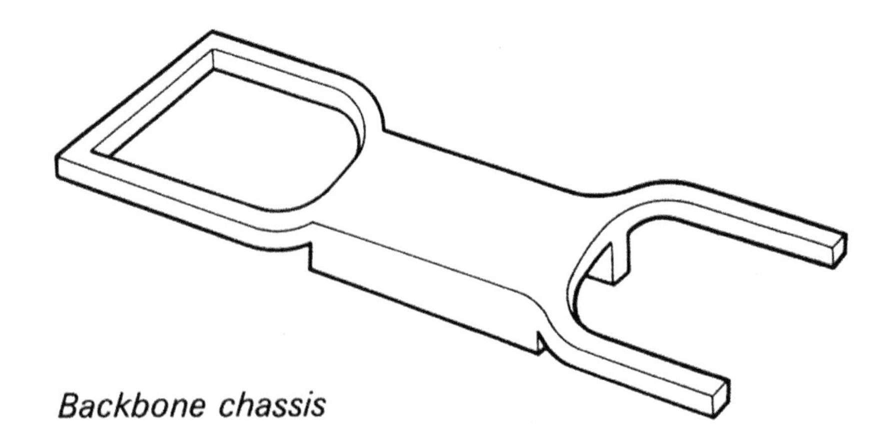

Figure 7.10 Backbone chassis

Integral construction

Integral construction is also known as **unitary construction or monocoque**. This is when the chassis and the body are made as one integral unit; that is as one piece from parts welded together, not as a separate body and chassis. These are often referred to as body shells. The floor, the sills, the roof and the quarter panels are all spot welded together to form an assembly to which the engine and the running gear are attached. The integral body/chassis is much lighter than using separate components; it is also very strong, especially in resistance to twisting, which make the car feel good to drive.

Tubs

Race car body and chassis units are often constructed as tubs. That is they are made from composite material – usually carbon fibre is laid over a honeycomb section board – to form an integral unit. The suspension and engine mounting points are built into the tub.

Nomenclature

Integral, unitary and monocoque, in relation to body/chassis, mean that it is made as one piece – that is the body and chassis is a two in one. The word tub is an abbreviation for bath tub, in other words the open racing car tub resembles a bath tub, or a hot tub.

Birdcages

Birdcage construction is where the chassis is made from tubular steel so that the engine, transmission and drive sit inside it. Then a non-stressed body shell is fitted over the birdcage to cover it.

Nomenclature

Although tubular usually refers to round sections, it can be round or square, as long as it is hollow.

Chassis sections

Different shapes of chassis sections are used for different purposes. The **round section** is the strongest, closely followed by the **square section**. The square section has the advantage of being easy to screw or weld component to. The round section is used for bicycle and motorcycle frames and on specialist racing cars like Thrust 2. Where the loads are limited to one direction the open 'U' section is idea; this is often made into a top hat section by the addition of extra bends, and welded to panels, such as the floor, to form a closed square section. The 'L' section is usually limited to load-carrying outriggers and trailer frames. The 'I' beam section is used for the longitudinal side rails on LGV and trailer chassis.

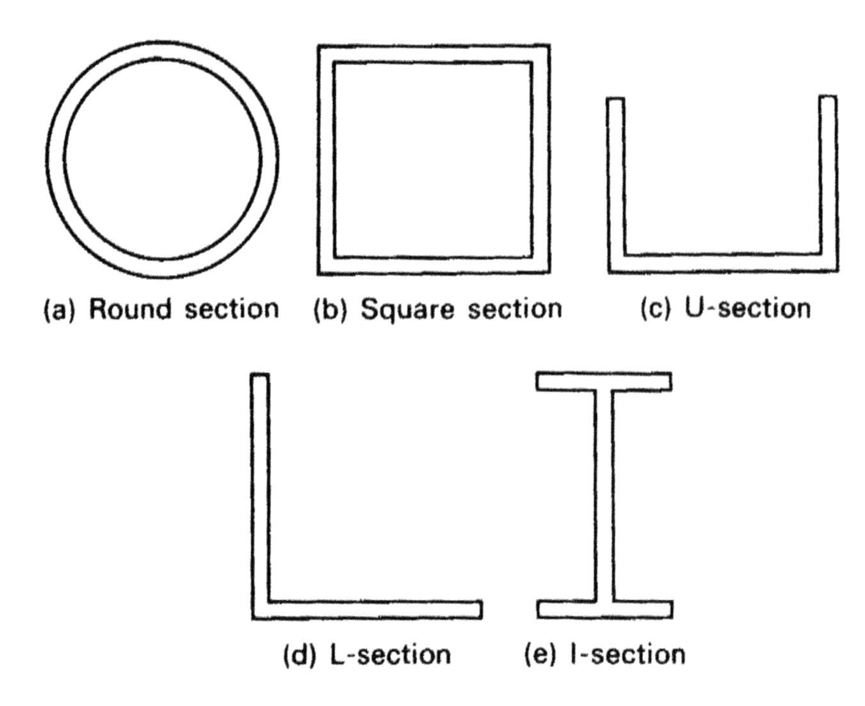

Figure 7.11 Chassis sections

Racer note

The tubular chassis of land speed record car Thrust 2 is made from Reynolds 531 alloy steel tube, just like good quality steel bicycle frames and some motorcycle frames.

Body/chassis mountings

The traditional method of bodybuilding, which is still used on some specialist cars like the Morgan, is to use sheets of steel or aluminium, on a frame of hard wood such as ash. The frame is attached to the chassis. Bodies may be made from a variety of materials, including steel, aluminium, GRP, carbon fibre, and a range of other plastics materials. Modern separate bodies do not usually have a wood, or any other frame, so the body is mounted to the chassis using some form of rubber mounting. The mounting may be of the spacer and filler type, or a rubber block bonded to two plates.

Sub-frames

To reduce **noise vibration and harshness** (NVH) and also to speed the vehicle assembly process, sub-frames are often used. The sub-frame typically may have the power unit and suspension attached to it, the sub-frame is then attached to the body shell using rubber mountings.

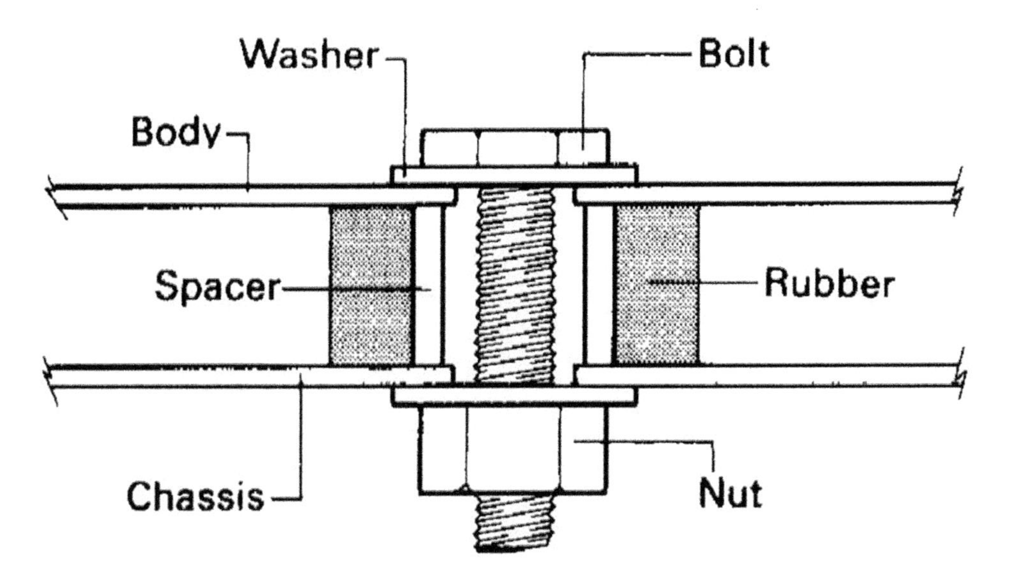

Figure 7.12 Body – chassis mountings

Body types

There are many different types of bodies in use; it is important to be able to identify them quickly and accurately. This next section looks at some typical styles.

Saloon – four doors, four or five seats, and a separate boot. This is sometimes called three-box construction, as from a side view you can see the passenger compartment, which looks like a box shape, with the engine compartment box to the front and the boot box at the rear. Saloons are very comfortable for carrying people separate from their luggage. Good examples are the Mercedes Benz 180 or 200 saloons.

Hatchback – many modern vehicle are hatchbacks, these combine the passenger comfort of the saloon with the load-carrying ability of an estate car. The rear seats fold down to enable the load to be carried. They have a shorter wheelbase than the estate car, making them easy to manoeuvre in traffic. The rear door opens upwards in the form of a hatch. They usually have three doors, that is the two at the side plus the hatch door. The VW Golf was the first popular hatchback.

Estate car – also called a **station wagon**, a **traveller** and a **shooting brake**. Four side doors, plus either an opening tail gate, or a pair of rear doors. Usually slightly longer than the saloon version of the same model. Ford models such as the Mondeo and Focus are available as estate cars.

Coupe – a two-door, two-seat car with a sloping roof line. Audi TT is a typical example.

Cabriolet – a car with four seats, two doors and a folding hood. Sometimes referred to as a convertible as they convert from closed to open. The old Ford Escort Cabriolet is a good example.

Fastback – a long, low sloping back with an upward opening hatch door. Aston Martin were the first to popularise this style.

Sports car – this is a two-door two-seat convertible. If there is a small space between the rear of the seats and the boot, then this is a **roadster**. This term is also misused for sports saloons and WRC/WRX cars. A sports saloon is a tuned version of a standard saloon. WRC (world rally car) and WRX (world rallycross) are cars made for special motorsport events as are Evolution versions. MG TF and Mazda MX5 are examples of sports cars.

Nomenclature

True sports cars are open two-seaters, like the Caterham 7; but in common language a BMW M3 may be referred to as a sports car.

Limousine – may have four, or six doors, and six or seven seats. Extended Limos are usually based on American Limousines, which have their wheelbase extended even more to add another row of seats, a television lounge, or a small swimming pool.

MPV – a multi-purpose vehicle or people carrier. Door arrangements vary; they may include a sliding door and an entry rear door. Usually there are three rows of seats giving a seating capacity of seven or eight, plus some space for luggage in the same area.

Minivan – a smaller version of the MPV, usually seating a maximum of six people.

Van – a vehicle for carrying two people and a load. There are many different types of vans. Those based on the shape of a car are called **derived vans**; vans which are box shaped, are called **panel vans**; the ones which have a load-carrying space above the driver are called **Luton vans.**

Off-road – for off-road use the body must be mounted high above the road to give a large amount of ground clearance; and have plenty of clearance between the large wheels and the wheel arches to reduce the risk of fouling by mud or snow. Off-road vehicles are usually four-wheel drive, that is 4x4 or AWD

Jacking points

Jacking points are specially **reinforced** areas of the underfloor area of the body, which can be used to raise the vehicle either with the tool-kit jack or a garage jack. This is needed to allow a wheel to be changed or other maintenance – for example the changing of brake discs – to be carried out. In the workshop the jacking points are often used to support the vehicle on a wheel-free device.

Safety note

The owner's handbook, or the workshop manual, will usually show you where to find the jacking points. Under no circumstances should you ever jack a car up under the engine, the gearbox, or the fuel tank; this is most likely to cause serious damage. You should never work underneath a car which is supported on a *hydraulic jack* or a *toolkit jack*; always use axle stands for safety.

Seats

There is a wide range of seats available. For the driver and the front seat passenger there is usually a pair of single seats; if these have a wrap-round shape they are referred to as bucket seats. The rear seat is usually a bench seat. Seats are attached to the vehicle floor using sub-frames. Of the two parts of the seat, the part which is normally sat on is properly called the cushion; and the backrest, or vertical part, is called the squab.

Racer note

The seat in a racing car is often moulded to match the contours of the driver's body. DIY kits are available if you wish to do this yourself on a limited budget.

Seat belts

By law, the driver and all passengers (with a few exceptions) must use seat belts. The most popular seat belts are three-point mounting. Front seat belts usually have an inertia reel devices to make them self-adjusting. For race and rally cars four- or five-point mounting seat harnesses are used.

Racer note

Correctly fitting harnesses are especially needed in race and rally cars. One-off hand-made harnesses are available from a small number of suppliers.

Corrosion prevention

The load-carrying part of car bodies are made from steel, if this is not protected from rain, or otherwise allowed to get water onto the bare metal, then it will rust. Rust is a soft brown coating, which will eventually work its way through the metal from one side to the other. Aluminium alloy components are also affected by water; these components turn into a white powder. The term corrosion applies to both steel and aluminium alloy.

To prevent corrosion, the surfaces of exposed metal components are treated with one of a number of materials (see Table 7.2).

Table 7.2 Corrosion prevention

Type of prevention	*Coating material*	*Example of use*
Painting	Acrylic	Visible body panels
Undersealing	Rubber or wax	Underbody
Chromium plating	Chromium or nickel	Bright trim
Surface coating	Rubber or plastic	Rubbing strips
Galvanising	Zinc	Suspension mountings

Interior trim

The interior of the vehicle is trimmed to make it comfortable to the touch for the driver and the passengers, and to reduce the level of noise. A lot of different materials are used. Top of the range cars use natural materials such as **leather** for the seats, **wood** for the facia and door capping, and wool carpets. Where loads are carried, hard-wearing, resistant materials are used, such as moulded plastic and nylon carpets. Rally and race cars are not usually trimmed.

SIPS

SIPS is the abbreviation for side impact protection system. SIPS is a set of reinforcing bars added to the sidedoors and vehicle side panels to protect the passengers if they are run into sideways. That is the type of accident that often happens at road junctions, when a vehicle going across a junction is hit in the side by another vehicle reaching the junction from another road. This type of accident is sometimes referred to as *T-boning*.

Crumple zone

The crumple zone is the part of the front wings that are designed to compress like springs during a head-on collision. The damage to the front wings protects the passengers by softening the crash.

Airbags

Airbags are like big floppy cloth balloons that are filled up during a crash to prevent the driver and passengers from hurting themselves on the steering wheel, windscreen and other hard parts of the car. The airbag is inflated with a small pyrotechnic device, in other words an explosive, in a time of about 2 milliseconds (0.002 second), faster than you can blink.

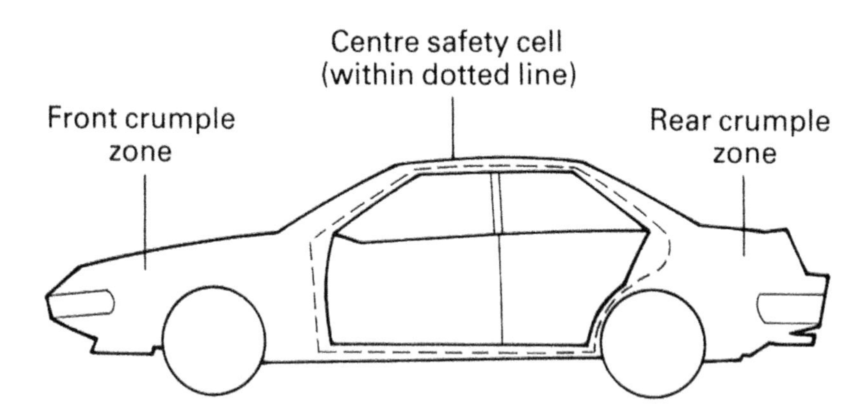

Figure 7.13 Crumple zones

Figure 7.14 V-Storm WR3 three-seat layout

Figure 7.15 V-Storm minimalist style

Figure 7.16 V-Storm aircraft-style interior

Figure 7.17 Built-in fire extinguisher

Figure 7.18 Roll cage

Figure 7.19 Vintage race car built using American Lockheed P38 bomber plane fuel belly tank

You must follow safety procedures when working with, carrying or just storing airbags. The pyrotechnic device is triggered by a small electrical current from an inertia switch situated underneath the dashboard. Therefore you must be very careful when working on anything electrical on a vehicle with airbags.

NB

Airbags contain a small explosive charge – pyrotechnic – which is usually set off by an electric current. However, undue jolting, and static electricity from your overalls, can trigger the charge. Remember:

- do not disconnect airbags, or anything electrical, in their vicinity unless you fully know what you are doing
- carry airbags carefully and with the moving part directed away from your body
- if you remove an airbag, store it in a locked cupboard for safety

Airbags are sometimes referred to as supplementary restraint systems (SRS); the primary restraint system is the seat belt.

Questions and skills

1 Draw a sketch to show how a particular car, of your choice, can be safely jacked up.
2 Look inside a car, of your choice, and describe how the seat adjusters work and/or how the seats move and fold, or how racing cars seats are made to fit the driver.
3 Make out a table to give examples of each type of vehicle layout.
4 Investigate how body tubs are made for open-wheel race cars.
5 Investigate the frame arrangement on a sports motorcycle and discuss how this is similar to, or different from, that on a clubman's type of racing car.

Chapter 8

Inspect, test and rebuild

Introduction

The choice of vehicle to inspect will depend on your specific area of interest: kart, single-seater, saloon or rally car; and the availability of vehicles. The wider the range of vehicles that you can use when training, the wider and more varied will be your experience and employability. Of course, you may choose to specialise at an early stage. Both approaches have advantages.

What is important is that you carry out vehicle inspections and record your findings. Carry out inspection with the IMI data sheets, which you can obtain direct from the IMI if you are registered. If you have not got access to these IMI sheets, then you may use MOT or service check sheets as guidance.

Figure 8.1 Scrutineering shed at Lydden race circuit

Safety first

When approaching a vehicle for the first time, especially a damaged motorsport vehicle, you must carry out a *risk assessment*. That is a mental assessment of the situation. If you are on duty at an event, maybe as a team mechanic or a marshal, you will be first on the scene. In this case you need to consider the following:

- Is it safe to get near the vehicle – think about location, other traffic, and other people – you must put your safety as first priority
- The next priorities are making the scene safe, calling for help, and the application of first aid and perhaps the paramedics

If you are approaching the vehicle after its recovery, even when it is in the pits or workshop, a mental risk assessment is needed.

Some typical examples of the dangers are:

- The rally mechanic who opened the bonnet of a vehicle at the end of a forest stage to be scalded by the coolant from a radiator fracture – the coolant was at about 130 °C
- The NASCAR mechanic who decided to touch the brake disc when the car came in the pits for a wheel change; he burnt his fingers. The disc were probably at about 800 °C. Remember that you need to wear gloves to even touch the tyres
- The technician who was feeling under the seat to plug in the diagnostic connector and had his skin pierced by a hypodermic needle left by a previous occupant

Use your senses

When inspecting a vehicle it is always good to use your senses, but do it with care.

Table 8.1 Using your senses

Sense	*Checks*	*Cautions*
Sight	How does this vehicle look? Is it level and square? Are there any leaks or stains? Signs of damage or misuse?	Use eye protection
Sound	How does this vehicle sound? Listen to the different systems or parts.	Use ear protectors
Smell	Is there a smell which might indicate a leakage or overheating?	Wear a mask
Taste		Not advised
Touch	Use your fingertips to check for damage or wear. Use a nail to check whether a blemish is raised or sunken.	Use hand protection
Kinaesthetic	Feel the operation of controls or mechanical linkages for smoothness.	Be prepared for the unexpected

Figure 8.2 Scrutineering underway at Lydden race circuit

Respect for vehicles

As a technician *you are responsible* for the vehicle that you are inspecting. Therefore you must *not* cause damage to the vehicle, even if the vehicle that you are inspecting is seriously damaged. You never know what repairs or salvage may take place. You are expected at all times to:

- Use seat covers
- Use floor mat protectors
- Use wing covers
- Protect from bad weather
- Jack up and support the vehicle safely using appropriate jacking points
- Ensure that the systems are treated with care
- Remove finger marks

The following tables (Tables 8.2–8.6) are to give you a practical guide to inspecting motorsport vehicles – you may choose to use them as a guide and to make up your own checklists.

Table 8.2 Tyre markings and wheel serviceability

No	*Task*	*Detail*	*Typical answers*	*Action points*
1	Use necessary PPE	Mechanic's gloves	Protects hands from cuts and burns when handling tyres	Look out for hot tyres and sharp edges
2	Correctly identify relevant tyre data for vehicle	Size, pressure, type, fitment	195 / 55 R 19 2 bar (30 psi) M&S Symmetrical	Use either vehicle mfg or tyre mfg data sheets
3	Locate and identify tyre markings	Tyre and wheel size, speed rating, load index, tread wear indicator, aspect ratio	195 / 55 R 19 82 H	Sketch tread wear indicator between tread ribs
4	Measure tread depth	Use MOT green depth gauge	Legal minimum 1.6 mm	Check across full width of tread and all circumference
5	Examine tyre condition	Look for damage to the tyre	Cuts, lumps, bulges, tears, abrasions, intrusions, movement on rim, concussion, tread separation	Check the tyre pressure
6	Examine tread wear patterns	Look for uneven wear (see 7), skid wear and patch wear on tread	Skidding, out of balance	
7	Identify reasons for abnormal wear patterns	State causes of edge or wear at one point	Incorrect tyre pressures, steering misalignment	
8	Examine wheel condition	Look for damage to wheel	Impact damage, cracks, distortion, run-out, security	
9	Check valve condition and alignment	Is it in straight?	May be damaged or bent after impact	
10	Record faults			List wheel and tyre faults
11	Complete data collection sheet			Record findings on data collection sheet

Figure 8.3 Compression gauge in use

Table 8.3 Pre-event set up – workshop

No	*Task*	*Detail*	*Typical answers*	*Action points*
1	Use necessary PPE	Mechanic's gloves, goggles		
2	Obtain vehicle set-up data	Check with appropriate authority	Use set-up management system	This may need a signature
3	Check previous data and driver comments		Use set-up data log	
4	Check battery serviceability	Use battery test procedure		System will not operate with our a sound power supply
5	Check oil and fluid levels	Note variance		
6	Check and adjust suspension and corner weights	Refer to data	Complete set-up documents showing changes	
7	Check and adjust steering geometry	Castor, camber, SAI, toe-out on turns	Complete set-up documents showing changes	

(*Continued*)

Table 8.3 (Continued)

No	*Task*	*Detail*	*Typical answers*	*Action points*
8	Check brake conditions	Measure pads, discs	Record on inspection sheet	
9	Check driver safety equipment	Seat belt harness, seats, roll cage, padding and mountings, fire extinguisher system		
10	Carry out spanner check	Check against checklist. Rectify as needed	Note any loose, damaged, or broken fastenings	
11	Record faults		Use set-up data file	Record any system faults
12	Complete data collection sheet			Record findings on data collection sheet

Table 8.4 Pre-event inspection – at event

No	*Task*	*Detail*	*Typical answers*	*Action points*
1	Use necessary PPE	Mechanic's gloves, goggles		
2	Check previous data and driver comments		Use set-up data log	Check previous data and driver comments
3	Check tyres	Fitment (direction/ corner), pressure, type, compound	Use set-up data log	
4	Check oil and fluid levels	Note variance	Check oil and fluid levels	Note variance
5	Check and adjust steering geometry	Castor, camber, SAI, toe-out on turns	Complete set-up documents showing changes	
6	Check security of bodywork	Panel attached, clips closed, screen clear, decal correctly located	Use checklist	May choose to photograph for record
7	Check seat belt harness with driver in place	May need adjusting		Note any pre-sets
8	Arm fire extinguisher system	Check setting, switches		
9	Record faults		Use data log	Record any system faults
10	Complete data collection sheet			Record findings on data collection sheet

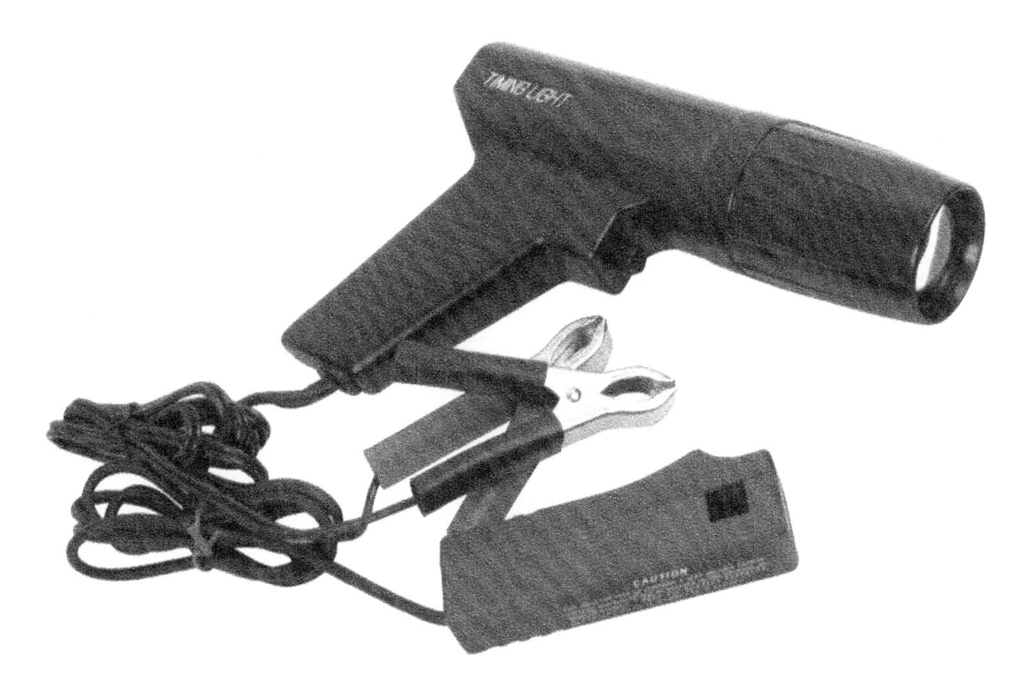

Figure 8.4 Strobe timing light

Table 8.5 On-event inspection

No	*Task*	*Detail*	*Typical answers*	*Action points*
1	Use necessary PPE	Mechanic's gloves, goggles		Remember vehicle components and systems remain hot for a while at the end of a stage or round
2	Check new data and driver comments	Down load data logger and record verbal comments	Use set-up data log	Check previous data and driver comments
3	Raise vehicle and remove wheels	Visual inspection		
4	Inspect wheels and tyres for wear/ damage		Record wear findings	
5	Carry out spanner check	Check against checklist. Rectify as needed	Note any loose, damaged, or broken fastenings	
6	Refit wheels	Check security and bearings		
7	Check fluid levels and pressures		Complete check sheet	

(Continued)

Table 8.5 (Continued)

No	Task	Detail	Typical answers	Action points
8	Check security of bodywork	Panel attached, clips closed, screen clear, decal correctly located	Use checklist	May choose to photograph for record
9	Refuel vehicle	Follow team procedure	Record fuel taken	Remember that petrol is highly flammable
10	Record faults		Save data file	Record any system faults
11	Complete data collection sheet			Record findings on data collection sheet

Table 8.6 Post-event inspection

No	Task	Detail	Typical answers	Action points
1	Use necessary PPE	Mechanic's gloves, goggles		Accurate recording and signing-off against names is recommended
2	Obtain and collate race data	Set-up sheets, data logging, fuel and tyre records		There may be both electronic and paper logging systems
3	Analyse race data	Record comments and make to-do list	Jobs to be completed before next event	
4	Check and correct log records	Team de-brief and post-event analysis	Include data check and other comments	
5	Clean vehicle and secure for transporting	Check for damage and fit transportation tyres	Photograph for record	
6	Drain fuel if appropriate	Use save storage procedure		Measure fuel left for fuel consumption calculations
7	Remove and charge battery			Gel batteries must be slow charged
8	Complete data records	Data logging and vehicle records	Complete all data logging and set-up sheets	Secure storage of data is essential – back up all files and keep in secure place
9	Complete data collection sheet			Record findings on data collection sheet

Test and rebuild

Before carrying out any overhaul work you should check the following:

- The customer fully understands the cost implications
- You have access to any special tools or equipment which is needed
- Spare parts are available
- Repair, set up and test data is available

Costs

The cost of overhauling a component – such as a gearbox – can, in some cases, be greater than that of a new one. On the other hand, given that not many parts require replacement, the overhaul may be one tenth of the new item. The biggest factor is often the labour cost, followed by the cost of obtaining the part needed. So the labour charge-out rate is a very big factor.

Let's look at a possible example using a labour rate of £85 per hour.

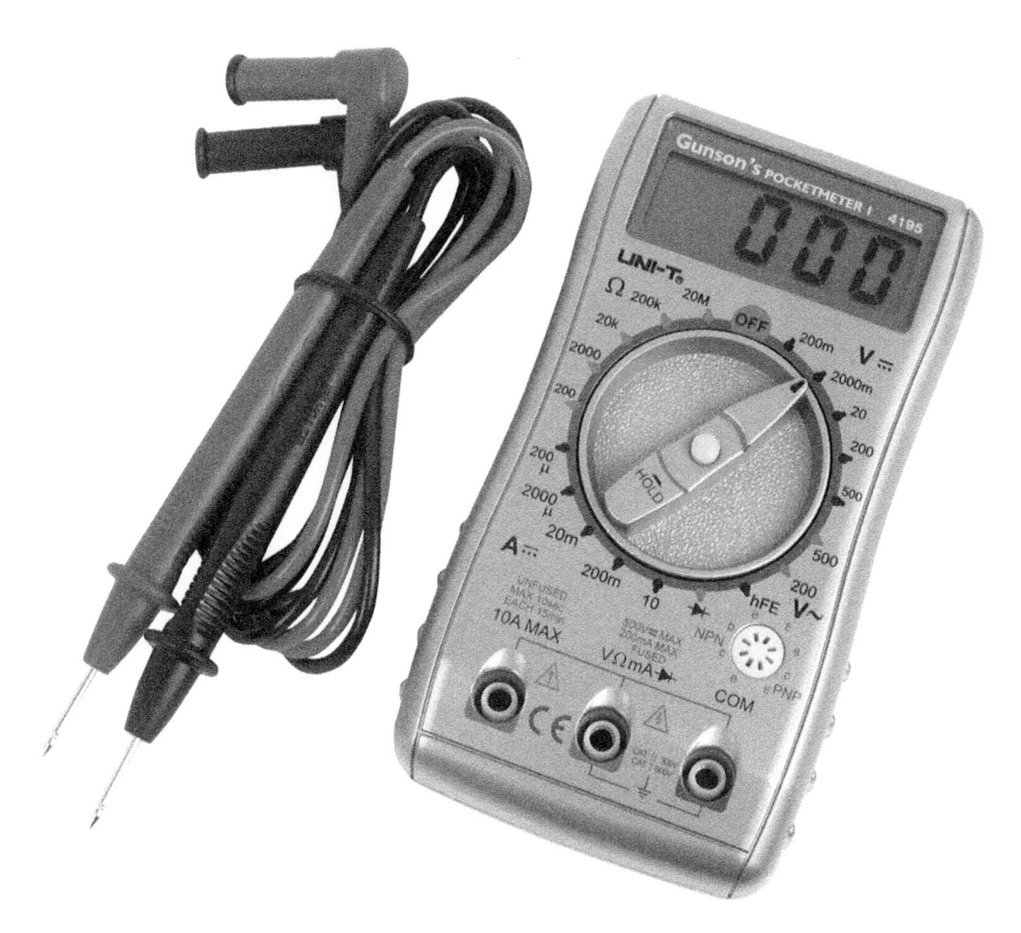

Figure 8.5 Multimeter

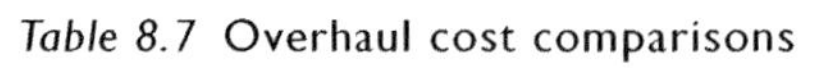

Table 8.7 Overhaul cost comparisons

Overhaul costs		*Replacement costs*	
Item	£	Item	£
Remove and refit gearbox 1.5 hours	127.50	Remove and refit gearbox 1.5 hours	127.50
Overhaul gearbox 5 hours	425	Replacement gearbox	500
Parts and oil	175	Oil	15
VAT	127.31	VAT	112.43
Total	854.81	Total	754.93

Figure 8.6 Impact sockets

Figure 8.7 Aircraft spanners

Figure 8.8 Socket set

In addition the replacement gearbox will have a period of warranty attached – giving the customer peace of mind.

When it comes to replacement parts such as engines and gearboxes the customer may wish to have the original part overhauled rather than fit a replacement simply to maintain the originality of the vehicle. Rather like the blacksmith who said, 'this is the original hammer that I've had since I was an apprentice – it has had five new heads and six new shafts.'

Beware, however, when buying or selling units, that there are many different definitions, or uses in this field. For example:

Original parts – OEM (original equipment manufacturer) – the units used on the vehicle when it was new. These are often not actually manufactured by the vehicle manufacturer – but sourced from an approved supplier. The boxes may have the logo of the vehicle manufacturer to show approval – but exactly the same item without the logo on the box may be obtainable from the local motor parts factors for 50% of the price. Also, be aware that many manufacturers – particularly German ones – often use recycled parts or products made from recycled materials. Such parts are still by definition OEM and will carry a price premium.

Figure 8.9 Torx drive set

Remanufactured units – particularly engines and gearboxes – the components are completely stripped, inspected, measured and crack tested; then all the machined faces are remachined and new wearing parts fitted – for example bearings and pistons.

Overhauled units – in this case the units are stripped and cleaned and inspected; but only the parts that are outside tolerance are replaced.

Pattern parts – these are lookalike items that fit exactly and do the same job as the OEM part but are made by another company. Sometimes the pattern parts perform better than the originals and often they are made by the same company that makes the originals but sold under a different brand name. Usually these units are 30% cheaper than the OEM equivalent.

Health, safety and the environment

Before carrying out any work whatsoever, you must ensure that you fully understand the health and safety issues and are able to comply with them along with the relevant environmental regulations and codes of practice. As a recap of your work at Level 2, you should consider:

Acts and regulations – Heath and Safety at Work Act, COSHH, EPA
Personal protection equipment – overalls, safety foot wear, gloves, goggles, masks
Vehicle protection equipment – wing covers, seat covers, floor mats
Identifying system hazards – such as fuel and high voltage sparks

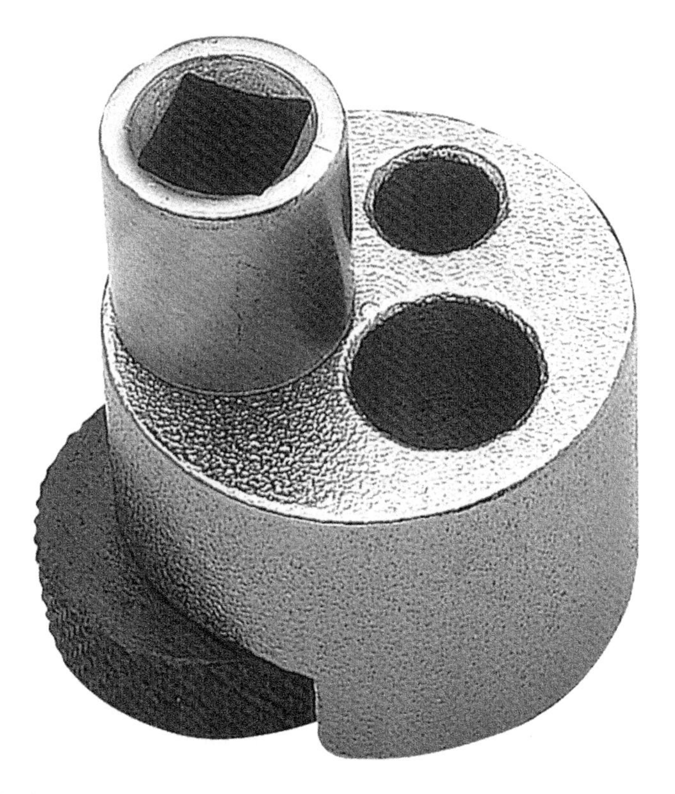

Figure 8.10 Stud remover

Safe disposal – using correct oil/fluid drainers, returning displaced units, disposing of electronic and pyrotechnic items correctly

Fire hazards and safety – knowing the muster point, especially if you are at a different site, such as the pit lane at a circuit. Evacuation drills and fire extinguisher identification

Procedures for dealing with accidents – knowing the contact persons and radio calls if at an event or on a different site

Risk assessments – before carrying out any task you must ensure that risk assessments are in place and the COSHH leaflets are available for all substances to be used

Personal conduct – you *must* conduct yourself professionally at all times. There is a code of conduct for this if you are a member of the Institute of the Motor Industry, see: www.motor.org.uk.

It must always be borne in mind that *motorsport can be dangerous* and that we are always under detailed scrutiny by other people.

Workshop

To be able to carry out any form of overhaul or repair – apart from the running repairs during competition – some form of workshop is needed. It sounds obvious, but often the functions of the workshop are forgotten about. Let's have a look at some of the reasons for a workshop:

- Provide safety, security and protection from the weather for the vehicle
- Provide safety, security and protection from the weather for the engineers and other staff
- Provide safe and secure storage for the tools, equipment and spares
- Provide somewhere to work on the vehicle
- Provide somewhere to overhaul and repair the units and components
- Provide an office facility

Workshops come in all sorts of shapes and sizes. I have worked in many, varying between a lock-up garage in a block without a permanent power supply and a Formula 1 team in a three-storey futuristic building with an underground entrance using solar power and heat-pumps to give a zero-carbon foot print. The ideal workshop is one where the race car can flow through in one direction. This should fit in with the loading and unloading of the car transporter, preferably having some form of docking-station so that the race car is kept both secure and dry on both its outwards and inwards journeys.

The race vehicle workshop is much more than a garage; it has a wide range of activities and a larger number of departments. One company I was involved with included a museum, car sales and mail order. We are not discussing these areas in this chapter. If you are employed in the motorsport industry, you may be involved in all aspect of motorsport vehicle overhaul, or just one tiny part. No matter how small it will be an important function and it will help if you have the bigger picture of what goes on in other departments. Some motorsport companies do just one of the functions covered in the next few pages.

Cleaning down area – when entering the building there will be a cleaning, or wash, area. Depending on the sort of competition involved, this will vary. Obviously a 4x4 off-road vehicle will have more mud to remove that a Formula Ford – unless maybe the FF has had an off. This area may also be used for removing body panels – especially damaged ones and other items such as sump guards and changing the wheels to ones more suited for the workshop.

Figure 8.11 Spring compressor

The Environmental Protection Act (EPA) and associated local authority (LA) regulations and building regulations (Building Regs) require that all buildings comply with certain criteria. The cleaning down area must have a drainage trap so that contaminated water, mainly that contaminated by oil and chemicals, is not allowed to enter the main drainage system. It is normal to recycle the cleaning down water, using a filtration and treatment system. This is environmentally clean and saves the water bill.

Preparation area – an area of bays where each vehicle can be worked on individually – the layout of the bays and the equipment in them will depend on the type of vehicle. For single-seaters it is normal to have stands or trestles so that the vehicle is at an ergonomically sound working height. In other cases wheel-free or four-post hoists (ramps) may be used. For a racing team it is usual practice to have a designated bay for each car and a designated technician too, in which case the bay will be kitted out with that technician's tool and the equipment and spares appropriate to the vehicle. In a jobbing shop – where cars are prepared for a variety of owners and/or drivers – the bay will be the sanctity of the technician – cars moving in and out as needed.

In the case of major unit overhaul, the bay may be used to remove the unit only, and then the vehicle moved to a compound area for secure storage until the unit is ready to be reinstalled.

Machine shop – this is where the machine tools and similar are laid out and used. Race cars, even touring car classes, tend to use a number of bespoke parts – ones made especially for that vehicle. So the machine shop is needed for both manufacture and overhaul. Also this may be where a number of special tools are kept – the ones that require floor mounting – as against those that can be taken to the benches in the preparation bays. Typically, a machine shop may contain some, or all, of the tools in Table 8.8.

Bench work area – the bench work area is often around the outside of the vehicle bays, using metal-topped benches with drawers and cupboards underneath. Vices and other tools may be mounted on the benches.

Where the work is solely unit based – such as overhauling gearboxes – the benches may be aligned in rows separate from the vehicles – with the use of stands, or rigs, for the gearboxes or other major components. The drawers, or open racks, will then contain the special tools needed for the job in hand.

Fabrication area – this is where items are fabricated and welded. Usually contains rollers, bending machine, croppers and MIG or TIG welding equipment. Specialist trained engineers in this area will provide these services to enable you to carry out your overhaul and repair tasks.

Table 8.8 Machine shop equipment

No	*Item*	*Specification*	*Purpose*	*Comment*
1	Small lathe	6–8 inch (150–200 mm) swing with 18–24 inch (450–600 mm) between centres	Making small items such as spacers, and cleaning up round parts	This will need a range of tools and chucks
2	Off-hand grinder – small	Approximately 6 inch (150 mm) diameter wheels – 1 fine, 1 coarse	General sharpening and cleaning	
3	Off-hand grinder – large	Approximately 10 inch (250 mm) diameter wheels	Sharpening drills and tools	Keep flat for accurate work
4	Pillar drill	5/8 inch (15 mm) chuck and variable speeds	Variety of drilling	Need variety of vices, or clamps, and drills and countersinks
5	Band saw	Approximately 4–6 inch (100–150 mm) cut	Cutting up steel stock	
6	Hydraulic press	20 ton (20 tonne)	Removing and replacing bearings and pins	
7	Milling machine	5-axis CNC milling centre	Manufacturing small parts	Used in conjunction with CAD system
8	Buffing – polishing	Floor-mounted buffing and polishing heads	Finishing parts	
9	Parts cleaning bath	Chemical cleaning bath with pressure spray	Cleaning parts	

Composites shop – where composite components are manufactured or repaired – a specialist clean area. This shop is staffed by specialist technicians who will make new parts or carry out specialist repairs to enable you to overhaul the motorsport vehicle.

Design studio – where vehicles and components are designed and modified. Usually using computer aided engineering (CAE) – that is the computer aided design (CAD) is connected to the computer aided manufacture (CAM) machine tools such as the five-axis milling centre. You will find the design staff supportive in providing technical data for your overhaul procedures.

Model shop – named because they create models, or macquettes, of vehicles for design and testing purposes – including scale models for the wind tunnel – as well as specialist full-scale parts such as aerofoil wings. The model shop is both a source of data and specialist parts and skills. The model shop in larger, or older, firms may incorporate clay and/or wood handling equipment and skills.

Paint shop – where the vehicles are painted. Again this shop has specialist staff and equipment. The overhauled vehicle, or its panels, will be refinished in this shop. An interesting move in refinishing is to use vinyl film instead of paint; the US army dragster, which is probably the fastest ever race car ever made, uses vinyl film. They applied this in their workshop at the Indianapolis raceway, known as the *Brickyard.* The comment of the technician applying the material – which is printed and cut on site –was, 'It's lighter that paint.' When it is on, it is almost impossible to tell that it is not paint.

Parts and storage – the safe and secure storage of parts, both new ones and ones waiting for completion of the overhaul, is very important. They can easily go missing. Small parts have an attraction for the floor (it's called gravity – 9.81 m/s^2) then they roll behind the largest possible box so that you can't find them. That aside, race car parts cost from 10 to 100 times the equivalent of the road-going equivalent. And some people just want to have the damaged piston out of the number 12 car that didn't win last race of the season; or similar.

Let's have a look at this in a bit more detail. You need a secure storage area large enough to store the large parts and a set of drawers and trays for storing the smaller parts. When you are stripping a component – such as a gearbox – then you need a tray system laid out to keep all the small parts – nuts, bolts, washers, spacers and so on – in order so that they can be assembled in the reverse order. If you are doing repetitive work – such as overhauling gearboxes – then you will probably be provided with suitably marked out trays. You will also be able to identify each component without any thought.

To ensure absolute recognition, there are a number of procedures. Two frequently used ones are:

- Plastic bags – like food storage bags – put each component into a bag and label the bag with the part name, part number, customer details (car number or VIN) and any other details – such as *left rear*. The bags are then placed in a tray which is also marked with customer, car and unit identification.
- Photographs – take photographs of both the assembled and the stripped parts. Record the photograph numbers – most digital cameras do this automatically – and reference the numbers to your notes.

These two techniques are used extensively in specialist firms – such as when repairing, or overhauling, Ferraris. Cars like Ferraris may fall into a model category, such as F430; but customers may have upgraded the seats, engine specification and other details, or the originals may have been replaced following an accident with used parts from another model. As it is unlikely that a workshop manual exists, you'll rely on the parts you have – so look after them, make notes and take photographs.

You will also need a storage system for every day consumables, that is items which are used on a day-to-day basis in the repair or overall of vehicles or units. Such items may include:

- Cleaning cloths
- Polishes and detergents
- Hand cleaning materials
- Specialist cleaning solutions
- Screws, nuts and bolts
- Specialist fixings – such as toggle fasters
- Washers, spacers and shims
- Locking wire
- Wiring cable
- Electrical connectors and fasteners
- Tape and adhesives
- Gaskets and seals

These items may be charged out against jobs in two ways:

1 Individually against the job number
2 On each customer's bill as either a percentage of the bill or pro rata against the number of hours work

Dynamometer and test shop – if working on complete vehicles you will need to have access to a rolling road dynamometer – referred to as the dyno, or the rolling road, by most staff. This allows the road wheels to sit on the dyno rollers and the power and torque measurements taken. Running any competition engine makes a lot of noise – so this is usually situated separate from the main building, or suitably noise-insulated from it.

If it is just the engine that is being tested, then a test cell is used. A test cell, of course, requires much less space than a rolling road. This takes the form of the engine mounted on a frame that is attached to the dynamometer (dyno). Because of their exceptional noise, aircraft engines are tested underground; some race teams do this too.

Checking and loading area – this is a secure area for the prepared race vehicle to be kept safely prior to, and during, loading them onto the transporter.

Office – office space is needed for a number of functions in motorsport. These include:

- Reception – to meet and greet customers and to record customer information and carry out tracking of customers' repairs
- Finance and administration – to control the flow of cash and communications with customers and suppliers
- Meeting area for the team and directors – often this take the form of a boardroom with a big table and chairs that can be used for a number of functions
- Engineering office area – with a racing team the race engineers will need an area to work on data from testing and racing, that is to analyse the data and work out strategies for

future developments. This usually takes the form of bar-type desking where the laptop computer can be docked into the team network – usually still hardwired for security, though high level encryption may be used with a suitable wire-free system (WiFi). This is a separate area to that of the design team who are working on upgrades and new designs for future vehicles and components

In addition to the functional part of the activities in the office area, this area can be used for displaying trophies – *silverware* – and photographs. Therefore the security of the office must be considered with appropriate locks and an alarm system. Interestingly many of the Formula 1 teams have private museums at their headquarter (HQ) offices – those of McLaren and Williams are outstanding, forming a detailed history with examples of each type of car used.

Tools and equipment

Table 8.9 lists the equipment needed for the machine shop. In this section we will look at the fuller range of equipment that is used in vehicle and unit overhaul.

Table 8.9 Overhaul equipment

No	*Item*	*Purpose*	*Note*
1	Four-post hoist – with wheel-free adaptor	Raise vehicle and independently raise one corner	MOT compliant
2	Two-post wheel-free hoist	Lift vehicle under chassis so all wheels are free	
3	Body scanner	Scanning bodywork to produce CAD drawings	Good when rebuilding historic vehicles
4	Cam profiler	Regrinding camshafts	Make cam profiles to your design
5	Castor, camber and SAI (KPI) gauges	Checking steering geometry	
6	CMM machine	Accurately measuring components for CAD drawings, or reverse engineering	
7	Coil spring gauge	Testing coil spring rate	Check all four springs
8	Compression gauge	Testing engine compression pressure	Compare reading of each cylinder
9	Corner weights	Checking weight distribution at each corner of the vehicle	Adjust suspension and redistribute weight as needed
10	Crankshaft grinder	Regrinding crankshafts	
11	Dial test indicator (dial gauge)	Measuring movement – such as for valve lift	
12	Durometer	Measuring the hardness of tyre treads	Check temperature first
13	Engine boring equipment	Reboring cylinder blocks	

(Continued)

Table 8.9 (Continued)

No	*Item*	*Purpose*	*Note*
14	Granite table	Providing a smooth and level surface on which to set up the vehicle	Cost is up to £1m, used for world class vehicles
15	Horizontal milling machine	Milling surfaces – such as cylinder head faces	
16	Laser, or light, suspension aligning gauges	Checking steering and suspension alignment	The manual system can achieve the same results
17	Mercer gauge	Measuring the diameter of a cylinder bore	
18	Micrometres – range, internal and external	Measuring inside or outside surfaces – such as cylinder bores and crankshaft bearings	
19	Pressure washer	Cleaning mud off the vehicle	Ensure EPA compliance when used
20	Surface grinder	Grinding surfaces such as cylinder head faces	
22	Turn tables	Measuring toe-out on turns and carrying out steering checks	Used in conjunction with No 5
23	Tyre machine	Removing and refitting tyres	Special machine needed with aluminium alloy rims
24	Tyre pressure gauge	Measuring tyre pressures	Need accurate gauge on motorsport vehicles – check temperature before adjusting
25	Tyre temperature gauge	Measuring tyre temperature	Take three measurements on each tyre – inside, middle, outside of tread
26	Vertical milling machine	Milling tasks such as when enlarging inlet ports	
27	Welding equipment (MIG or TIG)	All kinds of joints and repairs	
28	Wheel balancer	Static and dynamic wheel balancing off the vehicle	Use only approved weights – usually stick-on on inside of rim

Data

Sources of data are very, very important. The obtaining and storage of data is very much a profession in itself. Think of the word library – it does not just refer to books, but all forms of stored information – including photographs, films, tapes, posters, CDs, DVDs and other electronic storage media.

As a technician in motorsport you will have to collect, collate, store, use and communicate data.

There is a saying that knowledge, another word for data, is power. If that data means 0.1 of a second off each lap, that is very powerful, and therefore very crucial and valuable, data. As a technician obtaining, using and storing data are high priorities. Table 8.10 looks at the main sources of data.

Most manuals and other commercial sources of information are available on CD/DVD, and online through a subscription agreement, as well as in paper form.

Before you carry out any overhaul work you will need to check that the data in Table 8.11 is available.

Parts and fixings

On racing and competition vehicles of all types fixings are very important. It is good when working on a car to remember a bit of engineering science. To make these formulas more graphic, thinking of them in these different ways may help:

1 Force – if you want to fix a post into the ground take a big sledge hammer (14 lb / 6.3 kg mass) and swing it quickly holding the end of the shaft for rapid acceleration; now in terms of a car it is the weight (mass) multiplied by how fast it is accelerating. If you move the formula round you can see that to make it accelerate faster you can either fit a more powerful engine – increasing the force, or make it lighter.
2 Stress – as you can see this is inversely proportional to the area taking the stress – so usually the smaller the part the more stressed it will become.

Figure 8.12 Open block

Table 8.10 Data sources

No	*Source*	*Typical content*	*Accuracy – limits*	*Comment*
1	Autodata manual	Technical service data	Detailed and checked	
2	Autotrader	Vehicle sales		
3	Competition vehicle log	Set-up data for vehicle, work done, parts fitted	If correctly maintained	Must be maintained in detail to ensure continuity of operation and save time
4	Customer record/file	Name, address, vehicles owned, contact details	If correctly maintained	Highly confidential
5	Data logger	Vehicle operation data including acceleration, braking, LAG, gear, throttle position	Digital data – as good as system will allow	PI
6	ECU data	Codes for system operation set-up and fault codes	Digital data	Each system may have a separate ECU – a reader will be needed to access data
7	Glasses guide	Vehicle model guide	Detailed and checked	
8	Haynes (or similar) workshop manual	Vehicle and units service and repair procedures	Detailed and checked	Haynes offer an excellent range of manuals at competitive prices
9	Lap/section time from circuit system	Speed and timing of laps and sections	Digital data – as good as instruments will allow	Data available from race control
10	Manufacturer's workshop manual parts manual	Vehicle and units service and repair procedures	Detailed and checked	Often to be read in conjunction with training materials
11	Parker's price guide	Vehicle identification and pricing	Detailed and checked	
12	Stack system	Speed and timing of laps and sections	Digital data – made to high level of accuracy	In-car timing system – very useful for testing
13	Vehicle service book	Records of service and mileage	If correctly maintained	
14	Your company records	Vehicle changes an modifications	If correctly maintained	Annotated

Bearing these factors in mind – if we make the vehicle lighter and faster we are tending to make the parts more stressed, so we need to be sure that the parts used will cope with the situation. Therefore the parts and fixings tend to be made out of materials that are both light and strong; typically these are high strength steel (HSS), aircraft-grade aluminium alloy (7001 or similar series) and titanium. These materials are much more expensive than those used

Table 8.11 Pre-overhaul data check

No	*Data*	*Source*	*Comment*
1	Test data	Report on test	This should be added to vehicle log
2	Unit removal data	Workshop manual	
3	Unit test data	Unit manufacturer	
4	Unit stripping data	Unit manufacturer	
5	Equipment operation data	Equipment manufacturer	
6	Replacement parts	Parts manual	Parts used should be recorded
7	Unit assembly and test data	Unit manufacturer	This should be added to vehicle log
8	Set-up data	Vehicle set-up log and/ or unit manufacturer	This should be added to vehicle log

in standard cars; also the parts are made in smaller numbers and are therefore much more expensive to produce.

Aluminium alloy and titanium are both easily damaged, and their appearance can soon be marked, so ensure that they are handled with care and that the correct tools are used when working on them.

Procedures

When carrying out any overhaul work you must follow the procedures set out in the appropriate manuals and data sheets. Technical explanations for some tasks are covered in the appropriate chapters.

In all cases follow these basic steps:

- Obtain the necessary data
- Clean the vehicle in the vicinity of the unit to be removed
- In a clean work area remove the unit to be overhauled – following the appropriate safety sequences
- Take the unit to the bench, or mount on a stand
- Strip the unit noting the position of parts and using appropriate parts storage trays
- Repair and rebuild as needed
- Replace the unit and set-up as per data
- Thoroughly test and recheck setting against data

Products

The use of products – oils, greases and other chemicals – in the motor industry is big business. The correct use of products can often save time and money for the customer and increase sales revenue for you.

When you have carried out an overhaul task you will probably need some form of product for lubrication or cooling – the customer should be told of this, and if possible given an after-care leaflet, or card, with information about the product and its use to prolong the life of the overhauled unit.

Example of after-care leaflet

BrooklandsGreen – brake calliper care

1. Your brake callipers have been overhauled and should give you a high standard of braking.
2. You are advised to use ***AP Racing DOT 5.1*** brake fluid which complies with SAE J1703 – this is ideal for high-performance road, competition and track day use.
3. To prevent corrosion the brake fluid should be **changed frequently** – we recommend at least each racing season.
4. To prevent brake squeal use a **copper-based grease** between the pad back plate and the calliper assembly.
5. Be sure to **fully vent** (bleed) your brakes after fitting, before driving.

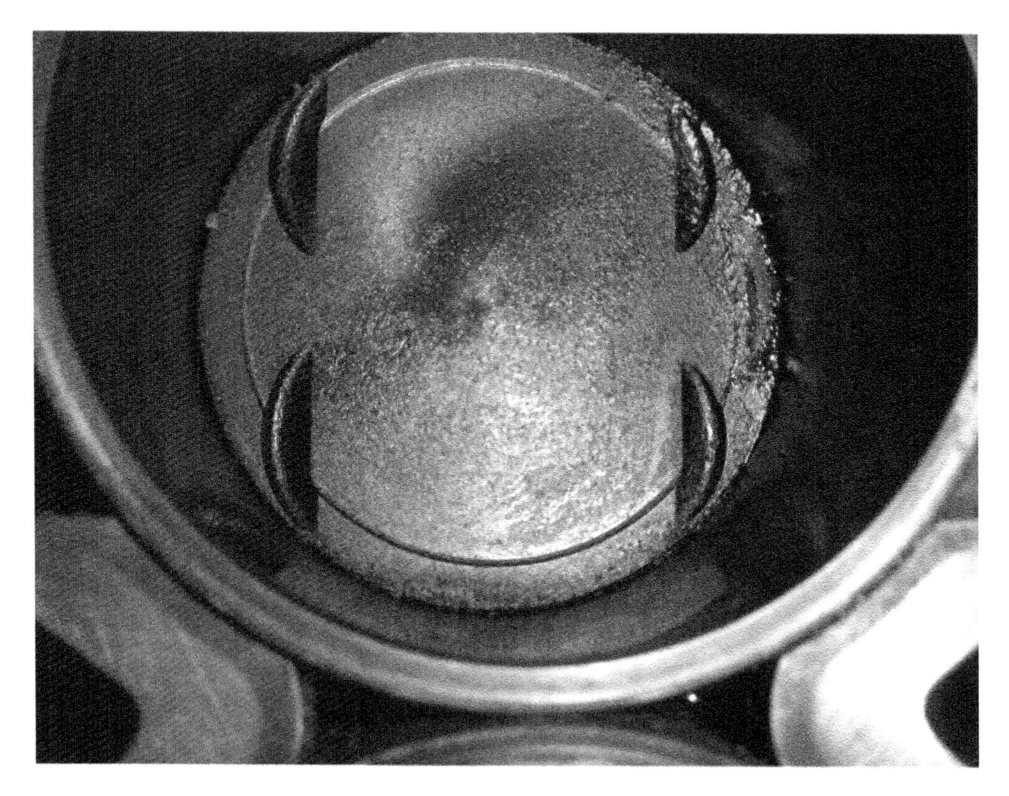

Figure 8.13 Wet liner close up

Clusters – motorsport valley

The implications of parts supply, knowledge and data, tools and equipment and transportation all revolve around cost. In motorsport there tends to be a team working spirit – yes, even between competing teams. After all, there would be no motorsport if there weren't any other teams. So, rather paradoxically, although it is all out on the track, or rally stage, you need to work with your competitors to achieve your aims of winning. Because of this, companies tend to work in clusters – small groups based at circuits, in trading estates, or sites with good linkages to each other. In the UK the bulk of the major motorsport companies are located in Motorsport Valley. This is geographically follows the Thames Valley from Oxfordshire, through the Home Counties, down to Kent. The communication routes use the M4 and M40 leading into the M25 and exiting with the M2.

Networking between the motorsport companies and the competitors – lots of the companies solely exist to be able to race, whilst others operate in other branches of performance engineering, such as aerospace and super yachts to provide an extra income stream – takes place in the Motorsport Valley. Organisations like the Motorsport Industry Association (MIA), the Institute of the Motor Industry (IMI), the Institution of Mechanical Engineers (IMechE), the Motorsport Institute (MI) and *Autosport* magazine organise some of these events. Often they are funded by government bodies such as UKTI and local development agencies. In the USA the magazine and show organiser Performance Racing Industry (PRI) is one of the main networking organisations for the industry; they have links with the British institutions as well as the Society of Automotive Engineers (SAE).

Questions and skills

1. Obtain data sheets for the vehicles that interest you and make these into a data booklet.
2. Look at the websites for the organisations that are mentioned in this chapter – or elsewhere in this book.
3. Make a spares list for a competition car that interests you.
4. Organise your workshop to suit your needs for the car or motorcycle that interests you.
5. Work on whatever vehicles you can; the more practical experience that you can get the better. Approach people and teams and ask if you can help. People involved in motorsport are always enthusiasts; they are usually keen to talk to others about their sport – but not in the pit lane on the last lap!

Glossary

This section defines a number of the words and phrases used in motorsport engineering, including some of the specialist racer and enthusiast vocabulary and jargon.

'An off' when you come off the circuit unintentionally into an area where you are not supposed to – if you are lucky it is just grass
'Esses' one bend followed by another
'O' rings rubber sealing rings
¼ mile quarter-mile drag racing strip
1/8 mile eighth-mile drag racing strip
24-hour a race lasting 24 hours, the winner is the one covering the greatest distance, usually there are different capacity-based classes. For low budget racers, there is an event at Snetterton for Citroen CVs and one in Sussex, near a pub, for lawn mowers
Acceleration rate of increase in velocity
Accessories anything added, which is not on standard vehicle
Ackermann steering set up to prevent tyre scrub on corners
Add-ons something added after vehicle is made
Adhesion how the vehicle holds the road
Airbags SRS – bags that inflate in an accident
Alignment position of one item against another
Alloy mixture of two or more materials; may refer to aluminium alloy, an alloy of steel and another metal such as chromium
Ally aluminium alloy
Anti-roll bar suspension component to make car stiffer on corners
Atom single particle of an element
Backfire when the engine fires before the inlet valves are closed – sending flames and gas out of the inlet
Barrel cylinder barrel – usually refers to motorcycle engine
BDC bottom dead centre
Bench working surface; also flow bench and test bench
Beta version test version of software or product
Birdcage tubular chassis frame which resembles a bird cage
Block and tackle used to lift engines
Block cylinder block – a number of cylinder in one piece
Bonnet engine cover (front engined car)

Bore internal diameter of cylinder barrel
Brooklands first purpose built racing circuit at Weybridge in Surrey with banking and bridge
Cabriolet four-seater convertible body
Carbon fibre like glass fibre but used very strong carbon-based material
Cetane resistance to knock of diesel fuel
Chicane sharp pair of bends – often in the middle of a straight
Chocks tapered block placed on each side of the wheel to stop the car rolling
Circuit race circuit
Clerk of Course most senior officer at a motorsport event – person whose decision is final, although there may be a later appeal to the MSA or FIA
Code reader reads fault codes in the ECU of a particular system
Composite material made in two or more layers – usually refers to carbon fibre, may include an honey comb layer
Compression ratio ratio of combustion chamber size to cylinder bore
Con rod connecting rod
Condensation changes from gas to liquid
Contraction decreases in size
Corrosion there are many different types of corrosion – oxidation / rusting are the most obvious
Cubes Cubic inches – USA term for size of engine, the saying is, 'there is no substitute for cubes.'
Cushion section of seat to sit on
Dash board instrument panel
Density relative density also called specific gravity
Diagnostic machine equipment connected to the vehicle to find faults
Dive front part of the vehicle goes down when under heavy braking
DOHC double overhead cam
Double 12 a 24-hour event divided into two 12 hours – day time only with parc fermé in the evening
Double 6 another name for a V12 engine
Drag racing two cars racing by accelerating from rest on a narrow drag strip
Engine cover cover over engine (usually refers to rear engined car)
Epoxy resin material use with glass fibre materials
Evaporation changes from liquid to gas
Event organiser person who organises the race or other event
Expansion increases in size
Fast back long sloping rear panels
Fender US term for mudguard or wing
FIA international motorsport governing body
Flag marshal marshal with a flag
Flag chequered flag, black flag, and red flag
Flow bench used to measure the rate of flow of inlet and exhaust gas through a cylinder head
Foam material used to make seats and other items
Force mass × acceleration

Formula car car built for a specific formula and usually refers to open wheeled single seaters such as Formula Ford or Formula Renault
Friction resistance of one material to slide over another
Frontal area (projected) area of front of vehicle
Gelcoat a resin applied when glass fibre parts are being made – it gives the smooth shiny finish
Glass fibre light weight mixture of glass material and resin to make vehicle body
GT Grand Turisimo (Italian for Grand Touring) first used on a Ferrari with two passenger doors and a rear luggage opening
Hatch back four-seater with rear upward opening rear
Heat a form of energy
Hill climb individually timed event of climbing a hill
Hoist used to lift vehicles, may be two-post or four-post
Hood US term for bonnet
Inboard something mounted on the inside of the drive shafts such as inboard brakes, which usually lowers un-sprung weight
Inertia resistance to change in the state of motion – see Newton's Laws, inertia of motion and inertia of rest
Intercooler air cooler between turbo charger and inlet manifold – to cool incoming air for maximum density
Kevlar fibre super strong material, often used as a composite with carbon
Le Mans Series 24-hour sports car races across the world – many in USA, organised by ACO
Le Mans 24-hour race for sportscars, prizes for furthest distance covered and best fuel consumption organised by Automobile Club de l'Ouest (ACO) (West France Auto Club)
Marshal person who helps to control an event
Mass molecular size, for most purpose the same as weight
Metal fatigue metal is worn out
Molecule smallest particle of a material
Monza World's second banked race track in Italy, copy of Brooklands
Motor-cross off-road circuit event (motor cycles)
MSA Motor Sport Association
Newton's Laws First Law – A body continues to maintain its state of rest or of uniform motion unless acted upon by an external unbalanced force; Second Law – the force on an object is equal to the mass of the object multiplied by its acceleration (F = Ma) Third law – to every action there is an equal and opposite reaction
Nose cone detachable front body section covering the front of chassis, which may include a foam filler for impact protection
Octane rating resistance of petrol knock
Off-roader vehicle for going off-road; or off-road event
OHV overhead valve
Open wheeler race (circuit) car with no wheel covering
Original finish original paint work, usually with reference to historic or vintage cars
Outboard something mounted on the outside of the drive shafts such as brakes, which usually increases un-sprung weight
Oxidation material attacked oxygen from the atmosphere; aluminium turns into a white powdery finish

Paddock where teams and vehicles are based when not racing
Parc fermé area where competition vehicles are left and cannot be worked on or prepared, usually between race rounds or rally stages
Parent metal main metal in an item
Pit garage garage in a pit lane; also just garage
Pit lane lane off the circuit leading to the pits
Pit wall protecting the pit lane from the circuit
Pit place for preparing / repairing / re-fuelling the vehicle at the side of the circuit
Pot another name for cylinder
Power work done per unit time, HP, BHP, CV, PS, kW
Prepping preparing the vehicle for an event
Prototype first one made before full production
Rallycross off-road circuit event (rally cars)
Regs racing regulations
Ride height height of vehicle off the road, usually measured from the hub centre to the edge of wheel arch
Rings piston rings
Roll bar frame inside vehicle which is resistant to bending when vehicle rolls over – safety protection for occupants
Rust oxidation of iron or steel, becoming reddish
Saloon standard four-seater car body
Scrutineer person who checks that a vehicle compiles with the racing regulations; usually when scrutineered whether the vehicle has a tag or sticker attached
Side valve when the valves are at the side of the engine (old engines and lawn mowers)
Single seater see formula car
SIPS side impact protection system – door bars (extra bars inside doors)
Skid vehicle goes sideways – without road wheels turning
Speed event any event where cars run individually against the clock
Spine backbone-like structure
Sprinting individually timed event starting from rest over a fixed distance
Sprung weight weight below suspension spring
Squab back of seat – upright part
Squat vehicle goes down a back under heavy acceleration
Squeal high-pitch noise
Squish movement of air fuel mixture to give better combustion
SRS supplementary restraint system – airbags
Stage rally when the event is broken into a number of individually timed stages, the vehicles start each stage at pre-set intervals (typically 2 minutes)
Stall involuntary stopping of engine
Steward a senior officer in the organisation of the motorsport event
Straw bails straw bails on the side of a track for a soft cushion in case of an off
Stress force divided by cross-sectional area
Strip drag racing strip
Stripping pulling apart
Stroke distance piston moved between TDC and BDC
Swage raised section of body panel
Swage line raised design line on body panel

TDC top dead centre
Temperature degree of hotness or coldness of a body
Test bench test equipment mounted on a base unit
Test hill a hill with a gradient that increases as it goes along, that is it gets steeper as you approach the top. Originally it was a test to see which car could get the furthest up the hill – climb the steepest gradient. There is one in use at Brooklands.
Tin top closed car with roof
Torque turning moment about a point (torque = force × radius)
Transporter vehicle to transport competition vehicles to events
Tub race car body / chassis unit made from composite material
Tunnel inverted 'U' section on vehicle floor; on front-engined rear wheel drive cars it houses the propeller shaft – propeller shaft tunnel
Turret US term for vehicle roof
Tyre wall wall on the side of a track built from tyres – giving a soft cushion in case of an off
Un-sprung weight weight below the suspension spring
Velocity vector quality of position change, for most purposes it is the same as speed

Index

Note: Page numbers in *italics* indicate figures and in **bold** indicate table on the corresponding pages.